D1703482

Automated Sample Preparation

Automated Sample Preparation

Methods for GC-MS and LC-MS

Hans-Joachim Hübschmann

Author

Dr. Hans-Joachim Hübschmann
Urbane Namba 1213
2-1-34 Minatomachi, Naniwa-ku
Osaka 556-0017
Japan

Cover Design: © ADAM DESIGN, Weinheim, Germany
Cover Image: © CTC Analytics AG

All books published by **WILEY-VCH** are carefully produced. Nevertheless, authors, editors, and publisher do not warrant the information contained in these books, including this book, to be free of errors. Readers are advised to keep in mind that statements, data, illustrations, procedural details or other items may inadvertently be inaccurate.

Library of Congress Card No.: applied for

British Library Cataloguing-in-Publication Data
A catalogue record for this book is available from the British Library.

Bibliographic information published by the Deutsche Nationalbibliothek
The Deutsche Nationalbibliothek lists this publication in the Deutsche Nationalbibliografie; detailed bibliographic data are available on the Internet at <http://dnb.d-nb.de>.

© 2022 WILEY-VCH GmbH, Boschstr. 12, 69469 Weinheim, Germany

All rights reserved (including those of translation into other languages). No part of this book may be reproduced in any form – by photoprinting, microfilm, or any other means – nor transmitted or translated into a machine language without written permission from the publishers. Registered names, trademarks, etc. used in this book, even when not specifically marked as such, are not to be considered unprotected by law.

Print ISBN: 978-3-527-34507-6
ePDF ISBN: 978-3-527-81750-4
ePub ISBN: 978-3-527-81752-8
oBook ISBN: 978-3-527-81751-1

Typesetting Straive, Chennai, India
Printing and Binding CPI Group (UK) Ltd, Croydon, CR0 4YY

Printed on acid-free paper

C087155_131012

For Paulina
In Memory of Amalia

Contents

Foreword *xiii*
Preface *xv*

1 Introduction *1*
1.1 A Perspective on Human Performance *2*
References *5*

2 The Analytical Process *7*
2.1 Laboratory Logistics *7*
2.1.1 Analytical Benefits of Instrumental Workflows *9*
2.1.1.1 Data Quality *11*
2.1.1.2 Turnkey Operation *11*
2.1.1.3 Green Analytical Chemistry *11*
2.1.1.4 Productivity *12*
2.1.2 Standard Operation Procedure *13*
2.1.3 Economical Aspects *15*
References *16*

3 Workflow Concepts *19*
3.1 Sample Preparation Workflow Design *19*
3.1.1 Transfer of Standard Methods to Automated Workflows *20*
3.1.2 Method Translation *21*
3.1.2.1 Sketching the Automated Workflow *22*
3.1.2.2 Robotic System Configuration *22*
3.1.3 Online or Offline Configuration *25*
3.2 Instrumental Concepts *25*
3.2.1 Workstations *25*
3.2.2 Revolving Tray Autosamplers *26*
3.2.3 Selective Compliance Articulated Robots *28*
3.2.4 Cartesian Robots *28*
3.2.5 Multiple Axis Robots *32*
3.2.6 Collaborative Robots *33*

3.3	Sample Processing	35
3.3.1	Sequential Sample Preparation	35
3.3.2	Prep-ahead Mode	35
3.3.3	Incubation Overlapping	36
3.3.4	Batch Processing	36
3.3.5	Parallel Processing Workflows	38
3.3.6	Sample Identification	38
3.3.6.1	Barcodes	38
3.3.6.2	Radio-Frequency Identification Chips	40
3.4	Tool Change	41
3.4.1	Manual Tool Change	41
3.4.2	Automated Tool Change	42
3.4.3	Tool Identification	44
3.5	Object Transport	46
3.5.1	Magnetic Transport	46
3.5.2	Gripper Transport	48
3.5.3	Needle Transport	50
3.6	Vial Decapping	50
	References	52
4	**Analytical Aspects**	**55**
4.1	Liquid Handling	55
4.1.1	About Drops and Droplets	55
4.1.2	Syringes	56
4.1.2.1	Precision and Accuracy	57
4.1.2.2	Syringe Needles	58
4.1.2.3	Syringe Needle Point Styles	59
4.1.2.4	Syringe Plunger Types	60
4.1.2.5	Syringe Termination	61
4.1.2.6	Operational Parameters	62
4.1.3	Vial Bottom Sensing	66
4.1.4	Pipetting	67
4.1.4.1	Air Displacement Pipettes	68
4.1.4.2	Positive Displacement Pipets	69
4.1.4.3	Pipetting Modes	69
4.1.4.4	Aspiration	71
4.1.4.5	Dispensing	73
4.1.4.6	Liquid-Level Detection	74
4.1.4.7	Liquid Classes	75
4.1.4.8	Pipet Tips	75
4.1.4.9	Functional Pipet Tips	78
4.1.4.10	Pipet Tip Materials	81
4.1.5	Dilutor/Dispenser Operation	82
4.1.6	Flow Cell Sampling	84
4.2	Solid Materials Handling	85

4.2.1	Workflows with Solid Materials	86
4.2.2	Automated Solids Dosing by Powder Dispensing	86
4.3	Weighing	88
4.4	Extraction	90
4.4.1	Liquid Extraction	91
4.4.2	Pressurized Fluid Extraction	92
4.4.2.1	Solvents and Extraction	93
4.4.2.2	Miniaturization and Automation	94
4.4.2.3	In-Cell Clean-Up	96
4.4.2.4	International Standard Methods	97
4.4.3	Liquid/Liquid Extraction	97
4.4.4	Dispersive Liquid/Liquid Micro-Extraction	100
4.4.4.1	Automated DLLME Workflows	103
4.4.4.2	DLLME for Soil and Urine	103
4.4.4.3	DLLME for Pesticides in Food	104
4.4.4.4	DLLME Hyphenation with LC	104
4.4.5	Sorptive Sample Preparation	104
4.4.5.1	Solid-Phase Micro-Extraction	105
4.4.5.2	SPME Fiber	109
4.4.5.3	SPME Arrow	112
4.4.5.4	Solid-Phase Micro-Extraction with Derivatization	116
4.4.5.5	Direct Solid-Phase Micro-Extraction Mass Spectrometry	119
4.4.5.6	Stir Bar Sorptive Extraction	121
4.4.5.7	Thin-Film Micro-Extraction	123
4.5	Clean-Up Procedures	124
4.5.1	Filtration	124
4.5.1.1	Filter Materials	125
4.5.1.2	Syringe Filter	126
4.5.1.3	Filter Vials	127
4.5.2	Solid-Phase Extraction	129
4.5.2.1	The General SPE Clean-Up Procedure	133
4.5.2.2	On-Line SPE	134
4.5.2.3	Micro-SPE Clean-Up	137
4.5.2.4	Syringe-Based Micro-SPE	141
4.5.3	Gel Permeation Chromatography	143
4.5.3.1	Standardized Methods	145
4.5.3.2	Workflow and Instrument Configuration	145
4.5.3.3	GPC-GC Online Coupling	146
4.5.3.4	Micro-GPC-GC Online Coupling	147
4.6	Centrifugation	148
4.7	Evaporation	150
4.8	Derivatization	153
4.8.1	For LC and LC-MS	154
4.8.1.1	Aromatic Acid Chlorides	154
4.8.1.2	Dansylchloride	155

Contents

4.8.1.3 Ninhydrin Reaction 155
4.8.1.4 FMOC Derivatization 155
4.8.2 For GC and GC-MS 156
4.8.2.1 Silylation 156
4.8.2.2 Acetylation 157
4.8.2.3 Methylation 157
4.8.2.4 Methoxyamination 158
4.8.2.5 Fluorinating Reagents 158
4.8.3 For GC and GC-MS In-Port Derivatization 159
4.9 Temperature Control 163
4.9.1 Heating 163
4.9.1.1 Incubation Overlapping 163
4.9.2 Cooling 164
4.10 Mixing 166
4.10.1 Vortexing 166
4.10.2 Agitation 167
4.10.3 Spinning 169
4.10.4 Mixing with Syringes 169
4.10.5 Cycloidal Mixing 169
References 171

5 Integration into Analysis Techniques 191
5.1 GC Volatiles Analysis 191
5.1.1 Static Headspace Analysis 192
5.1.1.1 Overcoming Matrix Effects 194
5.1.1.2 Measures to Increase Analyte Sensitivity 195
5.1.1.3 Static Headspace Injection Technique 195
5.1.2 Multiple Headspace Quantification 197
5.1.3 Dynamic Headspace Analysis 201
5.1.3.1 Purge and Trap 202
5.1.3.2 Dynamic Headspace Analysis with In-Tube Extraction 204
5.1.3.3 Dynamic Headspace Analysis Using Sorbent Tubes 207
5.1.3.4 Needle Trap Microextraction 208
5.1.4 Tube Adsorption 210
5.2 GC Liquid Injection 222
5.2.1 Sandwich Injection 222
5.2.2 Hot Needle Injection 222
5.2.3 Liquid Band Injection 224
5.2.4 Automated Liner Exchange 226
5.3 LC–GC Online Injection 230
5.4 LC Injection 233
5.4.1 Dynamic Load and Wash 234
5.4.2 Using LC Injection Ports with a Pipette Tool 235
References 237

6	**Solutions for Automated Analyses** *247*
	First About Safety *248*
6.1	Dilution *248*
6.1.1	Geometric Dilution of Reference Standards *248*
6.1.2	Dilution for Calibration Curves *251*
6.1.3	Preparation of Working Standards *256*
6.2	Derivatization *259*
6.2.1	Silylation *260*
6.2.2	SPME On-Fiber Derivatization *262*
6.2.3	Metabolite Profiling by Methoximation and Silylation *266*
6.3	Taste and Odor Compounds Trace Analysis *271*
6.4	Sulfur Compounds in Tropical Fruits *276*
6.5	Ethanol Residues in Halal Food *284*
6.6	Volatile Organic Compounds in Drinking Water *289*
6.7	Geosmin and 2-MIB *295*
6.8	Solvent Elution from Charcoal *301*
6.9	Semivolatile Organic Compounds in Water *304*
6.10	Polyaromatic Hydrocarbons in Drinking Water *315*
6.11	Fatty Acid Methylester *321*
6.11.1	Application *321*
6.12	MCPD and Glycidol in Vegetable Oils *328*
6.13	Mineral Oil Hydrocarbons MOSH/MOAH *339*
6.14	Pesticides Analysis – QuEChERS Extract Clean-Up *347*
6.15	Glyphosate, AMPA, and Glufosinate by Online SPE-LC-MS *362*
6.16	Pesticides, PPCPs, and PAHs by Online-SPE Water Analysis *368*
6.17	Residual Solvents *375*
6.18	Chemical Warfare Agents in Water and Soil *382*
6.19	Shale Aldehydes in Beer *390*
6.20	Phthalates in Polymers *394*
	References *400*
A	**Appendix** *413*
A.1	Robotic System Control *413*
A.1.1	Maestro Software *413*
A.1.2	Chronos Software *414*
A.1.3	Graphical Workflow Programming *415*
A.1.4	Sample Control Software *416*
A.1.5	Local System Control *417*
A.1.6	Script Control Language *418*
A.2	System Maintenance *418*
A.2.1	Syringes *418*
A.2.1.1	Manual Syringe Handling *418*
A.2.1.2	Syringe Cleaning *418*
A.2.1.3	Plunger Cleaning *419*
A.2.1.4	Needle Cleaning *419*

A.2.1.5	Confirming the Dispensed Volume of a Syringe 420
A.2.1.6	Sterilization 420
A.2.2	Pipettes 421
A.2.2.1	Calibration 421
A.2.2.2	Pipette Parts Maintenance 421
A.2.3	System Hardware Maintenance Schedule 422
A.3	Syringe Needle Gauge 423
A.4	Pressure Units Conversion 425
A.5	Solvents 425
A.5.1	Solvent Miscibility 425
A.5.2	Solvent Stability 428
A.5.2.1	Halogenated Solvents 428
A.5.2.2	Ethers 429
A.5.3	Solvent Viscosity 429
A.6	Material Resistance 429
A.6.1	Glass 432
A.6.2	Polymers 432
A.6.3	Stainless Steel 433
	References 437

Glossary 441
References 451

Index 453

Foreword

I feel honored by Hans-Joachim Hübschmann asking me to write this Foreword to his newest book, *Automated Sample Preparation: Methods for GC-MS and LC-MS*. I have read the near-final draft with pleasure, and I am impressed in all respects with the work, which was a great deal of work indeed! Hans-Joachim's choices of topic, title, chapters, sections, and their organization were splendid, and he presented the right amount of detailed yet concise information with much consideration and care.

Many books are written to follow trends that cover the same information as other books already available, but Hans-Joachim found a perfect niche by focusing on automation of sample preparation for instrumental chromatographic analysis. It is a topic that has rarely been published before, not even in journals due to the profitable advantages gained by laboratories that successfully implement automated methods.

Hans-Joachim and I met for the first time in 2016 at a food science conference in Singapore. We had an immediate connection in our work involving automated mini-cartridge solid-phase extraction, which has expanded to a larger connection after many conversations and email exchanges since then. I am grateful to Hans-Joachim for his valuable knowledge and input when I have requested it, and I am pleased to gain the perspective from the author of this work, as well as an 880-page 3rd Edition opus, *Handbook of GC-MS: Fundamentals and Applications*.

In my opinion, the three most important trends in sample preparation are streamlining, miniaturization, and automation. The fundamental chemistry of dissolution, precipitation, vaporization, partitioning, adsorption, hydration, chelation, and other phenomena that form the basis of all old and new sample preparation techniques have been known for centuries. However, the technical art of analytical chemistry requires the skill to manipulate those chemical properties in the most efficient way possible to still achieve accurate results for the purpose of the analysis.

Michelangelo Anastassiades and I did not invent any new chemistry when we developed the "quick, easy, cheap, rugged, and safe" (QuEChERS) approach for sample preparation in 2002, but we streamlined and miniaturized existing tools in an elegant solution at the right time when commercial GC-MS and LC-MS instruments were sensitive and universally selective enough to allow analysis of a wide scope of analytes. In this book, Hans-Joachim describes the next step for QuEChERS, which entails automation, and I've been calling it QuEChERSER (more than QuEChERS is also "elegant and robust").

Devising and implementing automated methods takes "brains," which constitutes the fourth important trend in sample preparation. Unlike the other trends that continually entail "more more more," required intelligence among laboratory workers is trending in both directions the same time. The growth in sheer knowledge and the complexity of problems, tools, and technology has required more brains to solve modern problems, but less brains are needed to perform routine operations, especially when using automated methods. For decades now, laboratories have saved money by hiring less educated, less talented, and less skilled technicians (thereby less expensive) to perform chemical analyses. In fact, many laboratory owners consider it a "no brainer" to fully automate as much of their operations as possible so they can hire as few staff as possible.

Despite what CSI and other fictional depictions of analytical chemistry may show, in which perfectly accurate results are beautifully displayed on colorful viewscreens in a matter of seconds, real-world analyses are not that fast and easy! Real-world analysts are not as smart as Abby Sciuto from NCIS either, and nobody is writing the plots and scripts for them leading to high certainty results that neatly solve the critical problem of the hour. With respect to reliability and data quality, brainless robots can outperform even Abby (without the personality quirks, need for sleep, or salary demands). The real-life Abby is a smart and savvy technical operator who writes the script and maintains the automated instruments.

Analytical chemists tend to be responsible, hard-working people who are motivated by laziness, which is a perfect combination to traits needed to implement laboratory automation. If it was easy, laboratory automation would have been implemented ages ago. Unfortunately, reliable, inexpensive, and user-friendly automated tools have not been commercially available until very recently, and the lack of technical know-how and abilities of laboratory staff remains the greatest obstacle.

This is how Hans-Joachim Hübschmann's book about the *Automated Sample Preparation: Methods for GC-MS and LC-MS* comes in handy. The smartest (and the dumbest) brains do not have to work hard to solve problems at all if the owners of these brains have knowledge that the same or similar problems have already been solved by others! Hans-Joachim and this book can help both the brainy and not so brainy analysts of the world, and their bosses, make the fictional world of CSI and NCIS become a step closer to reality.

Disclaimer

The opinions expressed in this foreword are the author's own and do not reflect the view of the USDA.

Wyndmoor, Pennsylvania; USA
June 25, 2020

Steven J. Lehotay, PhD (Lead Scientist)
USDA Agricultural Research Service
Eastern Regional Research Center

Preface

Although sample preparation is considered an enabling technology, it is among the most vital components of an analytical scheme.

Douglas E. Raynie, 2019 [1]

This textbook about *Automated Sample Preparation* originated from the frequent request of many analytical laboratories for integrated analytical sample preparation methods.

But, why the focus on "automated"? In fact, the motivation to employ instrumental workflow-based sample preparation is not "automation" by itself. In contrast, "automation" is the solution to several voiced drawbacks and bottlenecks in the laboratory. The traditional manual sample preparation is often slow and labor-intensive, not to mention the potential exposure of the laboratory staff to hazardous chemicals. A main aspect to be discussed is "manual," with all human impact on data quality, data comparability, and error. Multistep manual sample preparation can amount to 75% of the total method error [2]. Driven by the needs of international food, health care, and life science industries with regulations and standards, the data of chemical analyses are not for an only local approach anymore, but of global use and impact. Also, the potential environmental impact of the traditional analytical methods in use is increasingly scrutinized. This is due to the growing required number of samples for a steadily increasing number of contaminants and residues, and not to neglect, the environmental impact of the also growing usage of consumables and solvents, creating critical laboratory waste.

In the past, such topics did not receive much attention until quite recently with the overdue discussion and awareness of the impact of sample preparation on the error of analysis. On the recent survey question of the Analytical Methods Committee of the Analytical Division of the Royal Society of Chemistry "What caused your last poor proficiency test score?" the unequivocal response with 41% of impact was sample preparation, equipment problems, and human error, with "sample preparation" on top of all responses [3]. Figure 1 illustrates graphically the high man-made impact on analytical results. Significant improvements in standardized and automated sample preparation are required to keep pace with quality and productivity demands.

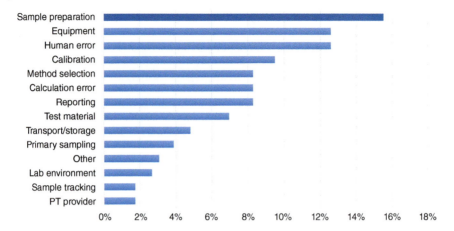

Figure 1 Causes of error in chemical analysis, after [3].

Considering all these, sample preparation has evolved as a separate discipline within the analytical and measurement sciences [2]. This compendium *Automated Sample Preparation: Methods for GC-MS and LC-MS* informs in detail on the available tools for the transfer of current manual preparation procedures to the automated level with x,y,z-robotic systems. The discussed concepts for automation try to answer on how to improve data quality, elaborate on the inherent potential for a green analytical chemistry, and finally conclude on improvements for sample throughput, method robustness, and productivity.

The objective of this textbook is to provide an overview of the current potential and proven examples for employing such robotic systems installed instrument-top or standalone with established tools, modules, and flexible workflows for a variety of sample preparation techniques.

The *Automated Sample Preparation: Methods for GC-MS and LC-MS* covers in the first part the technical concepts used in robotic systems for automated sample preparation workflows. This part enables the reader to create well-targeted and efficiently operating workflows with enough safety margin for robustness. It provides guidelines to transfer manual procedures to robotic systems. Besides the analytical expertise in the design of preparation methods, the knowledge of the technical possibilities and limitations, as well as a sense of logistics for the required sequential or parallel steps of operation, are discussed.

A second part goes into detail of popular analytical sample preparation workflows. It covers typical program-driven sample preparation examples from food, environmental, forensic, and pharmaceutical analyses. These examples are proven workflows found in many analytical laboratories and follow the published official guidelines. Each of them can serve as a template for individual use or additional modification.

My sincere thanks to many friends and colleagues for discussions, critical review, and contributions. I am very grateful to Steve Lehotay for the highly appreciated exchange on all aspects of laboratory sample processing, André Althoff for the valuable discussions, and Atsuko Bansho for her continued support, advice, and patience, my family and friends who did not get much of my time and attention during the preparation of this project.

The comprehensive compilation of technical material, images, and detailed operational and analytical background information would not be possible without the contribution by many companies actively involved in automated sample preparation. I am especially indebted to the following companies for kindly providing review, discussions, and approvals for the use of in-depth information, graphics, and image material for tools, modules, and data making the illustration of automated workflows possible: Agilent Technologies, namely, Paul Barboni and Eric Denoyer; Aisti Science Co. Ltd., namely, Sasano Sadato; Axel Semrau GmbH & Co KG, namely, Andreas Bruchmann; Tobias Uber, Brechbühler AG, namely, Philippe Mottay; CTC Analytics AG, namely, Jonathan Beck, Chiew Mei Chong, Stefan Cretnik, Toni Eberwein, Florian Gafner, Thomas Läubli, Gwen Lim, Jianxia Lv, Gerhard Nagel, Günter Böhm, Thomi Preiswerk, Kai Schüler, and Melchior Zumbach; CDS Analytical, namely, Roger Tank; Entech Instruments, namely, Dan Cardin and John Quintana; ePrep Pty Ltd., namely, Peter Dawes; Evosep, namely, Eric Verschuuren; GERSTEL GmbH & Co KG, namely, Ralf Bremer, Oliver Lerch, and Kaj Petersen; GL Sciences B.V., namely, Geert Alkema; Hamilton Bonaduz AG, namely, Beat Scheu; ITSP Solutions Inc., namely, Kim Gamble; LabTech Instruments Ltd., namely, Xue Liu; Leco Corporation Japan, namely, Michico Kanai; Markes International Ltd., namely, Massimo Santoro; Mettler Toledo AG, namely, Joanne Laukart; Plasmion GmbH, namely, Jan-Christoph and Thomas Wolf; SIM GmbH, namely, Rolf Eichelberg; Spark Holland B.V., namely, Florian van der Hoeven; Thermo Fisher Scientific, namely, Christina Jacob, Claudia Martins, and Fausto Pigozzo; Thomson Instrument Company, namely, Sam Ellis; Trajan Scientific and Medical, namely, Glenn Clivaz and Andrew Gooley; and Yaskawa Europe GmbH, namely, Richard Tontsch.

I am sure that there are many more colleagues contributing directly or indirectly with discussions during collaborations, conferences, and publications whom I have not mentioned and I apologize to them for any omissions.

Hans-Joachim Hübschmann
Mainz, July 2020

References

1 Raynie, D.E. (2019). The (mis)education of an analyst. *LCGC North America* 37 (11): 796–800.
2 Mitra, S. (ed.) (2003). *Sample Preparation Techniques in Analytical Chemistry*. Hoboken, New Jersey, USA: Wiley-Interscience.
3 Analytical Methods Committee, AMCTB No 56 (2013). What causes most errors in chemical analysis? *Analytical Methods* 5 (12): 2914–2915. https://doi.org/10.1039/c3ay90035e.

1

Introduction

Sample preparation remains the single most challenging aspect of chemical analysis

Mary Ellen P. McNally [1]

Samples arriving in the analytical laboratory usually cannot be applied directly to analytical instruments. After homogenization of the material, a more or less complex pre-treatment consisting of several steps is required for almost every sample before analysis. Each step is critical and can be a source of error and additional contamination. Often the used methods generate hazardous waste from solvents, chemicals, and consumables used. Most of the cost in chemical analysis is associated with the efforts in sample preparation procedures.

The goal of the sample preparation process is the extraction of the analytes from the incompatible matrix. Especially in trace analysis, the target analytes in low concentration are embedded in a high excess of a hard to eliminate matrix. The enrichment of the analytes to a suitable level of concentration for detection and quantification often also includes a necessary clean-up from the co-extracts. The sample preparation requires most of the time and the best skills in the analytical laboratory. Many of these tasks are still manual today. A recent survey revealed that two-thirds of the analysis time is spent on sample preparation [2]. Optimum and fit-for-purpose preparation methods for analytes in diverse matrices are subject to discussion in thousands of scientific publications every year. Google Scholar finds more than 2 million hits featuring specific sample preparations alone in the last 10 years! [3]. The demand for multi-analyte methods for larger groups of compounds with a potential occurrence in the same sample, targeted or non-targeted, like pesticides, mycotoxins, drugs, or personal care products, just to name a few areas, is increasingly addressed. The continuously growing number of analytes in different matrices is the main reason for steadily ongoing improvements.

These very first steps in the analytical workflow have the highest impact on the quality of the analytical data. Many laboratories use manual, time- and labor-intensive procedures. But the manual sample pre-processing is the biggest known source of error in the analytical sequence, no matter how precise and sensitive a mass spectrometer in the final step may be, it cannot correct for [4].

Automated Sample Preparation: Methods for GC-MS and LC-MS,
First Edition. Hans-Joachim Hübschmann.
© 2022 WILEY-VCH GmbH. Published 2022 by WILEY-VCH GmbH.

Errors in analytical measurements are the random variability, expressed as the precision of the method, systematic bias affecting the trueness of results, and gross mistakes, the handling errors. Method validation assesses precision and trueness of a method, while human spurious errors, reported as the greatest source of errors, cannot be corrected in the course of analysis [5]. It is reported from the survey that operator and sample processing errors amount up to 50% of all known potential sources of error. Evidently, measures in this area improve analytical quality significantly. Here is the standardization of sample preparation procedures a general goal. The introduction of instrumental and integrated sample preparation workflows addresses these weak points in the sample processing sequence of the otherwise excellently validated procedures.

Standardization of analytical methods including sample preparations for different kinds of samples is available for all routine areas with validated procedures published as European standards (EN), the methods of the International Organization for Standardization (ISO), the Association of Official Analytical Chemists (now AOAC International), American Oil Chemists' Society (AOCS), the Food and Drug Administrations (FDA), the Environmental Protection Agencies (EPA), Pharmacopias, and similar organizations. In these established methods, sample preparation is often the rate-determining step for sample throughput and too often the error-prone part of the analytical method. In this context, standardization calls for automation, not vice versa. Automation reduces the quite normal human manual variation and mistakes in the sample processing using adequately configured robots for the standard workflows.

In chemical analysis, the instrumentation for chromatographic separation and detection, in particular with the use of mass spectrometry, reached an operational and performance level of high technical maturity. Sensitivity, selectivity, separation power and mass resolution, speed of analysis, and data processing were the major instrumental developments with significant enhancements in recent years. Barely exploited are features for a major step ahead for an instrumental and robotic sample processing. The obvious gap is the missing use of such amazing instrument specifications for a greener sample preparation at the front end. Much smaller sample sizes are possible to reach legally required quantitation limits today. The miniaturization and standardization with automated robotic workflows for the analytical sample processing significantly release from the human impact on data quality. This handbook about *Automated Sample Processing* focuses on the tools, modules, and workflows for the next level of a greener analytical chemistry.

1.1 A Perspective on Human Performance

In contrast to the notable improvement in instrument performance and reliability, there was not much focus in the past on the instrumental integration of the traditional manually performed sample preparation workflows, knowing the significant impact of human error on reliable and true analysis results. A general industry view of human performance and root cause of events monitored is shown in Figure 1.1.

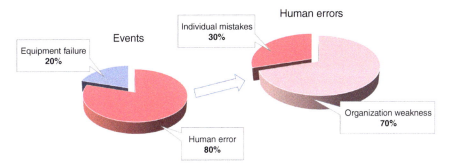

Figure 1.1 Human performance in analytical laboratories. Source: Adapted from U.S. Department of Energy [6].

Human error is not random. Mistakes are systematically connected to features of people's tools, the tasks they perform, and the operating environment in which they work [6]. About 80% of all events are attributed to human error. Further broken down, 70% are related to organizational weaknesses caused by humans in the past and 30% are directly related to human mistakes in the manual workflow.

Survey results confirm dramatically the sources of error in chemical analysis, illustrated in Figure 1.2 with the size of the square areas representing their impact. The major source of error is seen in such manual sample preparation steps with about 30% of all the potential causes. On top, operator-generated error and variation are estimated here to contribute to an additional 19%. The introduction of instrumental automated sample preparation workflows hence is expected to reduce such well-known human errors by half in routine analysis methods [5].

Samples to be analyzed usually require an often multifaceted preparation over several steps for extraction and concentration before application to an analytical instrument for analyte detection. More than 65% of the respondents to the recent survey reported the use of three or more sample preparation steps for one sample. The rate of more complex sample preparation went up significantly from previous

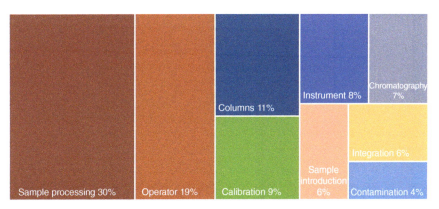

Figure 1.2 Survey results about sources of errors with impact in sample analysis (square areas in %). Source: Adapted from Majors [2].

years' responses, a number and apparent demand strongly increasing [7]. The degree of sample preparation depends mainly on the analytes, the physical status of the sample, the single or multi-compound approach, and finally the analytical method to be applied. It can cover a wide range from just dilute-and-shoot to multistep methods with extraction, concentration, and derivatization. The common goal is to reach just enough and fit-for-purpose sample clean-up to keep the analytical instrumentation in the validated status for large series of samples and reduce preventive maintenance downtime to the necessary. In the vast majority of analytical methods, the required sample preparation in the past and still today is manual. A large gap was left here, with only a few exceptions with some selected techniques, for the adaptation of analytical sample preparation methods on the instrument level. The solutions provided in this textbook address this notorious gap with the integration of robotic workflows for many routine tasks. A dedicated section covers automated turnkey solutions like the analysis of pesticides, off-odors, polyaromatic hydrocarbons (PAH), fatty acid methyl esters (FAMEs), mineral oil hydrocarbon contaminations (MOHs), or volatile organic compounds (VOCs), just to name a few that are presented in detail for reproduction.

The recent remarkable improvements in instrument sensitivity and selectivity, in particular in mass spectrometry using tandem instrumentation (MS/MS) and high mass resolution and accurate mass capabilities (HR/AM), greatly facilitated the miniaturization of such methods, allowing greener analytical chemistry with the reduction of sample sizes and solvent use. This sample volume reduction to the micro-scale opens up the way for compatible, instrument-top automation and further standardization of multi-methods. This welcome trend answers also the demand for improved data quality and higher sample throughput, especially in food, environmental, and pharmaceutical safety analyses.

Another important aspect and a strong improvement is seen in the hyphenation of instruments with workflows integrating the currently separated processes. The availability of online sample preparation, which finally includes the transfer and injection of the prepared extract to the analytical instrument, is today still the exception in analytical instrument design. Ongoing technical development starts covering some selected sample preparation procedures with benchtop or instrument top-mounted robotic preparation systems. The integrated software control for ease of use with just one sample acquisition sequence on screen is still a big but solvable gap to comply with operator demands and the required error-free operation. Several independent developments of integrated software control of the robotic sample processing systems with external devices and the hyphenated analytical instruments demonstrate successfully the benefits and feasibility.

And, even in light of the continuously increasing laboratory automation, a necessary practical remark at this point for the hands-on laboratory work, for safety and a green analytical chemistry: good laboratory practice dictates that all who handle solvents and chemicals should familiarize themselves with the compounds' material safety data sheets (MSDS) and manufacturer's recommendations for handling, use, storage, and disposal of the used chemicals.

References

1 McNally, M.E.P. (2013). Sample preparation: the state of the art. *LCGC Europe* 26 (2): 110–112.
2 Majors, R.E. (1991). Overview of sample preparation. *LC-GC The Magazine of Separation Science* 9 (1): 16–20.
3 Google Scholar (2018). scholar.google.com/, search for "sample preparation" in years "2008–2018" (20 April 2018).
4 Hein, H. and Kunze, W. (2004). *Umweltanalytik mit Spektroskopie und Chromatographie*. 3rd Ed., Weinheim, Germany: Wiley-VCH.
5 Lehotay, S.J., Mastovska, K. et al. (2008). Identification and confirmation of chemical residues in food by chromatography-mass spectrometry and other techniques. *TrAC Trends in Analytical Chemistry* 27 (11): 1070–1090. https://doi.org/10.1016/j.trac.2008.10.004.
6 U.S. Department of Energy (2009). *DOE Standard – Human Performance Improvement Handbook – Vol. 1 Performance and Principles*, vol. DOE-HDBK-1. Washington, DC, USA: www.hss.energy.gov/nuclearsafety/ns/techstds/.
7 Raynie, D.E. (2016). Trends in sample preparation. *LCGC Europe* 29 (3): 142–154.

2

The Analytical Process

The simplification of sample preparation and its integration with both sampling and convenient introduction of extracted components to analytical instruments is a significant challenge and an opportunity for the contemporary analytical chemist.

Janusz Pawliszyn [1]

2.1 Laboratory Logistics

All chemical analyses comprise a series of consecutive steps starting with the sample collection, transfer into the laboratory (if not analyzed onsite), registration, and allocation to one or more analytical testing groups as needed [2]. The first task, of utmost importance in the individual sample processing, is the preparation of a representative test portion of the sample [3]. The sample preparation follows with extraction, clean-up, and/or concentration for the instrumental measurement. Data analysis and reporting conclude the usual laboratory procedures (Table 2.1).

Looking at the sample processing steps in detail, we can distinguish those that are related to the bulk sample (lot, primary sample) typically with, for instance, freezing, cutting, cryogenic comminution, blending or milling, and finally dividing to achieve a homogeneous as possible laboratory test portion for analysis, the analytical sample [4–6]. It is mandatory to start with sufficiently large sample quantities in the gram to kilogram range (except for ab initio homogeneous samples water, beverages, blood, etc.), which are handled manually using appropriate homogenization equipment [7, 8]. The initial sample processing to achieve representative test portions of the original sample is of paramount importance, "or else the entire analysis is not just time, money, and effort wasted, it actually can provide false results to deceive and undermine the entire point of the analysis" [9]. The focus on careful sample processing prior to analysis is becoming even increasingly crucial for automated sample preparation and high-throughput analysis of only small test portions

2 The Analytical Process

Table 2.1 Steps in laboratory chemical analysis.

Process steps	Activities	
Sample collection	Manual or automated online sampling	
	Preservation	
	Sub-sampling	Manual sample handling
Sample processing	Generate a representative test portion	
	Milling, grinding, dissolution, comminution	
Sample preparation	Extraction	
	Clean-up	
	Concentration	
	Derivatization	
Chemical analysis	Separation (chromatography)	Robotic automation
	Detection (spectrometry)	
Data analysis	Identification	
	Calibration	
	Integration	
	Quantitation	
Reporting	Validation	Data system automation
	Summarize results	

Bold-faced steps are subject to automated instrumental workflows.

on the milligram level. Even the best recoveries and most precise analysis cannot correct for errors and bias occurring during the initial sample processing.

Most of the frequently applied sample preparation steps, many of them have to be performed in sequence, have the inherent potential to be transferred to integrated sample preparation platforms using appropriately programmed workflows and tools [10]. A recent survey identified the most frequently used sample processing steps in analytical laboratories [11]. The results are shown in Figure 2.1. The area in this graphics represents the frequency of use. The blue highlighted boxes express their potential for workflow automation on integrated sample preparation platforms.

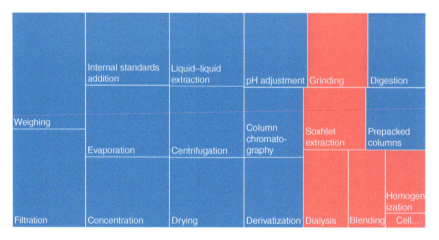

Figure 2.1 Frequently used sample preparation steps and their potential for workflow automation on integrated sample preparation platforms. Processes in blue can be automated more easily than those in red. Source: Adapted from Majors [12].

A constant trend with a continued increasing or decreasing frequency of application is reported over the last two decades for the ten most important sample preparation techniques [11, 13]:

Evaporation	↗
Centrifugation	↗
Filtration	↗
Sonication	~
Vortexing	~
pH Adjustment	~
Weighing	~
Column chromatography	↘
Dilution	↘
Liquid–liquid extraction	↘

(with ~ stable, ↗ increasing, and ↘ decreasing use)

Concerning the analysis techniques used in environmental and food safety analysis, Steven J. Lehotay and Yibai Chen analyzed the current trends based on the frequency of scientific publications during the last two decades [14]. Relating to the application area of publications, the subjects on food safety analyses outperform the former focus on environmental analysis. While gas chromatography (GC) slightly declined in attention over the years, a constant growth of liquid chromatography (LC) methods with classical detection was found. Not unexpectedly, mass spectrometry (MS) plays the most important role as the detection method for chromatographic analyses. A strong growth is noted with high-resolution accurate-mass instrumentation (HR/AM) using orbitrap or quadrupole/time-of-flight MS systems. The strength of HR/AM detection undoubtedly is the non-targeted analysis for unknown contaminations or metabolites. While the publication of new GC-MS methods reached a constant level, the LC-MS detection methods showed strong growth. Triple quadrupole MS instrumentation (MS/MS) found increasing application in the development of new methods, both for GC-MS/MS and even stronger for LC-MS/MS. This represents the dominant application of MS/MS instrumentation for the targeted analysis of regulated methods.

2.1.1 Analytical Benefits of Instrumental Workflows

Routine chemical analyses benefit from many undeniable analytical advantages of instrumental sample preparation workflows. Standardized methods can be implemented as an instrumental workflow. These standardized methods provide detailed descriptions of the applied materials, solvents, and chemicals and an optimized often validated set of parameters to be used. A time-consuming method development is avoided, targeting the comparability of the analytical data. But, even with standardized analytical methods, we know that human factors are involved leading to a wide range of possible results. Consistent inter-laboratory comparison of data is very often related to well-experienced technicians. Many publications about successful automation of analytical workflows report different aspects that were shown to be beneficial and incentivized the implementation of automated processes:

- Standard workflows for routine methods
- Turnkey configurations avoid lengthy method development
- Short familiarization time
- Quick implementation into production mode
- Standardization in different laboratory locations of international companies
- Consistent high-quality comparable data
- Less variation in results, independent of individual factors, change of operators
- Traceability
- Reduction in sample amount and solvents
- Green analytical chemistry by miniaturization, micro-extraction methods
- Facilitated sample transport
- Cost reduction related to expensive standards, reagents, consumables
- Increased sample throughput
- Proportionately scalable with increased sample volume
- 24/7 operation
- Shorter processing times, shorter reporting time
- Less contamination from labware
- Reduced human error
- Reduced rework and number of repeat analyses
- Safety, reduced human exposure to chemical reagents and ergonomic injuries
- Online operation with measurement systems
- Online sampling from air or water streams
- At-site unattended operation

A barrier in setting up automated sample processing workflows is often seen with the initial capital expense, missing technical expertise and new analytical challenges handling small sample amounts. The technical operation and logic of robotic workflows may appear complicated and requires a new learning experience. Regular system maintenance, in particular, preparation steps to avoid malfunctions and carry-over need to be implemented. The manufacturer-provided standard programs are typically prepared for specific single workflows with an inherent lack of flexibility and may require customization for the laboratory. Often, instrumental workflows for sample processing are then customized solutions for the methods of choice in collaboration with the vendor application teams. In many available turnkey solutions, the sample preparation methods are implemented by a suitable instrument workflow with an already predefined or range-limited set of parameters reducing significantly the required time to implement and train a method in the laboratory, also preventing invalid input for process safety. The automated sample preparation on the instrumental level has the potential to become routine widely independent from personal handling variation or interpretation, to provide comparable and predictable data.

2.1.1.1 Data Quality

Instrumental workflows deliver a consistent quality of data, independent of the human factor. Data generated are comparable and independent from an individual operator or location serving well-suited emerging international requirements related to, e.g. global trade or analytical standards harmonization.

On top of that, sample turnaround times are significantly reduced by the reduction of repeat sample measurements due to unavoidable mistakes in manual work. Automated workflows are well documented. The instrument operation and parameters used are traceable by accessing log files. Samples and related results can be tracked and identified by barcode labeling.

2.1.1.2 Turnkey Operation

Instrumental workflows establish the analytical sample preparation methods with an integrated or programmed operating sequence in the laboratory. Standard method workflows are increasingly also commercially available for robotic sample preparation systems in the application-specific configuration. No lengthy method development is required allowing a quick installation, short familiarization, and transfer into production. Typically, such workflows are documented as standard operation procedure (SOP) to become part of the laboratory internal quality management, keeping to international quality standards for increased regulatory compliance (see Section 2.1.2).

2.1.1.3 Green Analytical Chemistry

The concept of green analytical chemistry (GAC) covers the development of analytical procedures that replace or minimize the usage of hazardous reagents and solvents and maximize operator safety to establish environmentally friendly sample preparation techniques [15, 16]. The principles of GAC are summarized in the term SIGNIFICANCE and illustrated in Table 2.2 [17].

An important aspect in this context is the miniaturization and automation of sample preparation procedures, represented by the second "I" in Table 2.2. Traditional manual methods usually work on a larger deciliter scale as we still see in some established official methods. Manual extraction methods like the widely used liquid/liquid extractions are typical examples, e.g. the liquid extraction of organophosphorus pesticides from wastewater, sludge, or sediments using deciliter amounts of potentially hazardous dichloromethane (DCM) [18, 19]. Several extraction steps with following clean-up steps for purification and concentration of the extract before analysis of the analytes are required. An example of how new technologies improve analytical performance and at the same time significantly reduce critical and expensive solvent use is the increasing replacement of the classical Soxhlet extraction by pressurized or supercritical fluid extraction [20].

Miniaturization of the extraction process on the microliter level offers the potential for automation and to become compatible with the injection volumes typically applied with analytical instrumentation [21]. Downscaling of sample and solvent volumes is a significant achievement not only for GAC but also for the logistics and economy of the analytical laboratory. The increased usage of micro-methods

Table 2.2 Principles of green analytical chemistry [17].

S	Select direct analytical techniques
I	Integrate analytical processes and operations
G	Generate as little waste as possible
N	Never waste energy
I	Implement automation and miniaturized methods
F	Favor reagents from renewable sources
I	Increase safety for the operator
C	Carry out in situ measurements
A	Avoid derivatization
N	Number of samples and size should be minimal
C	Choose multi-analyte or multi-parameter methods
E	Eliminate and replace toxic solvents

Source: Adapted from Gałuszka et al. [17].

comes from their simple handling, rapid sampling, low cost, solventless nature, good recoveries, and large enrichment factors [22, 23]. Micro-methods like the solid-phase micro-extraction (SPME) achieve the same analytical results, avoiding solvents even completely, enabling the multiple use of the SPME device [24]. Another influential development is the replacement of the huge wastage of solid-phase extraction cartridges by replacement with online methods, for instance as already officially established for pesticides in drinking water with the US Environmental Protection Agency (EPA) Method 543 [25].

The big benefit of micro-methods is their compatibility with instrumental sample preparation concepts. Micro-methods work in the higher microliter range, which matches well the injection volumes of chromatographic analysis methods. In practice, the term "miniaturization" would fit better the automated approach of using a reduced test portion amount and limited solvent volumes. High extract volumes are not required for the miniaturized analysis scheme, only lead to unnecessary concentration steps. Working on the same volume level, even with many solutions without solvents at all, invites for automation of the sample preparation processes as further described in Chapter 3. This offers a great technical opportunity for the integration of the sample preparation processes with analytical instrument operation, another perfect fit for automated sample preparation methods with the green analytical chemistry concept.

2.1.1.4 Productivity

Productivity is the other important aspect of establishing instrumental sample processing workflows. Today less than 10% of the workload in analytical laboratories is dedicated only to instrumental analysis. More than two-thirds of the time in analytical laboratories is required for sample processing [26]. A productive use of such a big time slice would open up many improvement opportunities for extended services, higher sample throughput, and last not least reduced cost per samples in every analytical laboratory.

Instrumental workflows allow and are designed for unattended operation. Such workflows become the basis for an automated processing regime in the laboratory. Either a stand-alone or offline and the online coupling to analytical instrumentation can be realized, for instance, with the widespread use of GC-MS or LC-MS systems. A so-called "prep-ahead" functionality for executing a preparation workflow online during an ongoing analysis run offers a throughput optimized measurement cycle for a maximum duty cycle of the analytical instrumentation and a faster return of investment (see Section 3.3.2). For continuous monitoring from water streams or for reaction monitoring, the "prep-ahead" function facilitates high-frequency sampling intervals. A 24/7 operation can finally be achieved with sufficient sample, solvent, and reagent capacity made available from suitable reservoirs on the automated sample preparation equipment. The use of microplates in 96- or 384-well format is a good example of recent productivity developments for food safety and life science applications.

2.1.2 Standard Operation Procedure

In general, an SOP is a comprehensive workflow documentation within the quality assurance (QA) system of an analytical laboratory, but not limited to analytical purposes. The SOP provides approved information about the principle of an analytical procedure, the requirements for instruments, tools, chemicals, and the step-by-step guidance on how to execute this task. "The development and use of SOPs are an integral part of a successful quality system as it provides individuals with the information to perform a job properly, and facilitates consistency in the quality and integrity of a product or end-result" [27].

Why do we need SOPs? Laboratory staff follow well-established and approved sample preparation and analytical procedures with best practices. The guidance on the required work steps standardizes the workflow between different staff and locations, reducing the personal impact or interpretation, improves the comparability of data, and reduces errors. With these aspects, SOPs are also the basis for in-house training programs. International regulations for instance, those issued by the US EPA, the International Organization for Standardization/European Standard (ISO/EN), or the US Food and Drug Administration require the application of published SOPs for compliance.

SOPs are laboratory specific and need to be locally established even for the in-house or published methods representing, for instance, the locally available resources, consumables, and operation of the instrumentation to be used. It must be written in clear text with all necessary details to avoid any misunderstandings.

SOPs should be validated by a well-experienced person usually the lab manager and approved by a third person, e.g. the QA manager, for use. Document control with title, revision number, date, author, and lab manager approval is part of the QA plan. SOPs are version-controlled documents. The current SOP must be at hand for the operating staff and auditing purposes.

Required components of SOPs are:

- Purpose of the SOP
- Scope of application and limitations
  - Principle of the method
  - Reference to regulations or published procedures
  - Qualification of the operator
  - Safety, potential risk, and hazard information
  - Definition of terms and abbreviations used
- Safety information
  - Handling of dangerous substances
  - Workplace preparation
- Protective measures
  - Storage
- Equipment
  - Tools, glassware
  - Analytical instruments, setup, and method parameters
  - Calibration
- Chemicals and reagents
  - Analytical standards, quality control (QC) materials
  - Solvents, gases
  - Chemicals with CAS numbers, purity, sources
- Step-by-step workflow
  - Sample handling, collection, preservation
  - Standard preparations
  - Sample preparation, clean-up, concentration
  - Analysis and data acquisition
- QC
  - Blanks
  - Calibration samples
  - Spiked blanks
  - Matrix spike
  - Laboratory control samples
- Data analysis
  - Possible interferences
  - Data processing
  - Results calculation
  - Reporting
  - QC measures
- Waste management
- References

Robotic x,y,z-systems execute workflows and hence become an integral part of an SOP-guided sample processing and analysis process. These workflows follow strictly the given SOP in its step-by-step operation and the chosen parameter set.

More complex workflows are recommended to be accompanied by a checklist for the preparation of sample sequences.

2.1.3 Economical Aspects

Two major aspects drive the increased application of integrated and stand-alone robotic systems: first, about the way of implementation of such automated systems in a laboratory, and second, evenly important, about the quality of the analytical process.

The application concept of hyphenated robotic sample preparation systems cannot be limited to the hardware only. As an essential requirement, flexible and precise tools need to be made available on the system to perform an intended workflow. Equally important to the robotic hardware, the same as we know from laptops or any other computer system, is the program driving the hardware, using the installed tools and modules for the intended purpose. In the same way, without a suitable word processor, even a powerful laptop cannot be used for writing a single letter. Workflows for robotic systems are delivered as dedicated programs, for instance, as turnkey solutions, to execute sample preparation and injection tasks step-by-step as defined and documented in an SOP as outlined in Section 2.1.2. Typically, these programs offer a user interface with the main analytical parameters as defaults or for an advanced modified operation. For online analyses, the robotic systems are preferably integrated into the chromatography data systems (CDS) of the instrument manufacturer or get controlled by an "umbrella"-like software solution integrating the robotic workflow and involved analytical instruments [28, 29]. The selected workflow, sample indices, and injection parameter become part of a unified acquisition sequence table. In stand-alone installations, the robotic systems are operated via a local terminal or computer-connected.

In many cases, these workflow programs cover the regulated and standardized analytical methods. Also customized modifications according to local regulation needs or individual customizations are made available this way.

A successful implementation of a robotic sample preparation needs both the appropriate system tools and modules together with the analytical program to operate them for the desired SOP, respectively the intended analytical workflow. This concept allows an efficient implementation of a new or additional analytical workflow in the laboratory. The time- and labor-consuming process of developing a new method from scratch is avoided, and steps to productivity are significantly reduced.

Automation is the welcome side-effect of the programmed workflow control and CDS integration. The sample capacity of robotic systems often can be extended up to several hundreds of samples for 24/7 operation. CDS integration with analytical instrumentation then allows the productive execution of sample preparation steps to be performed during the ongoing analysis runs. This so-called "prep-ahead" mode significantly improves the duty cycle of costly analytical instrumentation for increased sample throughput with minimized instrument idle times (see Section 3.3.2). In many cases, the state-of-the-art network connectivity of the analytical instrumentation allows the remote access to monitor or control the unattended operation.

Much underestimated is the substantial impact of robotic systems on data quality and as such on the cost per sample, respectively the number of repeat analyses. Precision of e.g. volume dosing, time, and temperature control is unparalleled compared to manual operation. Inter-sample and inter-laboratory reproducibility is released from personal operator impact, a benefit highly appreciated by internationally operating companies with the demand for identical workflows at different locations such as production control laboratories in different geocenters. Complex multi-step sample preparation procedure, for instance, required for the manual determination of 3-chloropropane-1,2-diol in vegetable oils [30], polyaromatic hydrocarbons in wastewater [31], or the many kinds of derivatization reactions are naturally error-prone. The significantly reduced number of repeat analyses caused by manual handling errors further reduces the cost of analysis and improves laboratory reporting times to benefit customers. On top of that, the programmed workflow with all parameter settings is well documented and traceable for internal QA-regulated environments and greatly facilitates the audit procedures.

Most important cost savings are achieved by the significantly reduced consumption of solvents and chemicals in a laboratory using robotic sample preparation procedures. Comparing only the solvent cost for pesticides analysis using the traditional "Luke method" with liquid/liquid partitioning, consuming approximately 220 mL acetone, 250 mL petrol ether, and 350 mL DCM per sample (see Section 4.4.3), with the modern QuEChERS approach requiring only less than 60 mL acetonitrile (see Section 6.14), a total annual cost saving in the range of USD 250 000 can be achieved (estimated 25 samples/day, 220 days lab operation). On top up to 20% of solvent cost has to be calculated for the disposal cost of laboratory chemical waste. Additional growth in productivity adds on top. Such attainable cost savings speak by itself for the due implementation of automated sample preparation solutions.

References

1 Pawliszyn, J. (2009). *Handbook of Solid Phase Microextraction* (ed. J. Pawliszyn). Beijing, P.R. China: Chemical Industry Press.
2 Schatzlein, D. and Thomsen, V. (2008). "The chemical analysis process". Spectroscopy Oct 01, 23 (10), www.spectroscopyonline.com/chemical-analysis-process (accessed 29 September 2019).
3 Lehotay, S.J., Han, L., and Sapozhnikova, Y. (2018). Use of a quality control approach to assess measurement uncertainty in the comparison of sample processing techniques in the analysis of pesticide residues in fruits and vegetables. *Anal. Bioanal. Chem.* 410: 5465–5479. https://doi.org/10.1007/s00216-018-0905-1.
4 Gy, P.M. (1998). *Sampling for Analytical Purposes*. John Wiley & Sons, Ltd. ISBN: 978-0-471-97956-2. https://www.wiley.com/en-us/Sampling+for+Analytical+Purposes-p-9780471979562.

5 Pitard, F.F. (1996). *The sampling theory of Pierre Gy: Comparisons, implementation and applications for environmental sampling*, 3rd Ed. Boca Raton, Fl, USA: CRC Press, Inc. 113847648X.

6 Han, L., Lehotay, S.J., and Sapozhnikova, Y. (2018). Use of an efficient measurement uncertainty approach to compare room temperature and cryogenic sample processing in the analysis of chemical contaminants in foods. *J. Agri. Food Chem.* 66 (20): 4986–4996. https://doi.org/10.1021/acs.jafc.7b04359.

7 Anastassiades, M., A. Barth, D. Mack, et al. (2017). Sample Preparation and Processing and their Impact on Pesticides. Latin America Pesticides Workshop Presentation, San José, Costa Rica.

8 Fussel, R.J. (2016). Analytical Challenges for Pesticide Residue Analysis in Food: Sample Preparation, Processing, Extraction and Cleanup. Thermo Fisher Scientific, Hemel Hempstead, UK. White Paper 72048.

9 Lehotay, S.J. and Cook, J.M. (2015). Sampling and sample processing in pesticide residue analysis. *J. Agri. Food Chem.* 63 (18): 4395–4404. https://doi.org/10.1021/jf5056985.

10 Majors, R.E. (2013). Sample Preparation Fundamentals for Chromatography (eds. T. Robarge, N. Simpson, et al.). Wilmington, DE, USA: Agilent Technologies, Inc. https://doi.org/10.1515/9783110289169.

11 Majors, R.E. (2013). Trends in Sample Preparation, Chromatography Online, March 29.

12 Majors, R.E. (2015). Overview of sample preparation. *LCGC* 9 (1): 16–20.

13 Raynie, D.E. (2015). Trends in Sample Preparation. *LCGC Europe, March* 2015: 142–152.

14 Lehotay, S.J. and Chen, Y. (2018). Hits and misses in research trends to monitor contaminants in foods. *Analytical and Bioanalytical Chemistry* 410 (22): 5331–5351. https://doi.org/10.1007/s00216-018-1195-3.

15 De la Guardia, M. and Garrigues, S. (eds.) (2012). *Handbook of Green Analytical Chemistry*. John Wiley & Sons, Ltd. DOI: 10.1002/9781119940722.

16 Armenta, S., Garrigues, S., Esteve-Turrillas, F.A., and de la Guardia, M. (2019). Green extraction techniques in green analytical chemistry. *TrAC – Trends in Analytical Chemistry* https://doi.org/10.1016/j.trac.2019.03.016.

17 Gałuszka, A., Migaszewski, Z., and Namieśnik, J. (2013). The 12 principles of green analytical chemistry and the SIGNIFICANCE mnemonic of green analytical practices. *Trends Anal. Chem.* 50: 78–84.

18 US Environmental Protection Agency. (2000). "Method 1657 Rev. A – Organo-Phosphorous Pesticides in Wastewater, Soil, Sludge, Sediment, and Tissue by GC/FPD", September 2000.

19 Luke, M., I. Cassias, and Yee, S. (1999). Lab. Inform. Bull. No. 4178, Office of Regulatory Affairs, US Food and Drug Administration, Rockville, MD (1999).

20 De Koning, S., Janssen, H.-G., and Brinkman, U.A.T. (2009). Modern methods of sample preparation for GC analysis. *Chromatographia* 69: 33–78.

21 Foster F., O. Cabrices, Jackie Whitecavage, et al. (2014). Automating liquid-liquid extractions using a benchtop workstation, GERSTEL Inc. Linthicum, MD 21090, USA, Application Note 7/2014.

22 Costas-Rodriguez, M. and Pena-Pereira, F. (2014). Method development with miniaturized sample preparation techniques. In: *Miniaturization in Sample Preparation* (ed. F. Pena-Pereira), 463. Sciendo/de Gruyter, Warsaw, Poland. https://doi.org/10.2478/9783110410181.6.

23 Kloskowskia, A., Marcinkowskia, Ł., and Namieśnik, J. (2014). Green aspects of miniaturized sample preparation techniques. In: *Miniaturization in Sample Preparation* (ed. F. Pena-Pereira). Warsaw/Berlin: De Gruyter Open Ltd. www.degruyter.com/view/title/506137?tab_body=toc (accessed 16 July 2019).

24 Souza-Silva, É.A., Jiang, R., Rodríguez-Lafuente, A., and Gionfriddo, E. (2015). Trends in Analytical Chemistry. A critical review of the state of the art of solid-phase microextraction of complex matrices I. Environmental analysis. *Trends Anal. Chem.* 71: 224–235.

25 US Environmental Protection Agency (2015). Method 543. Determination of Selected Organic Chemicals in Drinking Water by On-Line Solid Phase Extraction Liquid Chromatography/Tandem Mass Spectrometry (On-Line SPE-LC/MS/MS). US EPA, Office of Research and Development, National Exposure Research, Cincinatti, OH, USA.

26 Majors, R.E. (1991). Overview of sample preparation. *LCGC* 9 (1): 1620.

27 US Environmental Protection Agency (2007). Guidance for preparing standard operating procedures (SOPs). EPA QA/G-6. Office of Environmental Information, Washington, April 2007.

28 GERSTEL (2013). Maestro Software. Product information, Mülheim an der Ruhr, Germany, GERSTEL GmbH & Co.KG.

29 Axel Semrau (2018). CHRONOS 4.9 Product information. Sprockhövel, Germany, Axel Semrau GmbH & Co KG.

30 American Oil Chemists' Society (2013). 2- and 3-MCPD fatty acid esters and glycidol fatty acid esters in edible oils and fats by acid transesterification. *AOCS Official Method Cd*: 29a–13a.

31 US Environmental Protection Agency (1984). Method 610 – Polynuclear Aromatic Hydrocarbons. In: *40 CFR Part 136, 43 344; Federal Register 49, No. 209.*

3

Workflow Concepts

Robots change the way people spend their time at work in a positive way, since the machines always get to do the most tedious, unpleasant and hazardous tasks.

Ray Perkins [1]

3.1 Sample Preparation Workflow Design

The steadily increasing number of samples in laboratories, requirements for reproducibility and traceability, and the challenge of lower detection limits recommend the automation of processes. In particular, the demand goes for integrated sample preparation with the measurement methods. Only a few automated measurement systems for integrated analytical measurement applications are currently available [2]. A serious impediment to the broad introduction of automated methods in laboratories is the status of the current standard and official methods. Many standard methods were launched even decades ago and do not reflect the recent progress in analytical instrumentation and their potential for automation. The methods are traditionally based on manual operation using standard labware and appropriate solvent use. An approach for green analytical chemistry was not yet on the horizon.

Automation with analytical instrumentation today is mostly limited to autosampler devices for a final extract injection to the measurement system (see, for example, Section 3.2.2). Integrated automated solutions with headspace analysis are still the exception of standard methods in the regulated market. Integrated sample preparation solutions are only very recent developments, mostly focused on particular applications or techniques [3]. A very recent development is, for instance, the fully automated online sample preparation and analysis of dried blood spots by Shimadzu Corporation with an online system that combines on-line supercritical fluid extraction (SFE) extraction with supercritical fluid chromatography (SFC) analysis in a single flow path [4]. Target compounds can automatically be extracted and transferred online to HPLC- or SFC-MS [5]. It is only a few years ago that Thermo Fisher Scientific launched the automated, high-throughput LC-MS solution for water and beverage analysis. The "EQuan™" called turnkey solution is based on online sample preparation, enrichment, and LC-MS analysis for trace contaminants in environmental water, drinking water, and beverage samples described for the analysis of, e.g. pesticides, pharmaceuticals, personal care products, endocrine disruptors, or perfluorinated compounds. The majority of the currently called "automated sample preparation" solutions are usually standalone and mostly mimic the manual working steps with similar sample size and solvents consumption, like the automated traditional solid-phase extraction (SPE) [6] or the popular QuEChERS extraction with dispersive solid-phase extraction (dSPE) clean-up [7]. Comprehensive, integrated solutions for the standard and official sample preparation methods, which are the major workload in routine laboratories, are rarely offered. The lack of integrated sample preparation solutions can be seen with the majority of the instrument manufacturer. This is in obvious contrast to the increasing customer demand as recently reported [8, 9]. Integrated automation concepts for increased sample throughput, reduced cost, and improved response time are often realized by laboratories themselves for particular and individual in-house solutions, or they are increasingly offered as highly customized applications by value-adding consulting companies [10–12].

The steps for chemical analysis as outlined in Table 2.1 starts with the representative sampling and comminution pretreatment, the necessary manual steps to provide a homogeneous subsample for processing [13, 14]. For homogeneous media like air and water also online sampling and online analysis can be realized. Sample preparation with extraction and purification are still manual process steps. Here the green concepts for hyphenation of the sample preparation with the analytical measurement can fill a current gap. In general, the preferred instrumental concepts for automated workflows follow well the green analytical chemistry requirements aiming for …

- preferred online analysis for reduced error, and most authentic measurements,
- same chemicals and reactions as the standard method,
- low solvent consumption,
- low consumption of expensive standard solutions,
- the use of inexpensive consumables,
- improved instrument duty cycle,
- reduced lab space requirements, installed preferably on instrument top,
- flexible configurations available for several workflows, and
- the use of open architecture equipment.

3.1.1 Transfer of Standard Methods to Automated Workflows

Transferring standard methods into automated workflows necessarily but also intentionally involves scaling down the suggested volumes [15]. The goal in the transcription into an automated workflow is to keep the "chemistry" of the method as described, but make use of the available modules, tools, and compatible vial sizes of robotic sample preparation equipment.

An instructive example is the widely used method for the analysis of the fatty acid composition of animal and vegetable fats and oils via conversion of the triglycerides to fatty acid methyl ester (FAMEs). Determination of fat content, fat composition, and fat quality is one of the most important methods in food analysis. A fast and flexible analysis method for various food commodities uses sodium methylate for transesterification of the triglycerides to FAMEs. The method was developed by the Official Food Control Authority of the Canton of Zurich in Switzerland in 1997 and accepted in the year 2000 by the Swiss Ministry of Health as the official method [16]. The method works even with aqueous media like milk and avoids prior fat extraction [17]. Some sample types require a pretreatment, which is then an initial manual task. For instance, nonhomogeneous and solid samples are homogenized using a blender, food powders must be prepared as a slurry with water, or cheese, meat products, or ready meals must be refluxed by heating in dimethylformamide (DMF) [17]. Aliquots of the pretreated samples are taken to the automated transesterification procedure with online GC and flame ionization detection (FID) or GC–MS analysis.

This transesterification method was standardized in May 2011 as the European standard method DIN ISO/EN 12966-2:2011-05 (DIN Deutsche Industrie Norm, National Standards Organisation of Germany) with an update in March 2017 as the ISO 12966-2:2017 method [18], and by the AOAC International in 2012 (AOAC, formerly in 1884 the Association of Official Agricultural Chemists, later the Association of Official Analytical Chemists) as the AOAC Official Method 2012.13 [19]. The rapid transesterification of triglycerides in alkaline conditions can replace the more laborious classical practice of lipid extraction followed by saponification and methylation of the free acids using methanolic boron trifluoride as catalyst [20], or the methanolic sulfuric acid method [21].

In contrast to the classical BF_3 catalyzed method, the automated transesterification not only reduces the exposure of hazardous chemicals like the aggressive BF_3, the concentrated acids and avoids the potentially dangerous diethylether, but also reduces the cost per sample by more than 50%. The short reaction time improves productivity and sample throughput significantly. The fully automated transesterification procedure works on the chromatographic time scale and can be accomplished by preparation of the sample for injection during the GC runtime of a previously injected sample in the "prep ahead" mode (see Section 3.3.2).

3.1.2 Method Translation

When translating a manual method to an equivalent automated and miniaturized workflow several decisions are to be met, all with the goal and consideration of keeping the "chemistry" of the method unchanged. Some basic requirements to be answered and steps to design the automated workflow are:

- Calculate the required sample amount vs. the required injection aliquot to the measurement system to meet the required method detection limit (MDL).

- Which injection volume for analysis is possible?
- Which injector, loop, or flow cell is available?
- Which specifics of the analysis instrument need to be taken into account?
- Which manual pretreatment is required for the bulk sample?
- How is the sample provided to the robot?
- Reduce the solvent volumes to fit the processing of the smaller sample amount.
- Decide on the vial size(s) to be used.
- Use a spreadsheet to calculate consumptions for n samples in a sequence.
- Decide on the reservoirs for the required volumes of solvents, standards, reagents, wash and waste liquids for n samples to run.
- Determine the required amount of solvents, standards, reagents, and consumables like pipette tips or filters, etc. for one and n samples.
- Is vial transport necessary? Are magnetic/nonmagnetic caps, a gripper tool required?
- Decide on the size and dimension of the sample racks for n number of samples.
- Decide on the tools and dimensions for liquid handling with syringes, pipettes, or dilutor.
- Decide on the kind and number of wash liquids for syringes.
- Decide on modules for incubation, agitation, mixing, etc.
- Decide on clean-up steps like filtration, SPE, centrifugation.
- Decide on concentration steps like evaporation, reconstitution.
- Which wait times are required? For reaction, incubation, sedimentation?
- Is a parallel operation possible for certain steps?
- Which online analysis is used? Define the injection parameter.
- Decide on the tool to be used for injection to the analytical instrument.
- For the prep-ahead mode, check the sample preparation and analysis cycle time.

3.1.2.1 Sketching the Automated Workflow

The transfer of a published manual method can be accomplished by identifying the sample and chemical processing steps with all details on the used solvents, reagents, vials, volumes, and times together with the described manual operation steps. A record in spreadsheet format as shown with Table 3.1 for the mentioned transesterification method facilitates the allocation of the proposed equivalent items in a workflow column, next to it the intended tool and module to be used. Additional workflow steps required for the sequential processing of a series of samples, not required and not mentioned in the manual method, can be inserted. For instance, the syringe rinsing with extract, several wash steps, or tool changes can then be introduced where necessary. From here the system configuration, volume requirements, and course of workflow can be taken and further refined.

3.1.2.2 Robotic System Configuration

The required configuration of an x,y,z-robotic system in open architecture for the workflow as developed in Table 3.1 is composed in a minimum arrangement as follows. The suggested tools could be made available for additional workflows, and as

Table 3.1 Transfer of the published method for FAMEs transesterification to an automated workflow.

Method tasks	Published manual method	Automated workflow	Workflow tool/module
Sample size	50–500 mg	15–50 mg	
Weigh accurately food or oil	50 mL Erlenmeyer flask	2 mL vial	Place sample vials into sample rack
Wait	—	*Ready* signal	GC
Prime dilutor w dioxane	—	Priming	Multichannel dilutor
Add dioxane with ISTD	5 mL	1500 µL	Multichannel dilutor
Add ISTD	—	15 µL	Liquid syringe (25 µL)
Clean dilutor	—	Wash step	Multichannel dilutor
Dissolve fat, mixing	Shake	Vortex	Vortexer
Transfer an aliquot	—	100 µL	Multichannel dilutor
Clean dilutor	—	Wash step	Multichannel dilutor
Add 5% methoxide in methanol	5 mL	100 µL	Multichannel dilutor
Vortex	3 s	3 s	Vortexer
Clean dilutor	—	Wash step	Multichannel dilutor
Wait	60–90 s	120 s	Vortexer
Add heptane	25 mL	1000 µL	Multichannel dilutor
Vortex	Vortex	Vortex	Vortexer
Clean dilutor	—	Wash step	Multichannel dilutor
Add 15% disodium hydrogen citrate	10 mL	300 µL	Multichannel dilutor
Clean dilutor	—	Wash step	Multichannel dilutor
Wait for phase separation	Wait	60–120 s	—
Transfer supernatant to a 2 mL vial	100 µL	—	—
Dilute, add heptane	1000 µL	—	—
Change tool	—	—	Injection syringe (10 µL)
Inject to GC	0.5 µL	1 µL	Injection syringe (10 µL)
Send	—	Start signal	GC
Clean syringe	—	Wash step	Injection syringe (10 µL)
Change tool	—	—	Multichannel dilutor
Proceed to next sample			

Grey, manual pretreatment; Blue, inserted workflow only tasks; Yellow, common manual/automated tasks; Red, not required in automated workflow.

of available space on the system bar, additional modules and tools can be present, or later installed for additional tasks:

- Basic x-rail — 80 or 120 cm length
- 1× Tool station — holds tools available for use
- 1× Tray holder — equipped with racks for vials
- 3× Vial racks — for 2 mL vials
- 1× Multichannel dilutor with tool — equipped with a 1000 µL syringe
- 2× Liquid syringe tool — for the 25 and 10 µL syringes
- 1× 25 µL GC syringe — used for ISTD addition
- 1× 10 µL GC syringe — used for GC injection
- 1× Vortex mixer — for reaction vial vortexing
- 1× Large solvent station — as the reagent reservoir
- 1× Fast wash station — for the various syringe wash steps
- GC Mounting kit — instrument top as of the GC model in use
- Consumables — 2 mL vials with magnetic caps, pre-cut septa

The required solvents get connected from their solvent bottles to the valve of the dilutor device, as shown in Figure 3.1. In this case, 250 mL bottles each for heptane, methanol, dioxane, water, and a tube to a waste canister are recommended.

The major difference of the automated workflow to the manual method is the reduced sample size and the reduced consumption of solvents, standards, and reagents. The chemicals, solvents, and standards used, as well as the sequence of steps and timing, are kept as suggested in the published method. The automated method is optimized for the transesterification reaction in 2 mL vials. Sample and solvent handling is achieved by using a dilutor tool. Internal standards are applied using a dedicated syringe for the best precision. The accurate handling of the reaction time for transesterification is critical as reported in the original publication. Precise timing is for sure the strength of an automated system for the overall method reproducibility and quantitative precision. A reaction timer can be started immediately after methylate addition and stopped with the citrate addition, which stops the reaction after the predefined time. After a completed transesterification reaction and liquid/liquid extraction (LLE) of the resulting FAMEs, the heptane extract is online injected for GC analysis. The GC injection is performed by using a dedicated injection syringe, not used for any other workflow step. A next sample can be prepared in time using the "prep-ahead" mode (see Section 3.3.2). Special care is taken in the automated workflow for priming and washing steps preventing

1 Dilutor tool
2 Waste
3 Water
4 Methanol
5 Dioxane
6 Heptane

Figure 3.1 Multidilutor valve port connections. Source: Courtesy CTC Analytics AG.

any potential carryover between samples. The required solvent volumes need to be calculated in the overall consumption sheet for the expected number of samples in the analysis sequence.

3.1.3 Online or Offline Configuration

Online hyphenation of sample preparation steps with the analytical measurement delivers unique advantages for routine analysis. Many standard methods can be implemented on a miniaturized level, saving cost, labor, and time. Especially with chromatographic methods, the preparation steps can go in parallel with the analysis run for improved data quality, time to report out, and sample throughput. Another important aspect is the identical preparation timeline for all samples until analysis. The "prep ahead" concept optimizes the instrument duty cycle and together with the unattended operation for an accelerated instrument amortization.

A batch processing of samples is always possible but leads back to the offline concept of analyses. However, automated offline sample preparation can be used as an option in routine laboratories. Offline concepts can serve several analytical instruments in a laboratory economically if the sample preparation time is short compared to a chromatographic analysis. Still, the offline scheme is a semi-automated approach available during the regular work hours of the laboratory, preparing the analytical instruments for the night shift. Extracts must be stable for a delayed analysis, a potential source of analytical uncertainty in many cases. In large sample series, the last sample suffers from the longest waiting time in the rack before analysis. Cooled trays can be an option to prevent reactions or solvent losses. Often, the parallel processing of samples via liquid handling is applied for photometric analyses employing microplate readers offline or online for biological screenings, for a sample transfer to matrix-assisted laser desorption targets (MALDI), or quantification with enzymatic assays, which tend to be conducted in batches all at once rather than sequentially as in chromatographic methods.

3.2 Instrumental Concepts

3.2.1 Workstations

The term "workstation" as an analytical device is lent from high-performance computer workstations that are dedicated to performing professional scientific, technical, or creative tasks by one individual. In the analytical realm, the term does not define the available extent of automation, technique, or workflow complexity, which can range from a simple dilution to a multistep and multitool sample preparation. An analytical workstation is widely regarded as an independent system specialized for a customized procedure performing repeatedly a set of different previously manual tasks in an automated fashion. Commonly, analytical workstations work as front-end standalone devices, but not connected or integrated to analytical instruments. Workstations use different technical realizations for automation. The standard processes are programmed and performed unattended.

The analytical processes are typically limited to liquid handling, for instance, with dilution, mixing, filtration, or extraction using a variety of syringe or (multi-) pipetting tools [22]. Open architecture workstations integrate devices such as incubators, cell counters, or centrifuges, allowing, for instance, nucleic acid purification, polymerase chain reaction (PCR) analyses, sequencing, microarray sample preparation, protein precipitation, immunoassay processing, or cell culture maintenance, just to mention a few popular applications [23].

3.2.2 Revolving Tray Autosamplers

Carousel autosamplers are the typical small automatic injection devices installed instrument top. They are most typically used in chromatography systems to deliver an individual sample from a series of sample vials to an analytical system. More than 80% of the analytical laboratories use this type of entry-level automation [24]. A popular design for GC is the arrangement of the sample and standards vials on a small eight or more position revolving carousel with a syringe in static x,y-position and only operating on the vertical axis. Rotating axes are easy and flexible in operation and control, and cost efficient by design. Low-cost simple liquid or headspace sampling systems use carousel sample trays. Often a gripper moves vials from adjacent sample racks or carousel tray of more than 100 vial capacity to a primary limited positions carousel, as shown in Figure 3.2. The concept can be extended with the installation of two injection towers on one GC system for parallel analysis. In a very similar design, also the tower rotates around an axis, making vial moves superfluous and enables access to vials on fixed positions and a large sample vial carousel. These

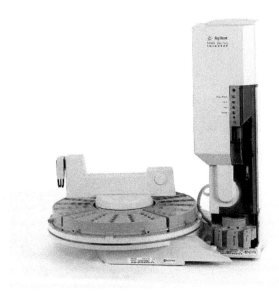

Figure 3.2 Typical carousel GC autosampler with a rotating large sample vial tray and vial transport with a gripper. Source: Courtesy Agilent Technologies Inc.

types of autosamplers are usually dedicated and limited only to liquid or headspace injection methods for gas or liquid chromatography analysis. This limited design does not allow workflow flexibility with different syringes, respectively tools and appropriate program control.

The modular extension of a carousel autosampler developed by Agilent Technologies is shown in Figure 3.3. This automatic liquid sampler comprises up to two injection towers with vial turrets and three attached x,y-oriented vial trays, delivering an increased total sample capacity of up to 150 vials. The sample trays can be heated or cooled using an external circulation thermostat. The two interchangeable towers, one used as vial transfer from the large sample rack, with 18 configurable sample, solvent, and waste turret positions, the second optional one used for up to 19 sample, solvent and waste turret positions. The vial transport to the revolving turrets is achieved by a gripper with an x,y,z-motion gantry. This two-tower concept offers a unique simultaneous injection of samples into two injectors for a parallel analysis in two independent channels for a duplicated sample throughput. Also, this concept allows several preinjection sample handling capabilities moving such typical manual tasks like dilution, mixing, heating, the reagent, or internal standard addition onto the automated platform, also offering the barcode reading and a handling of priority samples. Some sample preparation steps like liquid/liquid extraction or derivatization reactions can be realized by an icon oriented on-screen task organization. The applicable vial size is limited to the regular 2 mL autosampler vials. A flexible use of syringes up to 500 µL capacity, and programmable wash cycles, aspiration, and dispensing speeds, allows customization to different liquid handling applications. A special feature for sandwich injections addresses typical needs for the coinjection of standards, analyte protectants, or "keeper" solvents. Up to three

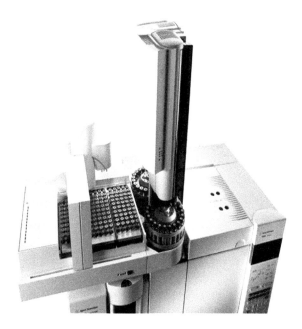

Figure 3.3 Current modular automatic liquid sampler comprising two GC injection towers with vial turrets and three attached x,y-oriented vial trays with a gantry-type vial transport. Source: Courtesy Agilent Technologies Inc.

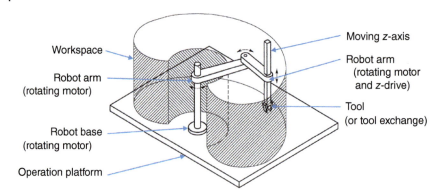

Figure 3.4 Circular workspace of a SCARA robot. Source: Redrawn from Siciliano et al. [26].

sample and solvent layers can be pulled into the syringe, separated by air gaps, for a combined injection (see also Section 5.2.1).

3.2.3 Selective Compliance Articulated Robots

Two or more revolving joints and a fixed base are the characteristics of the selective compliance articulated robot arm (SCARA) [25]. The SCARA serves a circular workspace with high motion speed on a small footprint. SCARA robots are based on serial architectures, which means that the first motor carries all other motors. Main applications are in production and assembly, but are found as automated analytical sample preparation robots as well. Although the speed of operation is a main attribute in high-throughput assembly lines, the low space requirement facilitates miniaturization of analytical sample preparation devices. Objects to be addressed are arranged in a circular area around the base on the same platform (Figure 3.4). Compared to Cartesian x,y,z-robots, the circular working space is limited to about 50% to 75% depending on the combined arm radius and size of the base mounted on the working platform.

A miniaturized automated solid phase extraction unit was developed by AiSTI SCIENCE CO. Ltd. in Japan [27]. Two robot arms with motors are operated from a base mounted centrally on the workspace platform. Each arm carries a motor. The second arm is equipped with a multifunctional tool for z-axis operation, rotation, and a coupling plate for tool exchange (Figure 3.5). Based on this operation principle, dedicated systems for the automated sample processing of pesticides extract clean-up, water, and metabolomics analysis with online GC-MS injection were derived.

3.2.4 Cartesian Robots

Robots are ideal for performing repetitive and high-volume tasks efficiently and accurately. Automated sample preparation units with capability for online injection or transfer into analytical instruments are mostly realized using x,y,z-robots, also classified as Cartesian robots. Every point in the Cartesian coordinate

Figure 3.5 Top view SCARA robot for the automated metabolomics sample processing by miniaturized SPE extraction, derivatization, and GC-MS injection by AiSTI SCIENCE CO. Ltd. Source: Photo by the author.

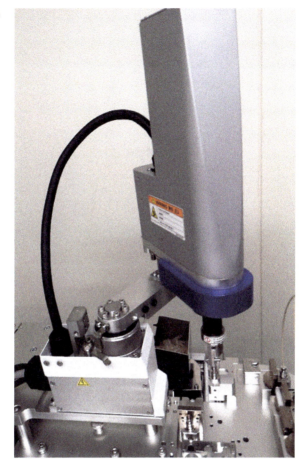

system (named after the French mathematician and philosopher René Descartes, 1596–1650) is defined by its x, y, and z-values (Figure 3.6). The principal working axes are in right angles to each other and controlled linear. The working space of Cartesian robots can be easily customized to the intended applications from just a small size to very wide dimensions. The accessible working area depends on the linear width of the three axes working together. Every point within this working space can be addressed, or excluded, by appropriate control programs. As a naming convention, the x-axis is the width of the unit from left to right, the y-axis described the depth from a front point to the back, and z is the vertical axis. A typical design of an x,y,z-robotic sample preparation system is shown in Figure 3.7. A range of tools and modules, e.g. for filtering, SPE, heating/cooling, are held available for programmable sample preparation steps. Samples are processed on racks using the typical autosampler compatible or customized vial sizes, from which extracts are transferred with suitable tools to the analytical instrument for analysis.

The x,y,z-robotic systems are used in an open-architecture design, as shown in Figures 3.7 and 3.8, and also in a boxed, closed design. The enclosed design of

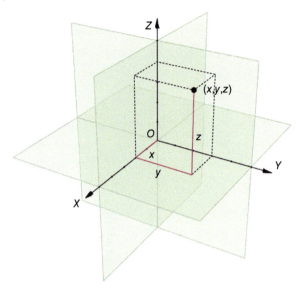

Figure 3.6 Cartesian system with linear x, y, and z axes and potential working space defined by maximum values x, y, and z. Source: Wikipedia [28].

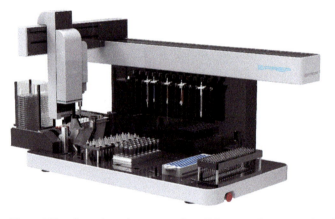

Figure 3.7 Cartesian robot system for off-line sample preparation. Source: Courtesy ePrep Pty Ltd.

Cartesian robots is the so-called gantry robot with two parallel top-mounted x-rails moving the y-beam, carrying the z-axis, in horizontal direction. This construction design is derived for laboratory benchtop stages from heavy industry portal or gantry cranes, known for instance from container terminals. In the laboratory, this structure offers a clear x,y-workspace free from obstacles for flexible positioning of any kind of objects. Processing tools are held and positioned by the z-axis for optimum utilization of the addressable workspace. Gantry robots are scalable, mostly designed as large parallel processing, and high-throughput liquid handling instruments. Built-in a closed box this compact design allows a sealed form for handling hazardous materials in low pressure with controlled filtered exhaust. Gantry systems are typically not used for sequential online injection to analytical instrumentation (if not installed inside the working space like photometers) but for

Figure 3.8 Cartesian robot system installed instrument top for online sample preparation and analyses. Source: Courtesy CTC Analytics AG.

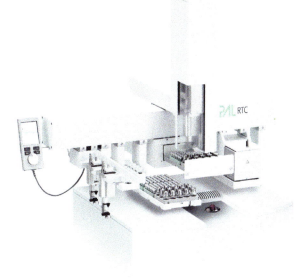

batch and parallel sample processing. Typical for x,y,z-robots is the high positional precision of less than a tenth of a millimeter in the complete working area, allowing precise handling of down-scaled objects like microvials, microplates, or MALDI targets, just to mention a few of the devices found in modern micromethods.

In the past, small x,y,z-samplers for online analysis were almost exclusively employed as liquid injection systems to serve the demand for processing high sample numbers. In recent years, the still called "autosampler" developed rapidly with extended functionality for sample preparation steps prior to analysis. Adding task flexibility for complete workflow operation with tool change capabilities and parallel operation of two towers moved this initial injection-only design into veritable x,y,z-robots. The principles of operation and use of tools and modules for analytical sample preparation applications are described in the following chapters.

With an instrument-top mounting the x,y,z-"autosamplers" quickly developed into the area of integrated robotic sample preparation devices online with analytical instrumentation. One strength is the perfect fit of the common operating timescales of sample preparation with chromatographic or direct analyses. A unique feature is the exclusive possibility for the mounting on top of analytical instruments, very often seen with a GC, as shown in Figure 3.8, or integrated with LC or other analytical systems, for online analysis. Bench space in laboratories is typically one of the most expensive in companies, and always limited. Instrument-top mounting hence is often preferred for space saving with substantial economic aspects.

A strong and constant growth in the use of x,y,z-robots has been reported. While in 1996 only 8% of respondents of a survey used automated samplers, the usage grew

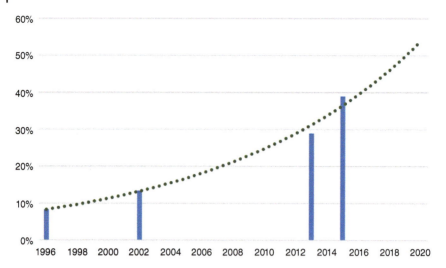

Figure 3.9 Use of automated sample preparation systems in analytical laboratories. Source: Adapted from Dahmer et al. [21].

in 2002 to 13%, and in 2013 to already to 17%, jumping in 2015 to 39% of laboratories using automated sample preparation systems [8, 9]. The currently observed strong growth, as illustrated in Figure 3.9 from the referenced surveys, allows an estimated use of automated sample preparation devices of more than 50% of the analytical laboratories in the coming years. Often, automated systems for only one dedicated preparation procedure entered the analytical industry market, for instance, the automated SPE using Cartesian robotic systems. In common here is the single application focus without a general workflow flexibility for different analytical sample preparation and analysis methods.

3.2.5 Multiple Axis Robots

One strategy for automated sample preparation workflows is the use of laboratory robots with computer-controlled arms and grippers for tools. Such articulated robots with two and more sequential rotary joints can be static or moving on rails and serve a defined operation area covering several benches, closed for human access during operation. Typical areas of application are the chemical synthesis for an increased compound variety in combinatorial chemistry, but are also considered for analytical sample preparation processes [29]. The aspect here is the automation of the continuous replication of the same preparation steps for a large number of samples. This mimics the manual process with a similar sample size and also similar consumables, glassware, or solvent volumes, as shown in Figure 3.10 [30]. After setup in the lab environment with instruments and the available tools, the robot tasks are selected from a job library to combine for the desired workflow. In this concept, the human hand is replaced and the robot should emulate the human skills (anthropomorphic) [31]. Ideal workflow tasks for articulated robots are demonstrated with

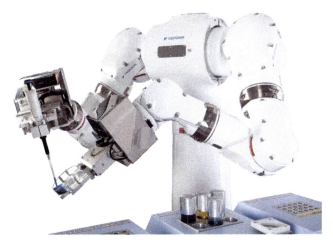

Figure 3.10 YASKAWA's dual-arm robot CSDA10F shown at the Analytica exhibition in Munich, Germany, 2014. Source: Courtesy Yaskawa Europe GmbH.

a general liquid handling, or the sample rack transport and loading to analytical instruments. High cost in the setup, programming, and method maintenance are typically involved and limit from this side the wide-ranging use in analytical laboratories, but benefit complex high-throughput environments. This solution for sure can replace human exposure in environments with chemical or biological hazards, which is also employed for sterile processing. Although working in constant quality, this concept is not a step ahead toward green analytical chemistry and the integration of miniaturized preparation processes for online analysis. Full-featured multiaxis laboratory robots are found in only few analytical laboratories and hence are not subject to further discussion in the context of this textbook.

3.2.6 Collaborative Robots

A new era started with the very recent advent of multiple axis machines as collaborative robots, in short, "cobots." Collaborative robots are designed to share the same workspace as humans for a human–robot collaboration. The collaborative approach is in contrast to the assignment of multiaxis articulated robots as autonomous systems as it is known from industrial uses, fenced to prevent any contact and accidents with workers close by. A patent application filed in 1997 by the Northwestern University in Evanston, Illinois, United States, was granted in 1999 to James E. Colgate and Michael A. Peshkin [32] describing cobots as "An apparatus and method for direct physical interaction between a person and a general-purpose manipulator controlled by a computer." Sensors make it possible, in particular for limitations in motion, speed, force, and torque. Cobots can respond immediately if they detect a human approach or contact, even more, stop movements when a distance is fallen short of a defined minimum [33]. Safety requirement features for collaborative robots are described with the international standards ISO 10218 [34] parts 1&2 and ISO/TS 15066 [35].

Technically, cobots fulfil many requirements for precise laboratory workflows and offer additional tough operations excluded for direct manual operation [36]:

- Movements and positioning with mechanical repeatability in the range of better than 0.1 mm (0.004 in).
- Tool change operation, e.g. grippers, pipetting.
- Prepared for remote control, e.g. glovebox operation in (bio) hazardous environments.
- Space-saving, lightweight design, and portable.
- Turnkey solutions concepts.
- Teaching concept for short and flexible set-up times.

Main aspects in the usage of cobots in analytical laboratories are the potential for increased precision and productivity by automation of many uniformly recurring manual tasks [37]. Most famous examples are the current applications as medical cobots in the healthcare area, for instance, as a remote operated surgical system. The ease of teaching and programming workflows by the operators at an affordable price point will facilitate the usage of cobots in research as well as routine laboratories in the near future. A current example to support sample preparation procedures is the automated powder dosing into multiple vials with subsequent automated sample processing, as shown in Figure 3.11.

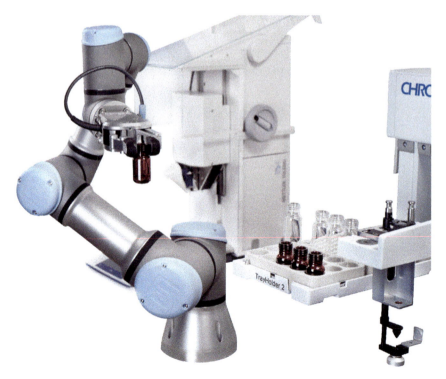

Figure 3.11 Six-axis Cobot working for automated powder dosing and online sample preparation. Source: Courtesy Axel Semrau GmbH & Co.

3.3 Sample Processing

A basic decision for every workflow concept is about the strategy to process samples in sequence, parallel, or batch-wise. Usually, the instrument hardware decides about the possibility of parallel sample processing steps. Special tools and modules, for instance, the multichannel pipettors for microplates processing, and related devices, are only available to a certain class of high-throughput preparation robots, outside the subject of this textbook. With the x,y,z-robotic sample preparation systems both, the sequential and batch-wise processing of samples can be realized, with the option for parallel tasks in a workflow with two operating heads and its hyphenation with analytical instrumentation.

With automated systems, the preparation of samples can be integrated with the following analysis on the analytical timeline. Many of the x,y,z-robots are installed dedicated to an analytical instrument for instance on GC-MS or LC-MS units. Installations can be space-saving instrument top or at-site as of most suitable space, or sample and reagent loading conditions.

3.3.1 Sequential Sample Preparation

The majority of analytical instruments work in sequential mode, sample by sample, see also Figure 3.12. The online hyphenation of sample preparation with instrumental analysis determines sequential workflow operation. From the analytical perspective, the sequential processing of samples is highly required for reproducibility. Most important here, each sample is treated on the same time scale before analysis. A concern with batch processing before analysis is avoided that the first and last samples undergo different conditions or reaction times until measurement.

3.3.2 Prep-ahead Mode

In all chromatographic analyses, there is ample time between sample injections. The instrument cycle time can be used for preparing the next sample for injection,

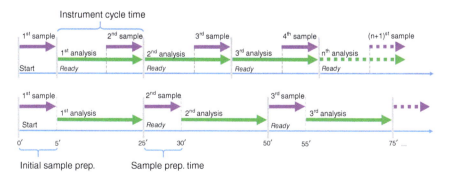

Figure 3.12 "Prep-ahead" principle for optimized instrument duty cycle. Top: Sample preparation during instrument cycle time. Bottom: Sequential sample preparation and analysis.

for even complex sample preparation steps. The principle of the preparation of a sample during the chromatographic runtime is shown in Figure 3.12, also known as the "prep-ahead" mode. With a sequential processing, as it is working with most external preparation devices, a *Ready* signal of the analysis instrument is expected to start the next sample preparation. This leaves the analysis waiting for the preparation workflow to complete. The available capacity for analysis runs is only used partially. With integrated robotic systems, the sample preparation can be executed during the analysis run. In practice, an initial sample preparation is performed with monitoring of the required processing time on the analytical time axis. With knowledge of the instrument cycle time (from *Ready* to the next *Ready* signal), the sample preparation can be scheduled to be completed at the calculated endpoint of the ongoing analysis. This ensures that all samples are treated on the same time scale without any uncontrolled waiting times for analysis. This "look-ahead" mode secures high reproducibility for all samples in an automated sequence. In addition, on an economical aspect, such prep-ahead automation leads to the most efficient instrument duty cycle for high sample throughput in the shortest time.

3.3.3 Incubation Overlapping

The productivity efficient incubation of a series of sample vials in an agitator/incubator with a limited number of positions requires a dedicated scheduler. The analytical requirement here is the constant and equal incubation time for all samples without waiting for injection. For a high sample throughput the optimum usage of the limited number of incubator vial positions is desired.

The overlapping of sample incubation in automated x,z,y-robotic systems is handled by a dedicated scheduler. When executed in a first run of a sample sequence, the scheduler monitors and records all activities in its time duration. Some activities have a set programmed duration, e.g. wait or incubation times, other activities have initially undetermined run times, e.g. the user-defined chromatographic separation or equilibration times.

All activity time slices together define the total cycle time of a sample run until a next *Ready* signal from a GC or LC system is received. An example of the timing information in action slices is shown in Figure 3.13. After the first run in a sample sequence is scheduled, the next samples in series are overlapped to get the completion of the incubation step aligned with the *Ready* signal of the GC or LC system, independent on the chromatographic separation conditions. Figure 3.14 illustrates how the scheduler uses the vial capacity of the incubator to its maximum depending on the required incubation time.

3.3.4 Batch Processing

The batch-wise processing of samples is mostly found with standalone robotic systems. Here a set of samples is processed and prepared for analysis. The final extract is provided in separate racks, which ideally can be transferred to an autosampler installed on top of the analytical instrument. The same sample preparation workflow

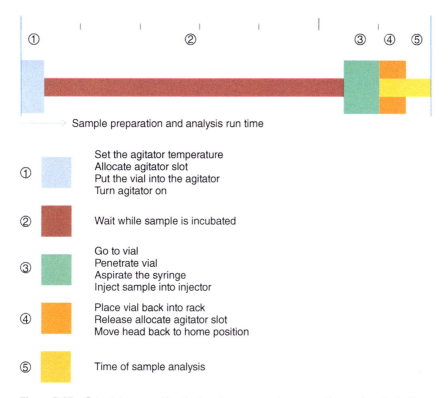

Figure 3.13 Scheduler operation for headspace sample preparation and analysis. Source: Courtesy CTC Analytics AG.

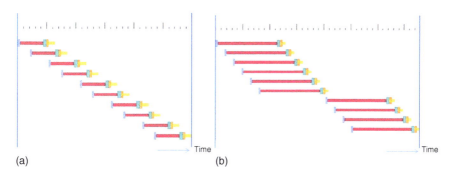

Figure 3.14 Overlapping with different incubation times. (a) Regular incubation times (red) in a continuous sequence, (b) Long incubation times (red), exceeding the capacity of 6 positions in the incubator uses a batch overlapping. Source: Courtesy CTC Analytics AG.

applies as with the sequential processes shown in Figure 3.12. The sample processing does not wait for the analytical instrument to be ready for analysis. A significantly higher sample throughput can be achieved for preparations. Such sample logistics is advantageous for laboratories with multiple instrument capacity using one or more dedicated preparation robots serving several analysis units for similar kinds of sample types.

3.3.5 Parallel Processing Workflows

Many x,y,z-robots are equipped with two heads for tasks with different tools (Figure 3.15). The heads share parts or the full length of the x-axis for operation. In this way, sample vials can be accessed from both heads treated with reagents, solvents, or transported to modules as the programmed workflow requires. Time intense steps like incubation, weighing, or centrifugation do not halt the processing and allow the parallel execution of tasks, even for the following sample. An important aspect for increased sample throughput is the reduced number, or for optimum workflow logistics, the prevention of repeated changes of the most engaging tools, and the reduction of travel distances with the heads in multimodule configurations.

Due to their extended working space, dual-head robotic systems facilitate the injection to multiple detection channels when installed over several analytical instruments for instance with several GC or GC-MS units. For LC analyses parallel and staggered LC, injections can be accomplished by a configuration with parallel injection ports on the instrument rail.

3.3.6 Sample Identification

3.3.6.1 Barcodes

Sample vials can be labeled with the barcodes provided at sample entry to the lab. This allows unequivocal identification and the traceability of the sample related analytical results. With appropriate scanners and data acquisition software, the monitored barcode is taken into the sequence table of the chromatography data system (CDS) and assigned to the acquired data. Usually, linear barcodes, also known as 1D barcodes (one-dimensional), are in use, which can be scanned by optical laser scanners (class 1 laser devices are not considered to be hazardous when used for their intended purpose). Often two scanners working simultaneously from two sides.

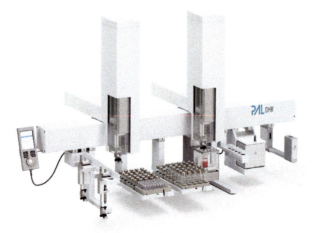

Figure 3.15 Dual-head x,y,z-robotic system for parallel workflow operation with shared tools and modules. Source: Courtesy CTC Analytics AG.

The barcode scanner developed by GERSTEL GmbH & Co KG reads 1D and 2D barcodes and uses a dual-camera setup with image analysis for the positive identification of samples [38].

Many different barcode types are in use for a variety of applications we know for instance with grocery product identification. Barcode types differ in length with start and stop characters. The supported types of barcodes compatible with the reader and decoder of a given robotic sampler and data system may differ. It is required to reference the manufacturer product specifications. In general, barcodes work with all types of vials in use, namely the standard 2, 10, and 20 mL vials. For laboratory use, the barcode labels should be made of thin polyester foil, and not paper. Polyester labels can withstand the high temperatures of the incubator/agitator with up to 200 °C. Thermo-transfer printers with a resolution of 300 dpi or higher have proven the most reliable. Caution is required for the thickness of the label material when inserting, e.g. into vial racks. Thick labels may keep vials stick to devices and hinder transport. In the decapper unit, multiple use of labeled vials, for instance, the labeling of standard and reference material vials, may affect label clarity for reading from the fixing claws and need to be checked regularly. For reliable barcode reading, the optimum position on vials is shown in Figure 3.16. The maximum tilt of the barcode label should not exceed ±20°. Labeling the vial horizontally around the vial body prevents most scanners to identify the barcode correctly, or requires an additional vial turning mechanism.

With barcodes or radio-frequency identification (RFID) vial labeling, there is future potential to choose or modify the sample preparation workflow for different sample or analysis methods. Such concepts lead to the next level of automated sample processing with upcoming "smart" decision-making data systems. For instance, a barcode-driven control of the sample analysis sequence is offered by GERSTEL Gmbh & Co KG for a robotic autosampler with 2D barcode reader. The barcode labeling enables an automated analysis by predefined methods and the automated generation of sequence tables [39]. Manual transfer errors are excluded, the manual

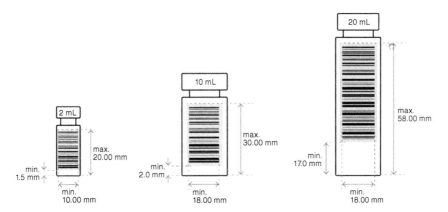

Figure 3.16 Recommended 1D barcode labeling on different vial sizes. Source: Courtesy CTC Analytics AG.

time consuming and error-prone sequence setup is even not necessary any more. The barcode of a sample vial can trigger at any time the start of a user-defined label-coded analysis workflow. This can activate a priority sample, blanks, or QC standards. The "Sequence by Barcode" feature works in batch (reading all vial barcodes at a time) or sequence mode (sequential barcode reading). Data import and export functions enable LIMS or database synchronization.

3.3.6.2 Radio-Frequency Identification Chips

Another means of providing a comprehensive sample identification and information is possible with the use of programmable RFID chips beside the standard barcodes as shown in Figure 3.17 for thermal desorption tubes. Markes International Ltd. introduced the tagging of sorbent tubes with RFID chips to be used for steel and glass tubes likewise [40]. Reading and writing information to/from tags can be customized. The sample information is recorded from the automated thermal desorption unit and can be stored in the sequence history. A tube-specific mode keeps the basic tube information such as sorbent materials and serial number. It allows tube specific information, such as sorbent type or the number of thermal cycles, to be associated with the tube for its lifetime and automatic tracking of sample information from sampling through to analysis. The use of RFID chips eliminates barcode reading errors and facilitates the chain of custody through field monitoring to lab

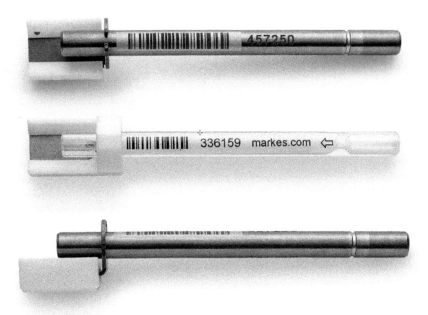

Figure 3.17 Thermal desorption tubes RFID and barcode labeled. Source: Courtesy Markes International Ltd.

analysis. The usage of tube ID tags is recommended by key standard methods such as US EPA Method 325.

3.4 Tool Change

Analytical flexibility for automated processing requires the frequent change of tools within a programmed workflow, in particular for sample preparation steps. Tools can serve many tasks comprising different volume syringes, pipettes or dilutors for liquid handling, or GC and LC injection. Also, grippers for transport, or the change of the analytical method for instance from a static to a dynamic analyte collection, are available tool solutions, just to mention the most applied ones. For analytical purposes, this request for tool flexibility to perform different sample preparation tasks was voiced for years.

3.4.1 Manual Tool Change

A tool exchange for an automated sampling system was first implemented in the analytical industry in 2009 with the Combi-PAL [41] by CTC Analytics AG for a manual exchange of a syringe holder, as shown in Figure 3.18. This design allows with only a few hand movements the change of the applied analytical method without the

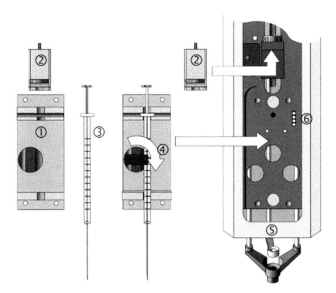

Figure 3.18 Syringe plate for manual syringe exchange in a programmable autosampler. 1 Syringe exchange plate, 2 Plunger button holder, 3 Syringe (different barrel OD requires suitable plate), 4 Syringe lock, 5 Head of robotic sampler, 6 Plate code reader. Source: Courtesy CTC Analytics AG.

installation of a new instrument. With such a feature, for instance, the change from liquid injection to SPME can be accomplished without much downtime. A syringe plate holds the required syringe by locking the barrel (Figure 3.19). The plunger tip is clamped into the system plunger drive. The syringe plate is manually placed (and exchanged) into the head of the robotic system. Optionally, the plate can be coded to the installed syringe for automatic recognition of a correctly installed tool as it is required by the program to be executed.

3.4.2 Automated Tool Change

The automatic tool change expands the system flexibility significantly for the execution of many different activities in one workflow. Complex analytical methods with programmed access to the required tools and tasks can be realized.

A robotic tool changer, also named as automatic tool changer (ATC), consists of two mating parts designed to couple and lock safely for high reliability, mechanical precision, and repeatability. Often necessary, electrical signals for a positive identification and operation are exchanged, media transfer for gas or liquids is provided, or any kind of mechanical actuation is enabled. Robotic tool changers are an installed standard today in particular on industrial robots to change grippers, welding devices, or other manipulators to perform multiple tasks. The coupling of tools for analytical purposes must maintain the high mechanical precision known from regular liquid autosamplers. An industry standard coupling mechanism between the tool and the robot head is based on ball-lock technology. Similar coupling mechanisms are used in heavy duty industrial robotic systems.

For x,y,z-robots, the tool connectivity is typically in z-orientation. A common connection part is used on the top face of the tools. The ball-lock mechanism

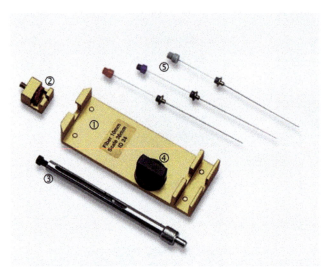

Figure 3.19 SPME syringe adapter for manual exchange. 1 Syringe exchange plate, 2 Plunger button holder, 3 SPME Tool, 4 Syringe lock, 5 SPME fiber assemblies. Source: Courtesy CTC Analytics AG.

with spring-loaded, pneumatically or actuator-driven actuation ideally allows the quick coupling and decoupling with the master connector located in the head of the robot. Several steel balls are positioned in a ring inside of the master coupling serving as tool locks. An actuator working in z-direction drives the coupling and decoupling of the tool. With the tool head in the coupling master, the actuator is moved upward, and the spring-loaded lock device forces the set of steel ball into the specially designed notch of the tool head (Figure 3.20). This mechanism maintains the precise centering of the tool during the coupling process for the required high positional x,y-accuracy for tool operation. It provides a highly resilient on/off connection without any play in conjunction with the tool. It is self-centering and of high lifetime [42].

The operation of syringes, pipettes, or other tools within a robotic system requires auxiliary connection points beyond the central tool connector (Figure 3.21). One additional z-axis actuator is required for the syringe plunger operation. A third one

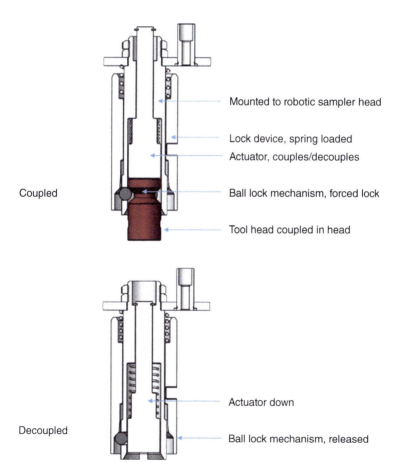

Figure 3.20 Ball-lock mechanism for tool change in a robotic autosampler. Source: Courtesy CTC Analytics AG.

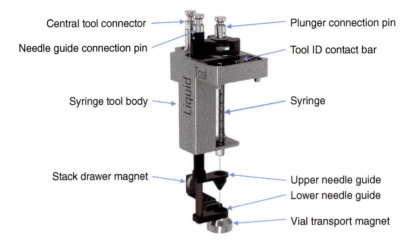

Figure 3.21 Exchangeable liquid tool for a robotic autosampler. Source: Courtesy CTC Analytics AG.

serves with syringes as a needle guide, or is used as the pipette tip stripper. Similar functions can be implemented, for instance, to manipulate the gripper claws. While the tool connection is static and keeps the tool in the robot head, the plunger and needle guide pins connect to a motor-driven vertical linear motion actuator each. Tools for several tasks are held ready in tool change or park stations for exchange when required by the programmed workflow.

3.4.3 Tool Identification

Robotic sample preparation systems use dedicated tools to accomplish workflow tasks. In many x,y,z-robots, the tools in use can be changed manually but most often they change automatically often even several times during a complex preparation workflow. It is necessary to identify the correct tool, but it is also required to provide necessary operation parameters of a tool to the robot. If there is a park station for tools involved, the location of the required tool needs to be transmitted. Active tool identification can be accomplished, as shown in Figure 3.21. A contact bar that allows the robot to read out the tool ID also serves to provide power to the tool and receive sensor feedback as needed for specific tool functions. A contact bar on the bottom side of the tool connection plate connects to an active tool park station and transmits ID and position when not coupled to the working head.

In general, a large number of attributes and important information is provided with the tool identifier. A unique identification number, the maintenance counters for tool couplings, and plunger movements of the actual syringe plunger are examples of stored and transmitted operation-related information. Also coded is the syringe class and dimensions with a syringe tool, e.g. the barrel diameter and needle length. In the case of a barrel diameter of 6.6 mm and needle length of 57 mm, the tool can be used for syringes with volumes of 1.2, 5, 10, 25, and 100 µL, all with a needle length of 57 mm. Based on this syringe and tool profile correct volumes during liquid handling can be aspirated and dispensed. Also, the precise movement calculation of differently equipped tools, e.g. vial, valve, or injector penetration is achieved.

A tool of the robot can for instance carry many different syringe types. An almost infinite variety of syringes is available and is also in use in automated systems. Besides the volume and needle lengths, more information about the plunger type, scale length, needle tip style, needle guide, maximum aspiration, dispense speeds and also the intended use for liquid handling, static or dynamic headspace, and more are required for correct operation in practical applications. The parameter set also serves as tool protection in case of out-of-range operation requests for instance with heated tools. The syringe specifications are provided either by manually logging in a syringe code provided with the particular syringe or is read from an embedded chip in the syringe head.

The so-called "smart syringe" concept introduced by CTC Analytics AG for robotic sample preparation systems reduces any potential input errors for enhanced process safety. A chip located in the syringe plunger coupling head shown in Figure 3.22 carries all syringe-related information. The complete individual syringe parameter set for operation is loaded automatically by the robot at the coupling of the liquid tool. The "smart syringe" concept extends the information exchange with the robot beyond syringes also for other devices like SPME Fiber, SPME Arrow, and static and dynamic headspace tools for further reduction of usage errors for improved operation safety. Figure 3.23 shows the color-coded SPME devices with embedded ID chip in the plunger coupling head. In these applications, the smart chip also logs the key parameters during workflow processing with a full operation tracking for preventive maintenance and support purposes. Parameters like a coupling or plunger movement counter for information about the frequency of usage are important in the automated environment to keep up with preventive maintenance procedures. Additionally, such operation related parameters can be monitored, updated, and stored with the tool for usage history and traceability. A selection of stored parameters with and without user access is listed in Table 3.2 highlighting, in particular, the process safety tool parameters and user information for liquid and headspace syringes.

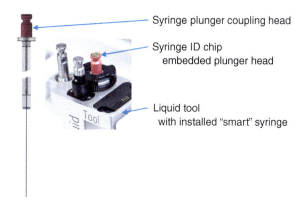

Figure 3.22 Smart syringe plunger head with identification chip. Source: Courtesy CTC Analytics AG.

Syringe plunger coupling head

Syringe ID chip embedded plunger head

Liquid tool with installed "smart" syringe

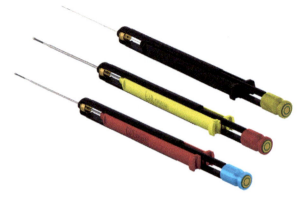

Figure 3.23 SPME Fiber and Arrow devices with "smart" ID chip. Color Code of SPME Holders: Black/Black: SPME Fiber; Black/Yellow: SPME Arrow 1.5 mm OD; Black/Red: SPME Arrow 1.1 mm OD; The colors of the SPME plunger heads indicate the type of sorbent phase material and thickness of the device: Red, PDMS 100 µm; Silver, Polyacrylate 100 µm; Blue, Carbon WR/PDMS 120 µm; Purple, DVB/PDMS 120 µm; Grey, DVB/Carbon WR/PDMS 120 µm; Black, PDMS 250 µm. Source: Courtesy CTC Analytics AG.

The usage parameters can be transmitted to the external control software or CDS for providing user guidance for compliance checking with the relevant SOP or maintenance schedule. This way a scheduled sample sequence can check the usage history and provide directions on-screen before the start if for instance the syringe or SPME device needs to be replaced in case it will exceed the permitted sample number during the planned analysis sequence. A high degree of inherent process safety for unattended automated procedures is realized with the appropriate usage of the tracking information by the controlling environment.

3.5 Object Transport

The activity "Transport Vial" is a very typical task for robotic systems. Vials are stored in dedicated racks on tray holders and need to be moved to various modules within the workflow for tasks like heating/cooling, equilibration, mixing, vortexing, weighing, barcode reading just to mention a few of the most used transport targets. Other objects needing transport within the workspace are well plates with a dedicated transport tool.

3.5.1 Magnetic Transport

The transport of vials is most simply accomplished using magnetic vial caps and tools furnished with magnets. The vial release is achieved by a sideward movement of the tool after placing the vial into its destination position. A common solution is the installation of rings with magnets to the lower needle guide of the tool in use. A

Table 3.2 Tool parameter information and tracking for process safety and selection.

Tool parameter	Display	User edits	User benefit
Syringe status	OK – Expired – Defective	Yes	Only use valid syringes
Syringe type	Type of syringe	No	Use of the correct syringe type
Strokes count	n	No	Direct PM measures
Max. strokes count	n	Yes	Quality assurance parameter
Remaining strokes	n	No	Check fitness for next sample series
First usage	dd/mm/yyyy	No	Quality assurance parameter
Last usage	dd/mm/yyyy	No	Traceability
Expiry duration (days)	n days	Yes	Quality assurance parameter
Duration 50 °C - 100 °C	s	No	Operation safety parameter
Duration 100 °C - 150 °C	s	No	Operation safety parameter
Duration >150 °C	s	No	Operation safety parameter
Syringe volume	µL	No	Quality assurance parameter
Needle length	mm	No	Operation safety parameter
Point style	Conical, LC, sideport	No	Operation safety parameter
Plunger type	Steel, Teflon, other	No	Operation safety parameter
Injection penetration count	n	No	Quality assurance parameter
User description 1	Free text	Yes	Provides user information on workflow use
User description 2	Free text	Yes	Provides user information on workflow use

Add. headspace only parameter

Main process safety parameter

Figure 3.24 Magnetic screw caps for 20 and 2 mL vials. Source: Courtesy CTC Analytics AG.

small ring of magnets in a dedicated lower needle guide is used for the classical 2 mL autosampler vials of different shapes. Larger magnetic rings enable the transport of 10 and 20 mL capped vials, the typical headspace vials, as shown with a syringe tool in Figure 3.21. This allows the transport of vials with magnetic screw or crimp caps into almost any destination without mechanical limitations (Figure 3.24). Magnetic screw caps are preferred over crimp caps for a rectangular flat vial top. This guarantees a reliable vertical position of the vial during transport and rack positioning. Crimp caps can easily be over-crimped and skewed with the risk of collisions when placing the vial into a target object. The magnetic rings can be removed or replaced with different dimensions manually or by a dedicated magnet stripping station to fit the intended workflow.

Limitations can arise from the magnetic force concerning the vial weight of a filled vial, e.g. with fully loaded headspace vials. Large and heavy vials cannot be transported magnetically. Also, some space as strip distance next to the vial is required after positioning the vial and release by a programmable horizontal x- or y-movement of the transporting tool. This movement needs to be considered when designing custom racks for different vial sizes. Vials that cannot be transported magnetically require the gripper transport.

3.5.2 Gripper Transport

Grippers are very versatile tools for the transport of different objects that do not necessarily need to or cannot be equipped with magnetic caps. Even larger, heavier, and longer objects, for instance, the tubes used in the benchtop nuclear magnetic resonance spectroscopy (NMR), flip tubes, or microplates can be handled reliably using grippers. Limitations in the operation can arise from the available jaws used and gripper dimensions for different vial sizes, the parallel or trapezoidal movement of the jaws, and the necessary access to space-limited modules or devices. Depending

Figure 3.25 Gripper tool with parallel jaw movement for the transport of objects. Source: Courtesy CTC Analytics AG.

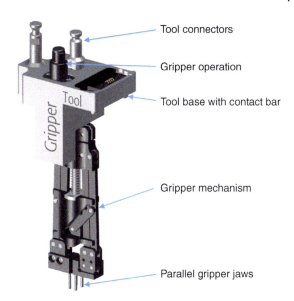

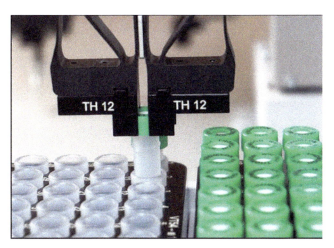

Figure 3.26 Customized Grabber with 12 mm jaws operating Thomson filters. Source: Courtesy Brechbühler AG.

on the type of jaws used, there is space required for the jaw opening movements. The recommended diameter of vials for transport is in the range of 8–20 mm for the gripper, as shown in Figure 3.25. The mechanical limitations need to be checked for the intended operation. In many cases, the jaws of a gripper tool can be exchanged to suit particular vial or object formats, as shown in Figure 3.26.

The handling of deep well plates requires special gripper dimensions for this standardized format, as shown in Figure 3.27. This allows the transport of any kind of well plates, in particular for the automated online processing of large life science sample series by LC–MS with integrated sample preparation steps by dilution, shaking, heating, or centrifugation.

Figure 3.27 Grabber for the transport of deep well or microplates. Source: Courtesy Brechbühler AG.

3.5.3 Needle Transport

It is possible to transport light objects after penetration hanging on a syringe needle. Additional tools other than the used syringe tool are not required. But, objects like vials, microvials, or cartridges need to be capped with a septum as used for the regular 2 mL autosampler vials. For transport, the syringe needle penetrates the septum and lifts the object with the syringe tool. If the weight is not too high, the object sticks well to the needle. Figure 3.28 illustrates the needle transport used for the move of a micro-SPE cartridge (µSPE) to an elution tray. The inner design of the cartridge with a narrow bore for the needle guarantees the upright orientation for insertion of the cartridge into racks. The shown needle guide pushes down for stripping off the cartridge from the syringe needle at the destination.

3.6 Vial Decapping

Many sampling or dispensing operations need to avoid the penetration of septa. Once septa are penetrated, they do not seal anymore well and pose a risk for loss

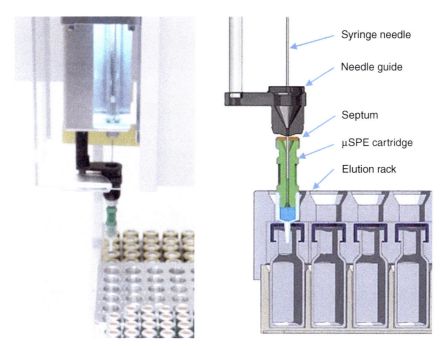

Figure 3.28 Needle transport for a micro-SPE cartridge. Source: Courtesy CTC Analytics AG.

of volatiles, solvents or become even a source of contamination with siloxanes from the soft septum material, in particular with the use of PTFE-coated septa. Also, applications working with a pipetting tool require the removal of the vial caps.

Technically, a decapper device for screw-capped vials comprises two parts, as shown in Figure 3.29. A bottom part keeps the vial with clamping jaws in position. The jaws in the bottom and top parts are moved radially and allow the handling of different vial and closure sizes. A top module attaches to the vial closure, clamping and turning the vial closure. During turning the closure, the top part moves upward, lifting the closure and keeping it inside. A further backward move allows access to the opened vial from the vertical axis while keeping in the bottom part. Torque control during the vial closure step is essential to guarantee the leak-free closure of different vial sizes. Especially for those vials for repeated decapper operation, the usage of metal screw caps with inside polymer coating of the thread is advantageous to avoid progressive particle shift from the glass thread of the vial.

The typical applications for the decapper in workflows are related to the handling of bulk reference standards, or stock solutions, in general for all activities involving dilutions, adding solvents, reagents, or internal standards. In most applications, the dilutor or pipetting tools are applied for liquid handling, especially syringes that work equally well for small volumes.

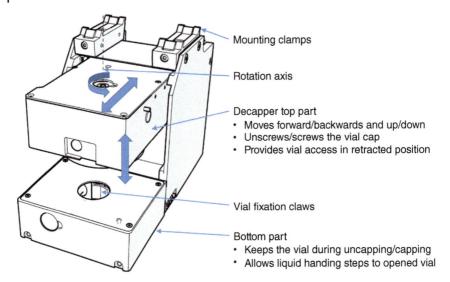

Figure 3.29 Decapper for screw cap vials. Source: Courtesy CTC Analytics AG.

References

1 Perkins, R. (2018). *Automation 101 – The Case for Automated Sample Preparation in Analytical Laboratories.* Environmental Technology www.envirotech-online.com.
2 Fleischer, H. and Thurow, K. (eds.) (2017). *Automation Solutions for Analytical Measurements.* Weinheim, Germany: Wiley-VCH Verlag GmbH & Co. KGaA https://doi.org/10.1002/9783527805297.
3 Pfannkoch, E. (2013). Sample preparation for chromatography: how much can be automated? *LCGC North America* 31 (3): 34–39. www.chromatographyonline.com/print/201119?page=full&id=&sk=&date=&pageID=2.
4 Shimadzu (2015). Supercritical Fluid Extraction/Chromatograph System Nexera UC Product Information, Shimadzu Corporation, Kyoto, Japan.
5 Liang, Y. and Zhou, T. (2019). Recent advances of online coupling of sample preparation techniques with ultra high performance liquid chromatography and supercritical fluid chromatography. *Journal of Separation Science* 42 (1): 226–242. https://doi.org/10.1002/jssc.201800721.
6 LCTech (2017). *Automated Sample Preparation.* Obertaufkirchen, Germany: Product Information, LCTech GmbH.
7 Teledyne Tekmar (2018). *AutoMate Q40.* Mason, OH: Teledyne Technologies Inc.
8 Majors, R.E. (2013). Trends in Sample Preparation, Chromatography Online, March 29.
9 Raynie, D.E. (2015). Trends in Sample Preparation. *LCGC Europe*, March 2015: 142–152.
10 Wittsiepe, J., Nestola, M. et al. (2014). Determination of polychlorinated biphenyls and organochlorine pesticides in small volumes of human blood

by high-throughput. *Journal of Chromatography B* 945–946: 217–224. https://doi.org/10.1016/j.jchromb.2013.11.059.

11 Zwagerman, R. and Overman, P. (2019). Optimized analysis of MCPD- and glycidyl esters in edible oils and fats using fast alkaline transesterification and ^{13}C-correction for glycidol overestimation: validation including interlaboratory comparison. *Europ. J. Lipid Sci. Tech.* 121 (4): 1800395. https://doi.org/10.1002/ejlt.201800395.

12 Zarate, E., Boyle, V. et al. (2016). Fully automated trimethylsilyl (TMS) derivatisation protocol for metabolite profiling by GC-MS. *Metabolites* 7 (1): 1. https://doi.org/10.3390/metabo7010001.

13 Meyer, V.R. (2019). Sampling: the ghost in front of the laboratory door. *LCGC North America* 1: 768–774.

14 Lehotay, S.J., Michlig, N., and Lightfield, A.R. (2020). Assessment of test portion sizes after sample comminution with liquid nitrogen in an improved high-throughput method for analysis of pesticide residues in fruits and vegetables. *J. Agric. Food Chem.* 68 (5): 1468–1479. https://doi.org/10.1021/acs.jafc.9b07685.

15 Pena-Pereira, F. (ed.) (2014). *Miniaturization in Sample Preparation*. Warsaw/Berlin: De Gruyter Open Ltd.

16 Bundesamt für Gesundheit (2000). Untersuchungsmethode 269.1. In: *Schweizer Lebensmittelbuch*, 1–11. Liebefeld, Switzerland: Bundesamt für Gesundheit.

17 Suter, B., Grob, K., and Pacciarelli, B. (1997). Determination of fat content and fatty acid composition through 1-min transesterification in the food sample; principles. *Z. Lebensm. Unters. Forsch. A* 204 (4): 252–258. https://doi.org/10.1007/s002170050073.

18 ISO 12966-2. (2017). Animal and vegetable fats and oils – gas chromatography of fatty acid methyl esters. Part 2: Preparation of methyl esters of fatty acids. www.iso.org/standard/72142.html (accessed 2 November 2019).

19 AOAC International. (2012). AOAC official method 2012.13: Determination of labeled fatty acids content in milk products and infant formula. *AOAC International* 2012: 1–9. http://stakeholder.aoac.org/SPIFAN/2012.13.pdf.

20 AOAC (2002). AOAC official method 996.06 fat (total, saturated, and unsaturated) in foods. *AOAC International* 41 (1): 28A.

21 Dahmer, M.L., Fleming, P.D. et al. (1989). A rapid screening technique for determining the lipid composition of soybean seeds. *JAOCS* 66 (4): 543–548.

22 EPrep (2018). *EPrep Sample Preparation Workstation*. Mulgrave, Australia: Eprep Pty Ltd.

23 Vorberg, E. (2015). Process automation for analytical measurements providing high precise sample preparation in life science applications. Thesis, Faculty of Computer Science and Electrical Engineering, University of Rostock, Germany.

24 Raynie, D.E. (2015). Trends in Sample Preparation. *LCGC Europe, March* 2015: 142–152.

25 Wikipedia. (2019). *SCARA*. https://en.wikipedia.org/wiki/SCARA (accessed 24 October 2019).

26 Siciliano, B., Sciavicco, L. et al. (2009). *Robotics*. London: Springer. ISBN: 978-1-84628-642-1.
27 AiSTI SCIENCE. (2017). *Fully Automatic Solid Phase Extraction System ST-L400*. Product Brochure, AiSTI SCIENCE CO., Wakayama, Japan.
28 Cartesian coordinate system. Wikipedia (accessed 20 April 2018) https://en.wikipedia.org/wiki/Cartesian_coordinate_system
29 Chu, X. et al. (2016). Automated sample preparation using a dual arm robot. *Am. Lab.* 2: 2016.
30 Yaskawa. (2018). See operational concept at www.yaskawa.eu.com/en/solutions/industry/labautomation (accessed 20 April 2018).
31 Fleischer, H. and Thurow, K. (2018). *Automation Solutions for Analytical Measurements*, 78. Weinheim, Germany: Wiley-VCH.
32 Colgate, J. E., and Peshkin, M.A. (1999). *Cobots*. US Patent 5,952,796, filed 28 October 1997.
33 Fraunhofer IFF (2016). *Mensch–Roboter–Kollaboration*. Germany, Fraunhofer IFF: Magdeburg. www.iff.fraunhofer.de.
34 ISO 10218-1. (2011). *Robots and robotic devices – Safety requirements for industrial robots – Part 1: Robots. Part 2: Robot systems and integration*. International Organization for Standardization.
35 ISO/TS 15066. (2016). *Robots and robotic devices – collaborative robots*. International Organization for Standardization.
36 Universal Robots (2019). e-Series from Universal Robot – World's #1 Collaborative Robot. Product Information EN/2314.11.2019, Universal Robots A/S, Odense, Denmark.
37 Tobe, F. (2017). 42 Companies empowering robots and humans to work side-by-side. *The Robot Report* 30.
38 GERSTEL (2016). *Barcode Reader Sample ID for MPS Robotic*. Mülheim an der Ruhr, Germany: GERSTEL GmbH & Co KG www.gerstel.com/pdf/Barcode_SID_robotic_Spec_en.pdf.
39 GERSTEL (2020). *Sequence by Barcode: Generating sequence tables from barcodes*. Product information. GERSTEL GmbH & Co KG, Mülheim an der Ruhr, Germany. http://www.gerstel.com/en/MAESTRO_Sequence-by-Barcode.htm (accessed 4 June 2020).
40 Markes (2019). "TubeTAG". Markes International Ltd., Product Information. www.markes.com (accessed 29 October 2019).
41 CTC Analytics (2019). PAL System History. CTC Analytics AG, Zwingen, Switzerland. https://www.ctc.ch/index.php?id=142 (accessed 12 July 2019).
42 Salois, G. (2013). Flexible manufacturing – utility couplers and tool changers safely save time. *FF Journal*. https://www.ffjournal.net/item/11315-flexible-manufacturing.html.

4
Analytical Aspects

To describe the relationship between analytical chemistry and automation is at the same time to describe the present position of both these fields, each of which strongly influences the other.

Hanns Malissa [1]
Austrian analytical chemist (1920–2010)

4.1 Liquid Handling

Syringe, dilutor and pipetting tools play a key role in liquid handling during automated sample preparation. The main aspect here is the flexible volume dosing with high precision for liquid samples, reagents, standards and solvents. Limitations arise from the materials in use, the reactivity of the agents, vapor pressure, or viscosity. Repeated usage requires appropriate cleaning steps, which have to be taken care of with suitable measures using appropriate wash stations and wash solvents in the design of the workflows.

4.1.1 About Drops and Droplets

In manual and even more in automated liquid handling, we need to avoid droplet formation during the dispensing process at the needle or pipet tip by suitable operation behavior, e.g. touching the vial wall with the pipet tip for dispensing, or setting an appropriate dispensing speed. Droplets (or pendants) hanging on a tip and not delivered into the target vial lead to lower accuracy and precision. Also, carryover due to smear contamination of septa can occur.

A drop or droplet is a small volume of liquid, bound to completely or almost completely to a surface. The volume of potential drops or droplets is not constant and depends on the diameter of the needle or tip orifice, the viscosity, density, the surface tension of the liquid transferred, and the strength of the gravitational field. The maximum mass of a pendant drop formed, then falling off as a drop, is given with Eq. (4.1).

$$m = \pi \cdot d \cdot \gamma \cdot g^{-1} \tag{4.1}$$

Automated Sample Preparation: Methods for GC-MS and LC-MS,
First Edition. Hans-Joachim Hübschmann.
© 2022 WILEY-VCH GmbH. Published 2022 by WILEY-VCH GmbH.

with

d = tube diameter
γ = surface tension of the liquid
g = gravity acceleration

A drop may form when liquid flows slowly from the lower end of a vertical tube of small diameter. The surface tension of the liquid causes the liquid to hang from the tube, forming a pendant, see Figure 4.1. When the drop exceeds a certain size, it is no longer stable and detaches itself [2]. Droplet volumes for water and hexane potentially forming from a syringe with different needle gauge values are given in Table 4.1. The difference in the maximum droplet volume is mainly caused by the very different surface tension than the needle gauge. Knowing about the formation of drops at syringe needles or pipetting tips allows setting the suitable speed parameter for dispensing and avoiding droplet formation. See the default and maximum syringe dispensing speeds in Table 4.2.

4.1.2 Syringes

Microliter syringes differ besides the volume in many aspects. The barrel, plunger and needle sizes, and materials are of particular importance for the many ways of analytical use. The syringe components are labeled and depicted in the schematics of Figure 4.2.

Figure 4.1 Droplet formation at the end of a tube. Source: Wikipedia [2].

Table 4.1 Droplet volumes from a syringe with different needle gauges.

Solvent	Needle [gauge]	ID [mm]	Droplet volume [µL]
Water	23s	0.116	0.27
	26s	0.127	0.30
Hexane	23s	0.116	0.10
	26s	0.127	0.11

4.1 Liquid Handling | 57

Table 4.2 Default and maximum syringe dispensing speeds.

Syringe volume [μL]	Needle [gauge]	Default setting [μL/s]	Maximum dispense speed [μL/s]
10	23s, 26s	20	100
100–250	22, 22s, 23s, 26s	100	900
500	23	100	1000
2500–5000	19	300	2000

Source: Courtesy CTC Analytics AG.

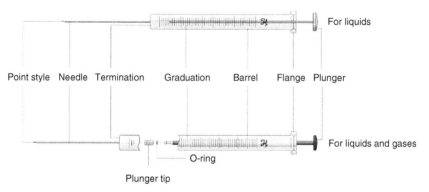

Figure 4.2 Microliter syringe schematics. Source: [3] Courtesy Hamilton Bonaduz AG.

Syringes are not only indispensable for the injection of micro-volumes to GC and LC instrumentation but also often used for liquid handling for solvents and liquid reagents. Microliter syringes are used for liquids only, especially for precise small volume dosing. Microliter syringes are available in a wide range from 1 to above 10 000 μL volume. Typical and practical working ranges are up from 10 to 100 μL volumes for dosing and injection purposes, and up to a few 1000 μL for liquid transfers. Large volume syringes are in use for gas sampling but avoided for liquid handling due to the extended aspiration times through low diameter needles, the high solvent need for washing steps, and a risk of dripping during movements due to the weight of the liquid. For fast and reliable large volume liquid handling, it is advisable to replace high volume syringes with a dilutor tool comprising different high volume syringes (see Section 4.1.5).

4.1.2.1 Precision and Accuracy

The precision of syringes varies slightly with the dosing volume. Syringe manufacturers advise not to use syringes for volumes below 10% of the nominal volume. "For accuracy and precision, the smallest dispensing volume for a given syringe should be greater than or equal to 10% of its total capacity. For example, the smallest dispensing volume recommended for a 10 μL syringe is 1 μL" [3].

In an automated workflow, this requirement for precision is managed by restricting the allowed working range of syringes to values from 10% to 100% of the nominal volume only. Typical operating volume precision and accuracy of microliter syringes are given with 1% measured at 80% of the nominal volume [3]. The repeatability of liquid volume dosing depends on the reproducible operation of the automated liquid handling system and is determined by gravimetric measurements with water. The precision of low volume dosing of 8 µL using a 10 µL syringe is given with ≤1% relative standard deviation (RSD), and larger volume dosing of 100 µL using a 100 µL syringe is ≤0.1% RSD [4]. The syringe precision with organic solvents differs from the specification experiments with water. Using for instance acetonitrile, it is observed that the precision of a 300 µL transfer with a large 1000 µL syringe delivers a well comparable precision to water of 0.04% RSD. With smaller volumes using a 100 µL syringe for dispensing 25 µL of acetonitrile, the precision drops by a factor of 10 to 0.4% RSD. In comparison, the manual pipetting of the same volumes of acetonitrile achieves a precision of 2.6% RSD for 100 µL dispensing, dropping to 3.1% RSD for 25 µL. Especially with micro-methods, the use of robotic systems achieve superior precision compared to manual operation. A slightly reduced precision for the transfer of small volumes of organic solvents compared to water needs to be considered (S.J. Lehotay, private communication).

Most critical for the accuracy of liquid transfers in automated syringe operation is the easily visible bubble formation by cavitation (refer also to Section 4.1.2.6). Such bubbles can safely be avoided by choosing appropriate slow aspiration speeds and a repeated quick dispensing of the liquid in the workflow programming. A proven procedure as well is the sandwich-like pull-up of a consistent air volume first, followed by the liquid (refer also to Section 5.2.1). The air plug serves as a pressure buffer during aspiration. Then, the entire liquid volume is dispensed through the needle pushed out by the air plug.

4.1.2.2 Syringe Needles

Syringe needles are available in a wide range of diameters and needle point styles. The needle "gauge" measure describes the inner and outer diameters of the syringe needle. Gauges are historical measures of thickness [5]. It was originally developed in the early nineteenth century in England for use in wire manufacture as the 'Stubs Iron Wire Gauge', also called the "Birmingham Wire Gauge." It began appearing for medical hypodermic needles in the early twentieth century. The current needle wire gauge was derived from the "Stubs Iron Wire Gauge," one of the many gauge systems that were in use at that time. The gauge measure system is expressed in gauge numbers starting with small numbers for large diameters and increasing numbers for thinner needles the syringe gauge chart. This number system reflects the wire (and needle) manufacturing process by diameter reduction through pulling the wire multiple times through increasingly smaller drawing dies. The gauge system starts at the lowest gauge number of 5 corresponding to the largest size of 0.500″ (12.7 mm) to the highest gauge number of 36, corresponding to the smallest diameter of 0.004″ (0.102 mm) [6], also refer to the needle gauge table in Appendix Table A.3.

The needle gauge system with its standardization is used today in a wide range of applications, well known still for medical equipment.

For analytical purposes, needle gauge numbers in the range of 10–34 are most found for manual and automated applications. Some needle gauge values are marked with "s" indicating special specifications, for instance, 22s, 23s, 25s, and 26s. These needles are of the same outer diameter as the regular gauge types 22, 23, 25, and 26, but of about half the inner diameter, providing a double wall thickness for increased strength, for details see Table A.3. The s-type needles offer distinct analytical advantages for automated operation with the improved ruggedness by the increased wall thickness, e.g. for penetrating vial or injector septa. The reduced inner diameter reduces the needle volume. More viscous media require a slower aspiration speed in the s-type needles to safely prevent bubble formation by cavitation compared to the faster flow of the regular gauge sizes.

Syringe needles can become bent or broken during automated operation. Root cause is mostly an incorrect teaching of vial or module positions. GC injector caps can be fastened too strong after septum change and heat up, making it impossible for the needle to penetrate the squeezed septum. These mechanical causes can be double-checked and solved by appropriate maintenance. Blocking a needle tip can occur frequently with particles from septum coring. Replacing the GC inlet septum regularly is part of the preventive maintenance schedule to prevent such and other effects like leakages and bleeding safely. The use of syringes with replaceable needles is depreciated for process safety as it requires special precautions due to additional risks due to leakage and improper installation.

4.1.2.3 Syringe Needle Point Styles

Several needle point styles are available depending on applications. The needle gauge sizes can differ (Figure 4.3).

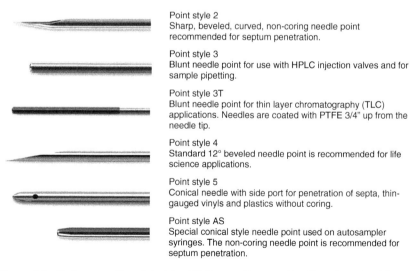

Point style 2
Sharp, beveled, curved, non-coring needle point recommended for septum penetration.

Point style 3
Blunt needle point for use with HPLC injection valves and for sample pipetting.

Point style 3T
Blunt needle point for thin layer chromatography (TLC) applications. Needles are coated with PTFE 3/4" up from the needle tip.

Point style 4
Standard 12° beveled needle point is recommended for life science applications.

Point style 5
Conical needle with side port for penetration of septa, thin-gauged vinyls and plastics without coring.

Point style AS
Special conical style needle point used on autosampler syringes. The non-coring needle point is recommended for septum penetration.

Figure 4.3 Syringe needle point styles [3]. Source: Courtesy Hamilton Bonaduz AG.

Several needle point styles are available and in use depending on applications. Also needle gauge sizes can differ. Figure 4.3 shows the most common point styles AS (AS stands for autosampler), HPLC, and side hole, also used in automated sample preparation and injection. Point style 4 must be avoided for sample preparation purposes. In practice, also the beveled point style 2 is completely replaced today by the conical AS type and depreciated for septum penetration, although advertised. In automated systems, the most used tasks like gas or liquid handling and the GC or LC injection are using syringes with dedicated needle point styles, which need to be considered when sketching a particular sample preparation workflow. The number of the point style type is widely harmonized between manufacturers; the available needle gauge can differ.

Guidance for needle selection in automated systems:

- For septum penetration (vials, injector, etc.), choose a gauge ≤26. Prefer s-type needles.
- Chose low needle volumes for low volume syringes.
- Chose a wide needle ID for viscous media to prevent bubble formation, or consider using a pipet tool.
- Avoid needle gauge <22 to prevent septum coring with the risk of needle clogging.
- Select the needle point style AS for septum penetration
- Use needle point style 3 for LC valve injections

4.1.2.4 Syringe Plunger Types

Metal Plungers Microliter syringes for low-volume liquid dispensing applications mostly comprise a stainless steel plunger. The metal plunger in the manufacturing process is made to fit exactly the glass barrel and cannot be interchanged from a different syringe. Syringes with metal plungers require priming and regular maintenance. Metal plunger syringes are not "gastight". A dry syringe should first be primed using the wash solvent. In an automated operation, the priming step typically is the initial full volume rinsing with wash solvent. The liquid film between plunger and glass barrel seals and facilitates plunger movement. Metal plungers should not be used with water or aqueous samples. It is observed that water causes a metal plunger to seize up in the glass barrel in a short time. A plunger with a polymer tip is recommended for automated use instead.

Low microliter "zero dead volume" syringes below 1–5 µL are the so-called nanoliter syringes with a zero needle dead volume for sub-microliter dispensing. They use a plunger-in-needle design. The plunger is extended into the needle up to the needle tip. Needle gauges are available from 22 to 32 size with ID in the range of 0.10–0.15 mm, cone or bevel needle tip. The sample is drawn only into the needle, not into the glass barrel. Due to limited volume precision of approximately ±2%, such types of syringes are reserved for special qualitative applications like the dosing of precious enzyme media or the sample preparation for matrix-assisted laser desorption ionization (MALDI) and are found only in highly specialized automated systems [7].

Polymer Tip Plungers Syringes with a polymer plunger tip are commonly called "gastight syringes." Gastight syringes are used for dispensing both liquids and gases. The plunger tip creates a leak-free seal with the barrel and does not need priming with solvent. Due to tight fit with the glass barrel, plungers with polymer tip have limited lifetime due to the wear on the inner barrel glass surface. The number of useful stokes is in the range of 50 000–100 000, which can be monitored and limited on robotic systems. Polymer tip plungers can be exchanged according to the manufacturer's specifications. Wetting the tip with a solvent upon insertion facilitates the plunger movement. Plunger tip materials use polyethylene (PE), or different fluorinated polymers, e.g. polytetrafluoroethylene (PTFE). Pure PTFE is soft and tends to abrasion causing a limited lifetime and a risk of valve clogging by particle accumulation. Modified fluorinated polymer or PE plunger tips should be employed for extended lifetime in automated systems.

The tight polymer tip fit to the glass barrel also provides analytical benefits concerning potential cross-contamination or deposit built-up. In contrast to a steel plunger, the polymer tip wipes the interior surface of the syringe barrel free of sample.

Often, a shrinking of the polymer tip after being exposed to higher temperatures can be observed. The following use at lower temperatures then can impair gas tightness. It is recommended with PTFE tip syringes to keep a constant temperature during syringe use in heated tools for volatile analysis. An almost invisible design detail in the manufacture of polymer tips for milliliter volume gastight syringes can compensate for shrinking caused by different temperature operation. A circular metal "spring" can be incorporated into the machined polymer tip providing a constant pressure of the polymer seal to the glass barrel, shown in Figure 4.4 for headspace analysis.

4.1.2.5 Syringe Termination

Several different configurations optimized for various applications connect the syringe barrel termination with the needle. In general, for liquid and gaseous samples handling, the fixed needle and removable needle types are distinguished besides other tailored designs, e.g. fixed dilutor, or the Luer termination, and more.

Fixed needle syringes are mostly used for microliter applications and recommended for repeat operation. Fixed needle syringes are ideally used in automated systems for long uninterrupted runtimes. The needle of a defined gauge value and point style is cemented into the glass barrel at a point corresponding to the zero

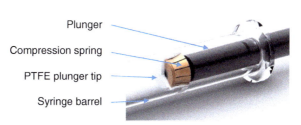

Figure 4.4 HD-Type plunger design. Source: Courtesy Hamilton Bonaduz AG.

graduation mark. The dead volume of the syringe is defined by the internal volume of the needle, see the needle gauge Table A.3 in Appendix A.3. The way of fixing the needle and technical implementation determines the temperature range for syringe operation. The maximum operating temperature specification is of importance for applications in heated tools to prevent condensation, e.g. in the analysis of volatile compounds. Cemented needle terminations are not autoclavable.

Removable needles are used in connection with syringe barrels or similar aspiration/dispensing devices. The exchange of needles allows the adaptation of the needle styles and gauge to different applications with a fixed syringe. Removable needles and corresponding syringe barrels are often autoclavable for biological applications. When using replaceable needles, special care is required for correct mounting with spring and checking for leak tightness.

Terminations for removable needles include the so-called Luer tip and a screw cap mount. The Luer tip or Luer taper is a standardized fitting system for leak-free connections between a male-taper fitting on the syringe barrel and its mating female part of the needle. It is named after the nineteenth-century German medical instrument maker Hermann Wülfing Luer and is today defined in the DIN and EN standard 1707:1996 and 20594-1:1993 [8]. The Luer termination does not position the needle to the zero graduation mark and increases the dead volume of the syringe significantly. The dead volume of the syringe is the volume of the needle installed plus the volume of the Luer termination at the barrel. Users can select the needle gauge, length, and point style to optimize the needle for custom applications. Luer tip syringes are in use in robotic systems for connecting to replaceable syringe filters, or applications using the push-through solid-phase extraction (SPE) cartridge format [9]. Luer tip needles cannot be used for injection to GC or LC analysis. Variations of the Luer design for a safe joint allow the needle to be screwed into a sleeve at the barrel, but is limited to manual use only, and not realized for automated systems.

4.1.2.6 Operational Parameters

Aspiration Speed With the use of syringes, the liquid aspiration speed becomes a critical parameter. The aspiration speed has to correspond to the chosen needle gauge and viscosity of the liquid to prevent the so-called cavitation bubbles in the liquid. Bubble formation leads to poor reproducibility and errors in volume dispensing. Cavitation occurs when the pressure in the liquid falls below the vapor pressure of the liquid by the rapid upward movement of the plunger and the flow restriction in the needle caused by the viscosity of the liquid. Minor bubbles move upwards and combine visibly below the plunger. A table with solvent viscosities relative to water is provided in the Appendix with Table A.6.

When liquid is withdrawn from an almost fully loaded and capped vial, the appropriate volume of air must be injected before aspiration to avoid the resulting pressure drop in the vial. This is mandatory for well-filled vials. If a vial is filled only half or less, the air in the vial will expand enough to replace the aspirated volume. In automated systems, this step of injecting air into a capped vial before liquid aspiration can become a general precaution for process safety independent of the vial fill grade.

Solvent vapor bubbles are removed manually by turning the syringe needle upwards, knocking gently the barrel to make bubbles move up to the needle, and then pushing the plunger in horizontal position a small step down to the desired volume. In automated systems with fixed top-down syringe orientation, cavitation needs to be avoided by appropriate needle dimensions and aspiration speeds. Slow aspiration does not cause any adverse effect other than being a little more time-consuming. As a safety measure, potentially occurring bubbles can be expelled by repeated rapid plunger down pushes three to five times combined with slow aspiration. Special care for syringe aspiration is needed for low boiling point solvents for instance with pentane. Such steps are typical safety measures recommended in automated workflows for precision dispensing.

Each liquid type has different maximum aspiration speed settings, dependent on liquid viscosity and the inner needle diameter. Practically, the best way to determine the correct speed setting for an unknown liquid is a comparison of the viscosity with water and according to experimentation.

Slow aspiration times for the prevention of cavitation can be a limiting factor in the use of large volume syringes. With milliliter syringes, the needle dimensions often cause the aspiration to become a time-critical step. Also, bubble elimination, as well as rinsing and cleaning steps, causes additional workflow delays, and on top of that requires large solvent volumes. Another downside of using large volume syringes in automated top-down orientation is their tendency of dripping low viscosity liquids due to the weight of the liquid in the barrel. The use of a dilutor tool as a replacement for milliliter syringes is recommended (see Section 4.1.5).

Dispensing Speed The syringe dispensing speed is usually not a critical parameter. Fast ejection speeds guarantee a precise volume dispensing by avoiding droplet formation at the needle tip. Only if the plunger speed is too high for a given needle ID and liquid viscosity, a significant pressure build-up can occur blocking motor movement with corresponding error situations. The maximum specified pressures of 1000 psi (69 bar) for 250 µL syringes, and 2000 psi (138 bar) for 10 and 100 µL syringes, 6000 psi (414 bar) for nanoliter syringes of 0.5–5.0 µL syringes allow high ejection speeds in the range of mL/s without damage to the syringe [3].

The speed of liquid, in volume per time, passing through a tube-like syringe needle is given by the Hagen–Poiseuille equation (4.2). The volume flow per time V/t that flows through a capillary tube is proportional in its fourth power (!) to the radius of the tube r, while the pressure P pushing the liquid through the needle is in just linear relationship. The length of the needle l (with syringe dispensing a constant) and finally the viscosity of the liquid η decrease the flow with increasing values.

$$V/t = \pi \cdot r^4 \cdot P / 8 \cdot \eta \cdot l \tag{4.2}$$

Priming Priming of syringes with a steel plunger is required to assure precision and accuracy. Syringes with steel plungers are not gastight. The liquid film between plunger and glass barrel seals and facilitates movement. It is necessary before installation to manually pump the plunger several times with the needle immersed in a wash solvent to be used later in the workflow. This will wet the plunger and expel any trapped air in the needle and syringe. The effect can be controlled visually with the pulled-up plunger. It must show the solvent plug in the barrel at the plunger tip without any bubbles. The volume of the visible solvent plug represents the sample plus needle dead volume.

In automated workflows, several pumping steps must be executed to expel air or cavitation bubbles. The solvent, sample, or liquid reagent must be slowly aspirated in a larger than later applied volume and ejected back in high speed. This removes bubbles safely from the barrel. The number of necessary pumping steps, usually three to five times, depends on the nature of the solvent. If cavitation occurs with low aspiration speeds, the type of needle needs to be exchanged to a larger inner diameter.

Priming often requires larger volumes of solvent than available as sample. Especially, dilutor devices require a larger solvent volume for priming syringe and tubings. A following necessary step after priming can be the rinsing with the sample.

Syringe Rinsing The term "rinsing" is reserved for priming a syringe with the liquid sample or extract. This step should not be confused with washing a syringe. Rinsing reduces the remaining solvents and may cover active sites within the wetted syringe surface. Rinsing with sample to waste is a recommended security step in automated workflows, typically done just once or twice, depending on the available sample volume.

Syringe Washing Syringe washing steps with one or more wash solvents is required before each workflow cycle, also as a priming step. Carryover is a main issue to treat carefully in automated processing. Several technical and programmable solutions are available on automated sample preparation platforms to address potential carryover, mainly occurring with the use of syringes.

With automated tool change systems, dedicated syringes can be employed in the workflow for tasks like standards or solvents dispensing, keeping the syringes reserved for known concentration levels, or the injection step. To prevent carryover between samples, the liquid handling and injection syringes must be flushed with suitable and different solvents repeatedly. It is known from analysis of the persistent organic pollutants (POPs) that the polychlorinated and polybrominated biphenyls, furans, and dioxins, as well as fluorinated compounds stick to glass walls, sealings, or other surfaces they get in contact to. A similar behavior is reported from many pesticides. Modern highly sensitive MS systems detect low percentage carryover even below the ppb range. Other compounds, mainly matrix or added analyte protectants, can affect the free plunger movement requiring a regular syringe maintenance (see Appendix Section A.2.1). Blank samples are required after calibration points and matrix spikes, also after suspected samples of high contamination to assure the reliable cleaning procedure.

Syringes for gas sampling, e.g. static and dynamic headspace, allow flushing the syringe body with clean (carrier) gas while at elevated temperature using a side hole for gas connection. In particular, in life science applications, flow-through cleaning solutions with dedicated tools that prevent the sample plug to get in contact with a syringe at all (refer to Figure 5.40) are also in use with active solvent cleaning.

Depending on the analyte, matrix, and application, three or a significantly higher number of solvent cleaning cycles for liquid syringes may be necessary. Dedicated cleaning cycles need to be introduced in workflows after each sample, standards, or reagent dispensing, and after and before starting a next work cycle. Syringe priming is necessary when changing to a new or a so far unused syringe, and also after a tool change.

Syringes are best cleaned with solvents of different polarity. The solvents should be most suitable for the processed sample matrix and analytes. Alcohols, acetone, or ethyl acetate, hexane are the most used cleaning solvents in GC. All cleaning solvents must be of highest purity (*pro analysis* [p.a.], American Chemical Society [ACS] quality grades). In LC applications, washing solvents should be stronger than the used mobile phase to remove any sample residues. Also, the cleaning solvent pH-value may be of importance. Suitable solvent reservoirs in sufficient volume for a complete sample sequence must be provided. Wash modules with an active solvent delivery from bulk reservoirs should be considered for complex and multistep workflows, see Figure 4.5. Upon insertion of the syringe needle, a liner is fed by a pump from the bottom with a clean solvent from the external reservoir. Unique with active syringe wash is the cleaning possibility of the outside of the needle from potential sample contamination as well, most appreciated in life science applications. A quick needle dip can also be programmed for the outside needle wash before an LC valve injection to avoid a continuous contamination.

Alkaline- and detergent-based agents are not recommended for regular cleaning procedures in washing workflows, but can be used with dismounted syringes according to the manufacturer guidance for the separate syringe maintenance.

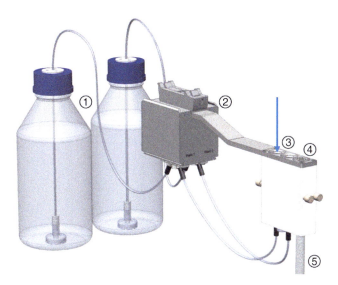

Figure 4.5 Active syringe wash module for two wash solvents. The arrow points to the first wash port. With (1) Wash solvents reservoirs; (2) Solvent pumps; (3) Two solvent wash positions; (4) Waste port; (5) Waste tube. Source: Courtesy CTC Analytics AG.

4.1.3 Vial Bottom Sensing

Sensing the bottom of a vial is a widely voiced demand for automated systems. This request comes from handling small volumes in trace analysis and the unique small volume samples from elaborate and often not repeatable sample preparation sequences. Also, samples after in-vial evaporation and concentration, or the safe access of a bottom phase from liquid/liquid extractions (LLE) require a precise adjustment of the needle penetration down close to the bottom of the sample vial. Well-known examples are applications in life sciences, POPs analysis, or the dispersive liquid/liquid micro-extraction (DLLME) methods. Knowing the required needle penetration depth can solve the issue by a low but constant penetration depth programming. But, often the inside vial bottom level shows height variations due to the manufacturing process of regular and tapered vials. In case a needle penetration depth is adjusted manually, the value is constant for all sample vials. Improper adjustment can lead to blocking the needle by pushing the opening against the vial bottom, so that no sample is aspirated. In the worst case, potential sample loss can occur by crashing the vial bottom with a further downward moving needle. Vial specifications are given only for the outside dimensions. The use of vial micro-inserts (Figure 4.6) or a usage of microplates (or "microtiter™ plates") for small volume handling are other present examples.

A special feature of some robotic systems for optional use detects the vial bottom or a valve seal in a pre-defined height distance. In most cases, tools with syringes are used for bottom sensing with tapered vials, but it works well with dispensing devices or pipets likewise. In case a syringe is penetrating a vial, a certain height distance is defined as the bottom search region. After sensing a bottom resistance, the downward movement stops. The needle then is pulled back a small defined distance, usually less than a millimeter, to aspirate liquid media, see Figure 4.7. Caution has to be taken using micro-inserts with polymer spring as the flexible spring interferes with the bottom sensing function. The spring needs to be removed, and inserts used without spring to allow correct bottom sensing and retraction.

Figure 4.6 Micro-vials and micro-inserts for autosampler use. (A) Standard 2 mL vial shape with built-in micro-vial; (B) micro-vial insert, ca. 300 µL volume; (C) micro-vial insert, ca. 300 µL volume, with polymer spring.

(a)   (b)   (c)

Figure 4.7 Tapered vial for low sample volume.

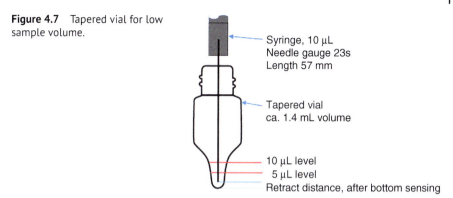

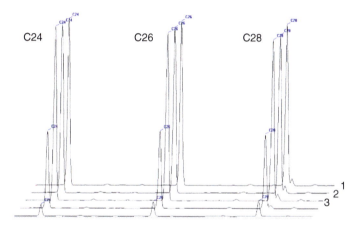

Figure 4.8 Bottom sensing allows 3 replicate 1 μL GC injections from a 5 μL sample, compounds C24, C26, and C28 alkane. Source: Image used with permission of Thermo Fisher Scientific Inc.

Typically, from a 5 μL sample in a tapered vial as shown in Figure 4.7, three injections of 1 μL are possible with a bottom sensing function, shown with an alkane mix with the chromatograms for C24 to C28 in Figure 4.8. Bottom sensing works well with syringe needles of gauges 26 and stronger using 57 mm needle length. Tapered bottom vials or micro-vials are recommended to prevent bending of the needle toward the sidewall of the vial. With the use of pipet tips for an LC valve injection, a pipet tip can be positioned well sealing into the valve port, making a tool change unnecessary for short workflows in LC-MS/MS screenings.

4.1.4 Pipetting

Pipets are very common tools for liquid handling in every laboratory. Pipets and pipet tips are available in multiple also customized designs for single and multiple use, for single or multi-channel, manual, or semi and automated operation. Capacities of up to 1000 μL are generally called micropipets.

The first micropipet was patented in 1957 by Dr. Heinrich Schnitger from Marburg, Germany. The founder of the company Eppendorf, Dr. Heinrich Netheler, inherited the rights and started the commercial production of micropipets in 1961 [10].

A major analytical benefit with the use of micropipets compared to syringes is the disposable pipet tip. The disposal of the tip safely avoids cross-contaminations and eliminates time- and solvent-consuming cleaning steps from the workflow, well-known and critical for potential cross-contamination with syringes. Working with pipet tips allows a metal contact-free liquid handling, for instance in life sciences, or for steel corroding media like hydrochloric acid, or the clean-up using magnetic beads.

Variations in pipetting volume as notorious from manual handling with different practice of aspiration and dispensing, caused e.g. by an increased aspiration volume due to non-vertical liquid aspiration, a known systematic pipetting error, are well eliminated in automated pipetting [11, 12]. Significant benefits with automated pipetting are the error-free, tireless, and consistent operation for large sample series. Robotic pipetting samplers with controlled and constant force operation also allow the injection with pipet tips into LC valves for LC-MS analyses.

4.1.4.1 Air Displacement Pipettes

The generally used pipets work by air displacement creating a vacuum above the liquid to be transferred, in contrast to the also available positive displacement technique. Micropipets are equipped with a movable piston for aspiration and dispensing, the same assembly for manual or automated operation. Two plunger stops define in the manual pipetting the aspiration volume (first stop) followed by a stop for dispensing (second stop). Pushing the plunger completely down, micropipets dispose the replaceable tip. Manual micropipets use fixed or variable volumes. For aspiration, the defined air volume is first pushed out by moving the piston down. Upon immersion of the pipet tip with piston down into the liquid, the vacuum caused by the upward moving piston draws up the liquid into the pipet tip. The liquid volume is equivalent to the previously displaced air volume. In automated operations, the transfer volume is limited to the capacity of a usually disposable pipet tip, requiring different pipet tools for different liquid volume ranges.

An important factor in pipetting accuracy with air displacement pipets is the sample temperature and liquid vapor pressure. If the sample liquid has a different temperature other than the pipet tool, the accuracy can change in the range of −5% to +5% for refrigerated samples from below 10 to 35 °C in a hot laboratory environment. If the temperature of the liquid, pipet, and air is the same, the accuracy is not affected. For best accuracy, the sample liquid should be close in temperature with the pipet, at room temperature. Rinsing the tip with sample one to three times with the liquid to be pipetted may improve accuracy, in particular for cooled stack samples and volatile solvents [13].

Pipets are usually calibrated with water. Hence, air displacement pipets work most precise with aqueous samples. Calibrations need to be repeated on a regular basis.

Liquids with higher vapor pressure than water will deliver a lower than the set volume. Expect reduced accuracy and precision with organic liquids [14].

4.1.4.2 Positive Displacement Pipets

Pipets using the positive displacement technique comprise a disposable plunger that reaches directly into the pipet tip. The plunger displaces the aspirated liquid in the dispensing step directly for the best volume accuracy. Positive displacement pipets are in use for accurate usually manual pipetting of viscous samples and volatile solvents, even for corrosive samples due to the disposable plunger. Also, for accurate low-volume pipetting in the range of below 1 µL positive displacement pipets are recommended as the air displacement technique in this range is sensitive to temperature effects in the laboratory. Air expands to a greater degree with temperatures than liquids.

4.1.4.3 Pipetting Modes

Several pipetting modes are possible. These modes are available on robotic systems by appropriate parameter settings. Depending on the sample media and tasks, pipetting modes like forward pipetting, reverse pipetting, dispensing, sequential dispensing, or diluting are used for optimized liquid transfers in numerous applications.

Two pipetting techniques are mainly applied in laboratories, the forward or standard mode, and a reverse mode. For pipetting of aqueous solutions, the standard forward mode should be used. It typically yields better accuracy and precision than the reverse mode, and it is used by manufacturers to calibrate their pipets. The reverse mode is recommended for viscous solutions only. The standard mode can result in under-delivery of viscous liquids [11].

In the *standard forward pipetting* mode, the plunger is pressed in manual operation to the first stop with the pipet tip outside of the sample. For taking up the sample, the tip is immersed into the liquid keeping the plunger down at the first stop. The plunger is then slowly released to aspirate the volume of the displaced air volume up to the first stop. At the destination vial, the entire contents are dispensed by pressing the plunger to the second stop, the blow-out step.

With viscous or foaming liquids or when dispensing very small sample volumes, the results with air displacement pipets may be improved by *reverse pipetting*. The robotic pipetting provides significant analytical benefits compared to manual operation for viscous media, e.g. like glycerol, with the slow and constant aspiration, minimizing the pressure difference inside of the pipet for the slowly flowing media. Pipet tips with larger orifice for viscous media are available commercially. The robotic systems allow the definition of "liquid classes" according to the sample viscosity (see Section 4.1.4.7 for details). The liquid class settings are an experimental constant parameter set useful for instance for biological materials, viscous media, samples that may form bubbles, or are adhesive to the tip surface. In general, positive displacement pipets are preferred, when available, for aspiration and dispensing of viscous media due to the direct contact with the plunger. A gravimetric control of the dispensed viscous media is recommended for both techniques.

Applying the reverse pipetting mode, the plunger is pressed in the manual operation down to the second stop before immersing into the sample. Upon release of the plunger, a larger than required liquid volume is pulled into the tip. The desired correct volume of the sample is then delivered by moving the plunger down only to the first stop. Some of the liquid remains in the tip and is subsequently discarded together with the tip [11]. In robotic systems, the discussed stops for depicting the principles are for sure not available. Instead, appropriate programming of the pipetting parameters in the workflow allows the implementation of the pipetting schemes illustrated in Figures 4.9 and 4.10.

The precision of automated standard pipetting is excellent. Accuracy values of better than 1% are achieved with dispensing volumes >20% of the nominal capacity for a small 200 μL tip, or for volumes >10% using a large 1000 μL tip, see Table 4.3.

Sequential dispensing distributes the aspirated liquid in defined volumes to several target vials. This technique is often used when processing longer test series or serving MTP cavities, see Figure 4.11. The repeatability of sequential pipetting is only slightly lower as for single pipetting steps. For a multiple 20% dispense volume of 4 × 40 μL with a 200 μL tip, a precision of better than 2% RSD is achieved, for a 1000 μL tip a precision of better than 1% RSD for a 20% dispense volume of 4 × 200 μL.

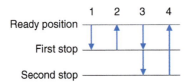

Figure 4.9 Forward pipetting technique (stops are related to manual pipet operation). (1) With tip attached, the pipet plunger is pushed to the *first* stop and then immersed into the solution. (2) With the tip kept immersed in the sample, the plunger is slowly released to the standby position aspirating liquid into the pipet tip. (3) The correct liquid volume is dispensed by pushing the plunger to the *second* stop. (4) Release of the plunger to the standby position.

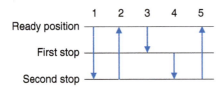

Figure 4.10 Reverse pipetting technique (stops are related to manual pipet operation). (1) With tip attached, the pipet plunger is pushed to the *second* stop and then immersed into the solution to a certain depth according to the volume set. (2) With the tip kept immersed in the sample, the plunger is slowly released to the standby position aspirating liquid into the pipet tip. (3) The correct liquid volume is dispensed by pushing the plunger to the *first* stop. Some liquid will remain in the tip. (4) The sample remaining in the tip can be moved to waste or discarded with the used pipet tip. (5) Release of the plunger to the standby position

Table 4.3 Precision of automated pipetting.

Pipette tip [μL]	Dispensed volume* [μL]	Repeatability RSD [%] n = 7	Accuracy n = 7	Linearity** R^2
20	10	<0.8%	<0.8%	
200	20	<1.5%	<2%	>0.999
200	40	<0.5%	<1%	
200	100	<0.5%	<1%	
200	200	<0.5%	<1%	
1000	100	<0.5%	<0.5%	>0.999
1000	200	<0.5%	<0.5%	
1000	500	<0.3%	<0.5%	
1000	1000	<0.3%	<0.5%	

*Using water. **Linearity for multiple dispensing.
Source: Courtesy CTC Analytics AG.

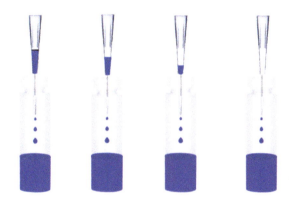

Figure 4.11 Sequential dispensing into several vials. Source: Courtesy CTC Analytics AG.

The *dilution pipetting* mode allows the aspiration of two liquids into one tip and transfers them together into one target. An initial liquid volume is first aspirated, followed by pulling-up a small air bubble before aspirating the second liquid volume. Both liquids are then dispensed in one step into the same target vial. This dispensing technique increases throughput by a reduced number of workflow steps and shorter processing time.

Table 4.4 provides pipetting recommendations for different kind of sample types for the use of air or positive displacement pipets, suitable pipet tips and techniques.

4.1.4.4 Aspiration

When aspirating the liquid, the tip should be immersed only a few millimeters into the medium. The pipet tip volume determines the recommended immersion depth as given in Table 4.5. The immersion depth for the most used pipet sizes from 1 to 1000 μL volume is generally 2–4 mm below the surface, and 3–6 mm for volumes

Table 4.4 Recommendations for pipetting.

Solution/compound	Examples	Pipette	Tip	Technique	Comments
Aqueous solution	Buffers, diluted salt solutions	Air displacement	Standard	Forward	
Viscous solution	Protein and nucleic solutions, glycerol, Tween 20/40/60/80	Air displacement Pos. displacement	Standard wide orifice Pos. displacement	Reverse	Pipette slowly to avoid bubble formation.
Volatile compounds	Methanol, hexane	Air displacement Pos. displacement	Filter Pos. displacement	Forward	Pipette rapidly to avoid evaporation. Carbon filter tips prevent vapors
Nucleotide solutions	Genomic DNA, PCR products	Air displacement Pos. displacement	Filter or wide orifice Pos. displacement	Forward	For genomic DNA wide orifice should be used
Radioactive compounds	(^{14}C)Carbonate, (^{3}H) thymidine	Air displacement Pos. displacement	Filter Pos. displacement	Forward	
Acid/Alkalis	H_2SO_4, HCl, NaOH	Air displacement	Filter	Forward	

Source: Redrawn from AccuTek Laboratories [15].

Table 4.5 Optimum pipet tip immersion depth.

Pipet tip volume [µL]	Optimum immersion depth [mm]
0.1–1	1
1–100	2–3
101–1 000	2–4
1 001–10 000	3–6

Source: Redrawn from Ewald [11].

larger than 1 mL. The optimum pipet position during aspiration is the vertical position, as inherently fixed by automated systems. Tilting in manual operation may cause a variation in the sample volume.

Rinsing the tip 2 or 3 times improves the volume accuracy and precision. It is in particular recommended for samples stored in a cooled stack. While in manual pipetting, the filled tip should be moved up against the wall of the vessel to avoid residues of liquid on the outside of the tip, in the automated mode, this is mechanically not possible. Instead, the immersion depth in the liquid is held constant to a low but safe immersion depth using the automated function of liquid-level tracking. The sample liquid should then be aspirated slowly and evenly. A short waiting time of one to three seconds depending on the sample viscosity should be allowed for the liquid to rise into the tip [11].

Liquid-level tracking means the pipet tool will move continuously up or down in vertical direction, following the falling, or rising liquid level during aspiration or dispensing in a vial or well plate cavity. This feature is important, if the best performance shall be achieved during aspiration or dispense of larger volumes (no contamination of the tip outside, accurate aspiration). By this automated function, the penetration depth can be kept constant during aspiration or dispensation. The liquid tracking only works if the liquid handling is used with liquid classes (see Section 4.1.4.7). The penetration depth is specified using the Liquid Offset parameter in the liquid class definition. For tracking the liquid level, e.g. in a vial, the system follows the volumetric geometry of the vial and adapts the height of the pipet tool simultaneously. The volumetric geometry describes the volume of a vial or a well-plate cavity as a function of the horizontal area at a specific height and allows volume calculations with the moving pipet tip.

4.1.4.5 Dispensing

Pipet tips, in manual operation, should be checked for droplets on the outside of the tip, with visible droplets removed very carefully with a lint-free cloth. This should only be done if absolutely necessary, however. Liquid can be wicked from the tip opening, causing sample loss and under-delivery. Once clean, the tip should be placed against the receptacle wall with a recommended 45° angled micropipet position and the plunger pressed to the second stop when using forward mode. The tip should then be slightly dragged up the receptacle wall to allow all liquid to be drawn from the tip. Repeated actions produce repeatable results. The plunger should always be pressed and released with consistent speed and pressure. For large sample series, manual pipetting can quickly become a tiring task, also raising the human error frequency.

In automated pipetting operation, the free dispense is used due to the mechanical limitations that do not allow a wall contact dispense as common in the manual mode. To avoid droplet formation, a sufficiently high dispense speed has to be set, depending on the tip dimensions and the liquid viscosity, e.g. with aqueous solutions in the range of 100–500 µL/s.

Automated pipetting operation often integrates a backlash compensation volume, a small variable volume that works as an overfill volume for increased reproducibility and precision. The volume is hardware related and used for backlash compensation for all activities that acquire a liquid volume with a tip overfill that is directly afterward dispensed again.

4.1.4.6 Liquid-Level Detection

Automated pipetting allows a liquid-level detection and liquid-level tracking as illustrated in Figure 4.12. The liquid-level tracking maintains a constant immersion depth of the pipet tip in the sample liquid for improved repeatability independent of the changing liquid volume aspirated or dispensed. A contamination of the outside surface of the tip is avoided, which would influence reproducibility depending on pipetted volumes, in particular for viscous liquids. The pipet tool moves continuously down for aspiration and up for dispensing following the falling or rising liquid level in a vial or well-plate cavity. This feature is important if the best performance shall be achieved during aspiration or dispense of larger volumes.

Achieving a constant immersion depth independent on a vial or well-plate cavity volume requires a liquid-level sensing. The liquid-level sensing with pipet tools is accomplished by different technical solutions. Capacitive level sensing or by using a highly sensitive low range pressure sensors are in use, both located in the pipetting tool of the robot. While a capacitive measurement is suitable for liquids with high dielectric constant like water only the pressure sensing works independently on the type of solvent. The most suitable solution for any liquid is employing a pressure sensor built inside the pipetting tool this way that a pressure change can be monitored when immersing the sample liquid with the pipetting tip. This allows the tool to

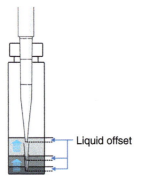

Figure 4.12 Liquid-Level Tracking keeps a constant tip immersion depth. Source: Courtesy CTC Analytics AG.

maintain a constant immersion depth with different vial filling levels. Also foaming samples or blocked tips can be monitored. A workflow limitation with liquid-level sensing can be seen in the restriction to the full pipet tip dispensing. The multiple dispensing mode is not available.

4.1.4.7 Liquid Classes

The automated handling of liquids with diverse physical properties like viscosity or specific gravity can be optimized by the creation of common parameter sets called liquid classes. These parameters influence the internal handling of the pipetting activities for improved precision and accuracy and allow flexible automated applications optimized for each sample or solvent in use.

Each liquid class can specify sets of parameters for several volume ranges (scopes), while each set of parameters is only valid for the scope it has been specified for. The scope refers to the sample volume to be aspirated or dispensed.

As general parameters for aspiration and dispensing the optional liquid-level detection, liquid offset and the liquid tracking parameters can be used. These parameters are unique to automated pipetting operations and assure the highest accuracy and repeatability of liquid dosing. The liquid detection allows the liquid level to be detected automatically. The liquid offset defines the penetration offset applied during the liquid tracking procedure that keeps a constant penetration depth of the pipet tip during aspiration or dispensation of a liquid. The penetration depth with this function is continuously adapted to the changing liquid level.

Single and multi-dispense parameters available for liquid class definitions are shown in Table 4.6 for "Water". The volume parameters used in the automated pipetting process are illustrated in Figure 4.13 at a pipet tip. The delay and speed parameters for aspiration and dispense are used to compensate for the viscosity of a liquid. The delay time specifies a waiting time after aspiration or dispensation to allow the viscous liquid to flow accurately. Liquids of higher viscosity require longer waiting times (see the Appendix Viscosity Table A.6). The correction factor is a calibration value multiplied to any liquid volume (slope). The volume offset is a calibration value as well and defines an added correction volume to the aspirated liquid volume (offset).

4.1.4.8 Pipet Tips

The air cushion between the pipet piston and the surface of the liquid should be as small as possible. The smaller the tip, the lower the air volume, and the greater the accuracy of the results will be. It is advisable to use the tips recommended by the pipet manufacturer. Racks are typically available in standard format (SBS format of the former Society for Biomolecular Screening, today American Standard Institute ANSI [16]) for holding pre-assembled tip racks with 96 tips. Every pipet shaft has a specific taper angle that is designed to fit optimally to the recommended tips. Tips have to form a tight seal with the pipet shaft, as a leaky seal contributes to errors in pipetted volumes. When using tips from other than the pipet's manufacturer, the user must carefully check the fit and confirm that there is no gap at either the top or bottom of the shaft and that the tip is properly aligned and fits snugly.

Table 4.6 Parameters for a liquid class definition with example "Water".

Parameter	Liquid class "Water"		Function
	Single dispense	Multiple dispense	
Airgap, rear	100 µL	0 µL	Blow out fluid residues for a complete dispensation. Usually used only in single dispense mode.
Volume, rear	0 µL	20 µL	Excess volume as reserve for the multi dispense mode. Not used for single dispense mode.
Sample volume	Variable	Variable	Net sample volume for single and multiple dispense mode. The sample volume is corrected using the correction factor and volume offset.
Volume, front	0 µL	2 µL	Excess volume for backlash compensation and priming. Released to waste directly after aspiration.
Airgap, front	0 µL	5 µL	Air gap to protect against droplet formation. Acquired again after each dispense step in the multiple dispense mode.
Aspirate flow rate	100 µL/s	100 µL/s	Compensates for the viscosity of a liquid
Aspirate delay	300 ms	300 ms	Wait time after aspiration. Compensates for the viscosity of a liquid.
Dispense flow rate	100 µL/s	500 µL/s	Compensates for the viscosity of a liquid
Dispense delay	100 ms	100 ms	Wait time after dispensation. Compensates for the viscosity of a liquid.
Correction factor	1.04	1.04	Calibration value. Defines a factor to the aspirated liquid volume.
Volume offset	1.6 µL	1.6 µL	Correction value. Defines a constant added volume to the aspirated liquid volume.

Source: Courtesy CTC Analytics AG.

Setup of pipet tips for automated operation requires in some cases the adaptation from the manufacturer default of the tip connection parameters to the used pipet tips, including the confirmation that a tip is connected, with the following parameters:

Pick up force is used by the robot arm while driving down with the tip adapter into the tip to pick it up.

Drop force is used for placing a tip back into a rack position and to detect the surface of the rack.

Strip force is used to strip off a tip from the tip adapter of the tool.

Detection force is used to check for a mounted tip. The system detects the resistance while pushing slightly against the tip. If no tip is mounted, no resistance

Figure 4.13 Liquid class parameter representation in the pipet tip. Source: Courtesy CTC Analytics AG.

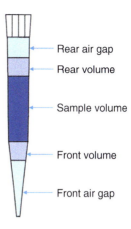

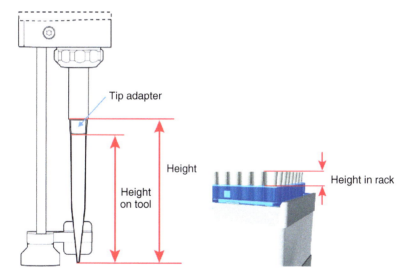

Figure 4.14 Height of pipet tips for automated operations. Source: Courtesy CTC Analytics AG.

is detected. The force value should be low to avoid stripping off the tip during the check.

Height in rack is the height of the upper part of the tip that protrudes the surface of the tip rack, see Figure 4.14.

Height on tool is the height of the lower part of the tip that does not overlap with the tool adapter, while mounted on the tool, see Figure 4.14.

After use, the operator can choose in automated operation if the pipet tip is set back into the rack, in case the same tip should be used several times, or is disposed to a waste bin. It should be checked for workflow safety that the tip is dropped off correctly.

4.1.4.9 Functional Pipet Tips

Besides liquid transfer, pipet tips can carry functional elements with new formats for sample preparation. Aside from below featured uses, micropipet tips have also been used to miniaturize dialysis and enzyme digestion. "An advantage of micropipet tip–based sample preparation is that it can be adapted easily to liquid-handling robotics." A comprehensive overview of functional pipet tips in bioanalysis is provided by Shukla and Majors [17].

Filter Tips Commercially available pipet tips may be equipped with a piece of filter material close to the connection point to the pipet shaft, shown in Figure 4.15. This version of pipet tips is also known as "barrier tips" describing well its function to protect the pipet tool from aerosols or foaming samples against cross-contamination between pipetting steps. Vice versa the sample is protected against potential contamination back from the pipet tool.

Automated Disposable Pipet Extraction Pipet tips containing sorbent material can be used for automated SPE. This patented solution employs disposable pipet tips "for the rapid, low-volume" SPE of analytes from a variety of sources. For this purpose, "the pipette tip contains a loosely confined stationary phase. The mobility of the stationary phase particles enables rapid mixing and equilibration with a sample solution during agitation. The analyte may thereby be extracted in less time with less solvent, removing the need for a separate concentration step" [18]. Disposable pipet extraction (DPX) is a dispersive SPE technique developed by Professor William E. Brewer from the University of Southern Carolina. The principle design of the extraction pipet tip is shown in Figure 4.16. The DPX technique has been automated and commercialized by GERSTEL GmbH & Co. KG, Mülheim, Germany, using a proprietary x,y,z-multipurpose robotic sampler.

Figure 4.16 shows a cross section of the disposable extraction pipet tip. The purpose of the bottom frit is to provide a permeable barrier, which permits the unrestricted passage of fluids in either direction but does not allow the stationary phase material to pass through. The purpose of the top frit is to prevent the passage of either solids or fluids. This ensures the retention of the stationary phase and the fluid and

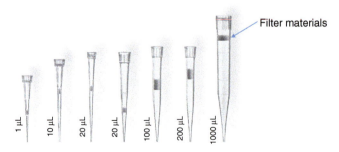

Figure 4.15 Pipet tips with filter as aerosol barrier against cross-contamination. Source: Courtesy BRAND GMBH + CO KG.

Figure 4.16 Design of the Disposable Pipet Extraction Tip. Source: Redrawn from Brewer [18].

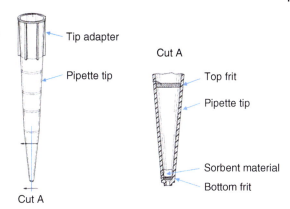

prevents contamination of the pipet tool by sample solution or solvents during the agitation step. The pipet tip volume serves as a mixing chamber with the adsorptive particles of the stationary phase [18].

Applications of the automated DPX method are as manifold as SPE is for clinical, forensic, and food safety applications. Commercial pipet tip solutions are provided as INTip™ sample preparation solutions by the patent holder Dr. William E. Brewer with the DPX Technologies company [19]. An important aspect is that the syringe and pipet tool is never in contact with the sample eliminating the potential for cross-contamination. The DPX pipet tip is disposed of after use.

A good example of an application is the analysis of drug residues, which is performed for a large number of samples in many forensic laboratories worldwide [20, 21]. Lehotay and Lighthouse applied the DPX clean-up for the multiclass and multi-residue monitoring of veterinary drug residues and aminoglycoside antibiotics [22]. DPX is also used for the automated clean-up of QuEChERS extracts for pesticides, drugs of abuse analysis in blood, or the drug and metabolites screening analysis, illustrating the analytical potential of the automated DPX sample preparation. The DPX method enables the full automation of all workflow steps using x,y,z-robots resulting in a significant reduction in processing time [23]. The workflow as shown schematically in Figure 4.17 can be performed in <10 minutes, ideally in a "prep-ahead" mode during the chromatographic run of the preceding sample.

Pipet Tips for Protein De-Salting In proteomics applications dedicated disposable pipet tips for de-salting become common practice for automated sample preparation prior to nano-flow LC-MS. Evotips™ provided by Evosep Biosystems are miniature, disposable C18 trap columns [24].

Samples are loaded onto Evotips for de-salting and cleaning before analysis. The loading and cleaning steps may be automated by an x,y,z-robot or done in parallel with multichannel pipets. The steps of manually eluting samples, drying down, and re-suspending are omitted. Racks of rinsed tips are placed into the robotic sampler which elutes the samples from each tip by two gradient forming low-pressure pumps and runs the separation in one integrated procedure illustrated in Figure 4.18.

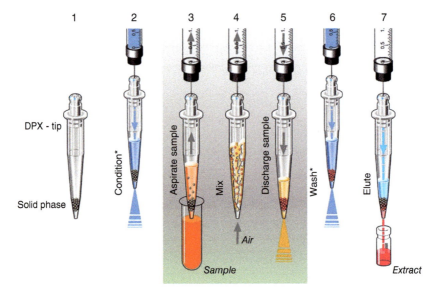

Figure 4.17 Workflow steps for the automated drug residue analysis using DPX. (1) The DPX pipet tip is taken up from the rack into the tool of the x,y,z-robot. (2) The solid-phase material is conditioned (optional). (3) The sample is pulled from the sample vial into the pipet tip. (4) The sample and SPE sorbent material are mixed by pulling air through the tip. (5) After extraction, the sample is discharged, typically after 30 seconds. (6) The sorbent material can optionally be washed to remove unwanted matrix. (7) Taking up a suitable solvent, the extracted analytes are eluted into a separate vial for analysis by LC-MS or GC-MS. Source: Bremer [22]. Courtesy GERSTEL GmbH & Co KG.

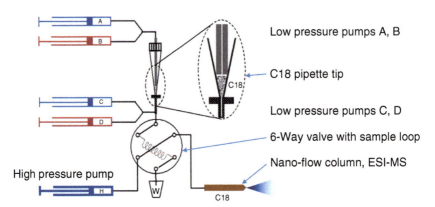

Figure 4.18 Configuration for automated C18 pipet tip de-salting and online nano-flow LC-MS [26]. Source: Courtesy Evosep Biosystems.

Each tip is eluted using "partial elution", that only runs the gradient up to a certain strength where contaminants such as polymers and large biomolecules remain on the tip, along with any particulate matter of the sample. By this means, the nano-flow separation column is protected for large automated sample series [25, 26].

4.1.4.10 Pipet Tip Materials

Pipet tips are usually made from polypropylene (PP), and in case they provide an aerosol filter, it is made of polyethylene (PE). The PP material allows special mechanical and surface characteristics for high precision liquid transfer.

Low Retention Tips The objective of a low aqueous sample retention is to obtain a highly "water repellent" tip surface that is hydrophobic enough to ensure that no liquid is retained. Although the used PP material is inherently hydrophobic, several special treatments are in use on the manufacturing side to improve the behavior of the so-called "Low Retention" tips. The low retention tips, also called "low binding" tips, are specially modified to reduce the adhesion of aqueous biological samples with for instance DNA, enzymes, proteins, cells, as well as other viscous materials to the tip surface. They are especially advantageous when pipetting detergent-containing solutions such as Triton™ X-100, Tween™ sodium dodecyl sulfate (SDS), or sodium lauryl sulfate (SLS), etc., which have a lower surface tension compared to water and are not completely repelled from the tip surface. A remaining film after dispensing can reduce accuracy and precision.

Three methods are in use for reducing liquid retention and offered with pipet tips for commercial use. In production, the mechanical diamond polishing of the tip molds provides an extremely smooth tip surface that prevents samples from adhering. No chemical additives are used in this purely mechanical manufacturing process.

But, also additives are in use either as additives to the polymer material or as a post-processing treatment. This creates the risk of such compounds leaching from the tip into the pipetted solutions with noticeable chemical contamination. A standard approach is the application of silicone or oleamide (e.g. 9-octadecenamide, CAS 301-02-0) surfaces [27]. The silicone and oleamide layers can be washed out with organic solvents and lead to background contamination in subsequent GC-MS detection [28].

Additives to PP before molding are found for instance with di(2-hydroxyethyl) methyl-dodecylammonium (DiHEMDA, CAS 22340-01-8) leaching into aqueous samples with effects on enzyme and receptor proteins, or get into solvents like methanol and dimethyl sulfoxide (DMSO) [29, 30].

On top of mechanical and additive measures, the goal to achieve the "Lotus Effect" with pipet tips leads to proprietary technologies with several manufacturers. Such modified pipet tips feature patented surface treatments, which makes them extremely liquid repellent. The physio-chemical processes are employed to avoid leachable coatings that might lead to sample contamination. Such tips maintain the high chemical resistance and are autoclavable [31]. Small volumes < 10 μL of aqueous liquids are not well suited for pipetting using low retention tips. The aspirated volumes may remain lower than targeted, or it may be impossible to aspirate the liquid at all [32].

4.1.5 Dilutor/Dispenser Operation

A diluter/dispenser component extends the liquid handling capabilities of automated systems significantly, adding solvent delivery as well as liquid transfer modes. A dilutor module is connected by a low volume flexible tubing to a dispenser tool mounted to the head of the x,y,z-robotic system. The dilutor comprises several ports for the choice of solvent sources, flex-tube connection to the dispenser tool, and a waste line for priming. The dispenser tool comprises a syringe needle to address vials also by septum penetration, shown in Figure 4.19.

Solvents or reagents from different sources can be precisely dispensed into vials in programmable volumes, which are typically beyond the practical capacity of liquid syringes. Also, the precise aspiration of liquid from a sample vial, for instance after LLE phase separation, can be performed for transfer into target vials or waste (Figure 4.20). The module features syringes of different volumes operated by a workflow programmable drive. Typical syringe sizes range from 100 µL to 10 mL. Excellent linearity of the dispensed volume can be achieved from 10 µL on as shown in Figure 4.21. Many automated systems allow the parallel operation of several dilutors modules. The solvent dispensing precision is typically better than 1%. Using a large volume syringe of 10 mL, a gravimetrical precision of 0.1% RSD can be achieved [33].

For serial dilutions in workflows, a dilutor module is the tool of choice. The source solvent reservoirs can be made to fit easily to the processing needs. Typical applications in sample preparation workflows are for instance the fatty acid methyl ester (FAME) analysis with the delivery of three different reagents sodium methylate, heptane, and citrate, with intermediate wash steps, as outlined in Table 3.1. A dilutor tool can also serve as an injector tool for large volume injections (LVIs). Any potential carryover is minimized. The sample does not contaminate a syringe. The sample

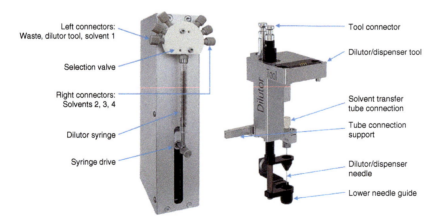

Figure 4.19 Dilutor module for multiple solvent sources with dispensing tool. Source: Courtesy CTC Analytics AG.

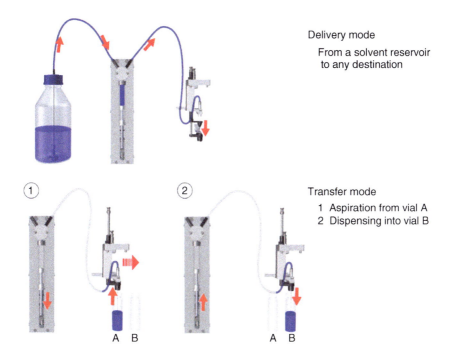

Delivery mode
From a solvent reservoir to any destination

Transfer mode
1 Aspiration from vial A
2 Dispensing into vial B

Figure 4.20 Dilutor operation in delivery and transfer mode. Source: Courtesy CTC Analytics AG.

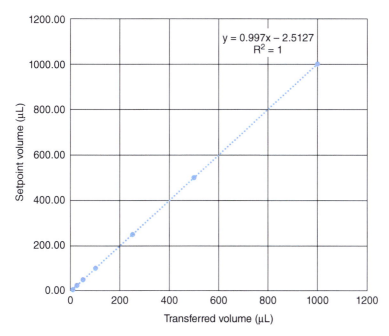

$y = 0.997x - 2.5127$
$R^2 = 1$

Figure 4.21 Dilutor module linearity for liquid delivery from 10 µL to 1000 µL (water, gravimetrical). Source: Courtesy CTC Analytics AG.

transfer tubing can be washed in backflush mode to waste using the connected wash solvents. An application using a dilutor tool for the online SPE with LC–MS/MS detection was published by S. Huntscha et al. [34].

Depending on the positioning of the dilutor device and the solvent reservoirs, a pressure drop in the solvent tubes to the reservoirs can occur. The reduced pressure during aspiration can cause cavitation with bubble formation of low boiling solvents. In this case, the degassing of the solvents and the positioning of the reservoirs close to or on equal height solves the cavitation effects.

4.1.6 Flow Cell Sampling

Many automated analysis applications require the sampling from continuous flows. This can be achieved from gaseous or liquid media. The media to be sampled are pumped or taken directly from pressurized lines and routed through a flow cell. If necessary pressure reduction, flow control, filtering, exhaust connection, or on/off devices in the supply line need to be considered separately. Flow cells in different designs and purposes are in use, mainly different for sampling from either pressurized lines equipped with a septum or for open access.

For the continuous monitoring of water samples, an open-access design as shown in Figure 4.22 facilitates sampling and maintenance. Several water streams can automatically be sampled and analyzed with a parallel flow cell setup. For sampling, flexible solutions can be used with tools for water withdrawal by a dilutor, syringe or pipet tool for sample transfer to vials. For the analysis of volatile organic compounds (VOCs), the water is transferred directly from the entry point into capped vials. After the withdrawal of a liquid sample aliquot, further analysis steps using LLE, DLLME, SPE/micro-SPE/online-SPE, or a derivatization can be implemented and become part of the workflow. Static or dynamic headspace analysis can follow, see Section 5.1. The analysis of dissolved analytes, the semi-volatile organic compounds (SVOCs), a direct on-line sampling using the solid-phase micro-extraction (SPME) direct immersion technique (DI-SPME) can be accomplished as shown in Figure 4.22, see also Section 4.4.5. An additional derivatization step for analysis of extracted polar compounds by GC-MS, or a liquid desorption to LC-MS, can be integrated for the online SPME-GC analysis [35].

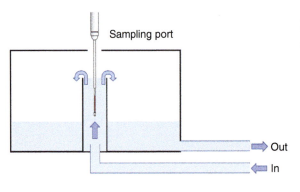

Figure 4.22 Flow cell for water analysis in open access design, shown with a SPME sampling device.

Figure 4.23 Flow cell with septum sampling point, shown with syringe sampling. Source: Courtesy CTC Analytics.

Typical applications are the analysis of wastewater, surface and drinking water. In continuous at-site process analysis, online sampling is used for instance the control of chemical or fermenter reactions, or exhaust gas analysis. The minimum frequency of sampling is determined by the analysis cycle time. In many cases, the prep-ahead function can be used to collect and prepare the sample in time for the next analysis, see Figure 4.74, for the usage of micro-SPE with prep-ahead function for sample clean-up.

A closed flow cell design is shown in Figure 4.23 connected to inlet and outlet piping. Multiple flow cells can be installed in parallel for sampling from different media sources. The closed flow cell has a low internal volume of 500 μL only and is equipped with a regular silicon rubber septum. The septum can be penetrated by a robotic sampler for liquid or gaseous media withdrawals with a syringe needle. The septum allows the direct connection of pressurized media. The flow cell shown does not provide a built-in heating, but can be accomplished separately for the in/outlet piping including the flow cell using a metal heating block design. Septum and cell maintenance is required depending on the nature of the media and sampling frequency.

4.2 Solid Materials Handling

The processing of solid materials requires special mechanical devices. Solids appear in analytical workflows routinely as the samples itself or are required as salts buffer or drying agent additions. High throughput preparative workflows as used for pharmaceutical formulations or combinatorial chemistry screening dominate the use of automated solids dispensing tools. Usually, the addition of different solid materials before further compounding and processing is required. The big challenge to overcome in automated solids dispensing is caused by the different material characteristics in various aspects covering mechanical, chemical, and physical properties. Another aspect addresses the handling of hazardous or active materials that have to be kept sealed, amongst others with the operation in exhaust or glove boxes. It is obvious that different technical solutions are necessary to cover this wide range of differing material properties, making dedicated and differently optimized dispensing heads necessary.

Instead of volume measurements used for liquids, the gravimetric control becomes indispensable for quantitative procedures from microgram range up. Accuracy and speed of dispensing are here the important characteristics of robotic dispensing systems. In the context of this textbook, the subject of solids powder dispensing is limited to aspects of applications in analytical workflows.

4.2.1 Workflows with Solid Materials

In general, sample preparation steps from solid samples received in a laboratory like plant materials, soil or food require individually optimized homogenization of a representative sample size by cutting, grinding, or milling, including cryo-milling for volatile analytes. This comminution of raw samples to achieve a representative aliquot as the test portion for analysis is mostly manual and not subject within the context of this textbook.

Many widespread analytical workflows require the addition of salts in solid form in particular for headspace extractions (see Section 5.1). Salts increase the ionic strength or are used in buffering the pH value. Popular examples are the VOC analysis with headspace extraction from water, or the pesticides analysis using the acetonitrile-based QuEChERS extraction. If robotic systems do not provide a possibility for powder dosing, the workflow needs to be designed by the addition of salts as a first manual step by adding prepared vials into the sample tray. This way also continuously working online systems are realized. In the example of the online control of the off-odors geosmin and 2-methylisoborneol (2-MIB) by headspace SPME, a sufficient number of prepared headspace vials, capped with 3 g of salt added, are provided with a rack capacity for at least 24 hours of operation [36], see also Section 6.7.

4.2.2 Automated Solids Dosing by Powder Dispensing

Manual dosing of solids is a time-consuming and repetitive process. Small amounts require tight tolerances, usually not easy to handle precisely. It takes a lot of manual time for weighing vials and recording the data. As in all manual operations, there is the risk of human error. There are as well safety concerns about hazardous materials and inconsistent samples. The demand for automation of such multiple times repeated handling steps is obvious. Automated powder dosing eliminates the manual variability and out-of-specification results, and protects users against exposure to potent substances. In a laboratory routine environment, the automated weighing and powder dosing is much faster than dosing manually and increases efficiency hence. Available solutions cover the range from dosing one powder to many vials (Figure 4.24), or extending the scope to the dispensing of many different powders to many vials (Figure 4.25). A cobot is used for the transfer of sample vials from a storage rack to the balance and back, as well as changing the dosing heads according to

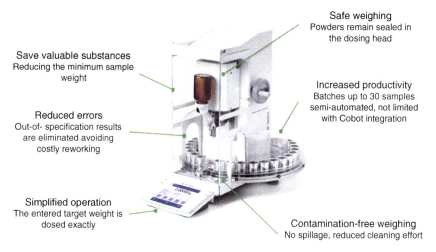

Figure 4.24 Automated powder dosing – One powder to many vials. Source: Courtesy Mettler Toledo AG.

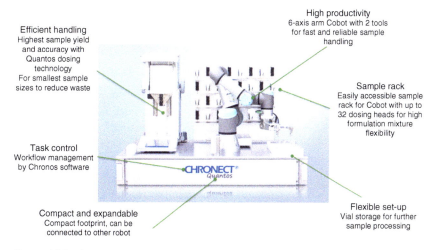

Figure 4.25 Automated powder dosing – many powders to many vials. Source: Courtesy Axel Semrau GmbH & Co KG.

the required chemicals, programmed as a workflow. Measures against electrostatic charging during powder dispensing must be in place. The dosing heads are RFID coded so that the automated processing also offers the required traceability for regulated environments. The vial storage rack is used in this setup not only as the reservoir of empty vials for weighing and powder dosing. The racks work as well as the transfer point for capping and decapping vials (see Section 3.6). Further sample processing is achieved by subsequent sample preparation workflows involving liquid handling, e.g. dissolution, dilution, liquid sample addition for extractions, reactions, chemical synthesis, and integrated online analysis (Figure 4.26).

Figure 4.26 Automated powder dosing and sample processing workstation with (1) Powder dosing head; (2) Cobot gripper, here taking a vial after powder dosing from the balance; (3) Analytical balance; (4) Rack of powder dosing heads for different chemicals; (5) 6-axis cobot; (6) RFID reader for dosing head tracking; (7) Vial rack transfer station; (8) Decapper/capper device; (9) x,y,z-Robotic sample processing system; (10) Vial racks for further processing. Source: Courtesy Axel Semrau GmbH & Co KG.

4.3 Weighing

Volumetric dosing is temperature-dependent. Automated sample preparation systems usually work in laboratories at different room temperatures, but not at the 20 °C level at which volumetric equipment is calibrated. The temperature-dependent media density defines the relationship between the mass and its volume. The marking of a volumetric flask for instance is typically calibrated at 20 °C and indicates the nominal volume at that temperature. For quantitative work and preparations at regular laboratory conditions, the determination of the mass of a compound or standard solution is compulsory. Weighing allows the correct dilution to a required analyte concentration. While the thermal expansion coefficient of borosilicate glass is negligible with only 9×10^{-6} °C^{-1}, aqueous solutions have a coefficient of volume thermal expansion of about 210×10^{-6} °C^{-1}, and organic solvents in the magnitude of $>1000 \times 10^{-6}$ °C^{-1} [37].

Beyond the accurate determination of mass, many practical aspects are solved by weighing. A typical example is the handling of viscous media. While pipetting with suitable tips works well (see Section 4.1.4), the mass determination via volume is critical. For viscous media, the weighing of the dispensed sample for further calculations is recommended. Also, dispensing of solid materials into processing vials requires the final weight confirmation. Typical laboratory implementations use cobots for handling the required powder dosing head and vial transport to and from the balance, see also Section 3.2.6.

Another application of general importance is the automated preparation of work standards or calibration dilutions, for instance for multi-compound pesticide analysis [38]. Using robotic systems, it is possible and recommended to keep the individual reference stock solutions in a temperature-controlled cooled tray, as it is the refrigerated storage in manual operation. These stock solution vials can be capped instead of septum sealed. For quality control, it is necessary to monitor the integrity of each reference by weighing and tracing the stock solution vial weight during its use. Differences in weight after last and before next use can indicate a leakage with loss of solvent, hence change in the expected concentration. Septum capping for needle penetration is depreciated for working standards preparation. Screwcap closures for automated de-capping and re-capping after withdrawal work reliably in the long run (see Section 3.6). Full traceability of the required standards preparation for quality control is achieved by barcode labeling and comprehensive documentation. For the weighing step, vials can be transferred by magnetic or gripper transport, as well as using cobots to and from a system integrated balance.

The integration of a laboratory balance to a robotic sample preparation system as shown in Figure 4.27 is the most flexible way. It allows the choice of balance type and measurement range best suited for the intended workflow. The balance must be mechanically fixed relative to the robotic system to maintain the taught x,y,z-position. The weighing pan uses a dedicated rack to accept the used vial sizes. Depending on the type of balance, the protecting doors are closed by the robot or the balance activates the door closure before starting the weighing process. A unique solution for gravimetrical solids dispensing is commercially available with an overhead balance integrated into the powder dosing head for simultaneous dispensing and weighing of solid materials [39]. The control of the balance and data transfer is achieved using the common communication standards via a bidirectional serial interface handshake, Ethernet/IP, or similar protocols.

Figure 4.27 Weighing step with magnetic vial transport using a system-integrated laboratory balance. Source: Courtesy Axel Semrau GmbH &Co KG.

4.4 Extraction

In trace analysis, the samples for food, soil, or tissue analysis cannot be applied directly to analytical instruments. In particular, in residue analysis, samples need to be homogenized from a larger batch. Analytes get extracted from a representative test portion and concentrated. Different extraction processes are in use depending on the physical sample properties. No single extraction technique serves all the needs. Many different techniques are applied depending on the range of analytes and sample matrix. All of them are initially developed for manual operation using a larger sub-sample, a large amount of organic solvent, the clean-up, and finally evaporation for concentration, and analysis. The classical Soxhlet extraction is a good example of a continuous liquid extractor. Invented in 1879 by the German chemist Franz Ritter von Soxhlet (1848–1926), Professor for agricultural chemistry at the Technical University of Munich, it is still in laboratory use today [40]. The solvent volume has to be at least three to four times of the sample volume in the extraction thimble, e.g. for samples up to 100 g a volume of more than 300 mL solvent is required (Figure 4.28). Automated Soxhlet apparatus can work with samples as low as 10 g and reduce the solvent need to 50–100 mL [41].

Micromethods gain ground [42]. With the miniaturization of the sample processing, the requirement of homogeneous samples is of special focus. The comminution and homogenization of a laboratory sample become the initial and essential part of the analytical process to obtain representative test portions for the automated sample preparation and analysis [43]. It is stated by the European Directorate-General for Health and Food Safety (SANCO) that "Sample comminution should ensure that the sample is homogeneous enough so that sub-sampling variability is acceptable. If this is not achievable, the use of larger test portions or replicate portions should be considered to obtain a better estimate of the true value." This SANCO document became complementary and integral to the requirements in ISO/IEC 17025 [44].

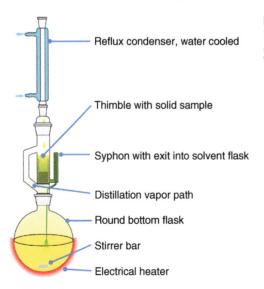

Figure 4.28 The classical manual Soxhlet continuous liquid extraction apparatus. Source: Redrawn from Wikipedia [39].

The micro-extraction techniques are the miniaturized versions of known traditional and mostly manual extraction techniques, providing the potential for automation with inherent significant advantages [45, 46]. Only microliters of potentially toxic organic solvents are used. They are easy to perform, offer high enrichment factors and extraction efficiencies. In the case of polar analytes, simultaneous derivatization and pre-concentration are possible avoiding additional evaporation. The extracts are cleaner with low matrix interference. And, the methods are compatible with different analytical instruments for online analysis.

4.4.1 Liquid Extraction

The classical extraction of solids and sediments is the ultrasound-assisted liquid extraction. Sonication is one of the basic extraction techniques concerning equipment and solvent choice. The first patents on the application of ultrasound for extraction purposes appeared already at the end of the nineteenth century. The sonication of a solvent/solids slurry intensifies significantly the solid/liquid transfer of analytes from the sample material. Ultrasound waves in the range above 20 kHz up to typically 40 kHz create smallest cavitation bubbles by the strong alternating stress in the liquid. The unstable bubbles implode after a short growth and create high shear forces at a boundary layer. This effect is noticeable with the typical noise of ultrasonic baths. The shear forces generated by cavitation occur predominantly at the solid/liquid boundary and lead to the crushing and disintegration of contiguous particles in the liquid media [47].

Very comprehensive methods for ultrasound extractions are laid down with US EPA methods. The EPA method 8270E for the analysis of more than 100 semi-volatile chemicals (SVOCs) from solid waste matrices, soils, air sampling media, and water samples also refers to the ultrasound-assisted liquid extraction [48] and lists the compounds extracted with this method. The EPA method 3550C describes in detail the procedure for extracting non-volatile and SVOCs from solids such as soils, sludges, and wastes [49]. As the ultrasonic extraction is not considered to be as rigorous as other extraction methods for soils and solids, a minimum power requirement of the ultrasound bath of 300 watts is required. The mentioned EPA methods are not miniaturized. For a sample of 30 g typically, 100 mL solvent is required, with two repeat extractions. After filtration the majority of the extraction solvent must be evaporated, the concentrated extract then must be subjected to clean-up. Gel permeation chromatography (GPC) must be performed for all soil/sediment extracts [50].

The EPA method 3550C restricts the ultrasound extraction to three solvent systems that may be employed for different groups of analytes. It is important to note that "the choice of extraction solvent depends on the analytes of interest. No single solvent is universally applicable to all analyte groups. The purpose of a water-miscible solvent like acetone is to facilitate the extraction of wet solids by allowing the mixed solvent to penetrate the layer of water of the surface of the solid particles. The water-immiscible solvent extracts organic compounds with similar polarities". Hexane is typically applied for the non-polar analytes, while dichloromethane (DCM) is used for the more polar compounds. Acetone/hexane

(1:1, v/v) or acetone/DCM (1:1, v/v) is used for the extraction of semi-volatile organics, organochlorine pesticides, and polychlorinated biphenyls (PCBs) [49]. For analysis, the extraction solvent can be exchanged with a compatible solvent for the following instrumental analysis.

Based on the mentioned EPA methods, developments for miniaturization were established. Mudiam et al. applied the sonication for assisted micro-extraction for the simultaneous determination of 10 phenolic endocrine disruptor chemicals (PEDCs) including alkylphenols, parabens, and bisphenols [51].

Robotic x,y,z-systems today can easily accommodate an ultrasound bath or module in the accessible working space. In many installations, a commercial ultrasound bath is equipped with a rack for 2, 10, or 20 mL vials. The on/off switching and sonication time are well controlled by the programmed workflow. Customized ultrasound modules are commercially available in different capacity. The shown ultrasonic module in Figure 4.29 developed by the German SIM GmbH offers vial cooling by a separate cooler for 4×20 mL and 5×2 mL vials. The lid of the ultrasonic bath is perforated at the vial positions and allows syringe access, while vials are kept inserted in the module. The vial transfer is achieved by magnetic or gripper transport. Beyond solid/liquid extractions, ultrasonic baths can serve and facilitate many additional workflow tasks and extraction methods like LLE, DLLME, dissolution, mixing, or reaction steps.

4.4.2 Pressurized Fluid Extraction

Pressurized fluid extraction (PFE) [52] is an instrumental analyte and matrix independent technique for preparative as well as trace analytical extractions. PFE is

Figure 4.29 Ultrasound module with a perforated cover for vial access used with x,y,z-robots. Source: Courtesy SIM GmbH.

fast, uses only small solvent volumes, and often provides cleaner extracts than the manual and time-consuming classical extraction procedures. PFE as the generic term is known with different branded names like accelerated solvent extraction (ASE) [53], pressurized solvent extraction (PSE) [54], or pressurised liquid extraction (PLE) [55]. The terms are used interchangeably in the literature. PFE as a method does not cover the high-pressure regime of the supercritical fluid extraction (SFE). The instrument configuration with solvents, pump, extraction oven with cell, and finally the extract collection vial is illustrated in Figure 4.30. The publication of PLE as a US Environmental Protection Agency (EPA) extraction method in 1995 contributed to its widespread acceptance as a green extraction technique with high efficiency for a wide range of matrices [56].

Extraction cell volumes can vary between 10 and 100 mL. In trace analysis, typically 10 g of sample in a 10 mL extraction cell is applied. Solid samples, and with auxiliary sorbents also liquid samples, can efficiently be extracted with pressurized solvents at elevated temperatures. Samples with high moisture or water content should be dried before extraction, for instance by freeze-drying. Samples must be fine-ground and well dispersed for example in sand or hydromatrix, a diatomaceous earth material. Glass or cellulose filters are used on the bottom of the extraction cell.

4.4.2.1 Solvents and Extraction

The PLE extraction solvent or a mix of solvents is provided by a high-pressure pump. A static extraction phase is used to permeate the sample material and desorb the analytes from the matrix, followed by a dynamic step flushing the extract into a collection vial. Solvents known from the traditional or manual extraction method are usually a good starting point for method optimization. Mixing solvents of different polarities can extract a broader range of target compounds. Organic solvents, as

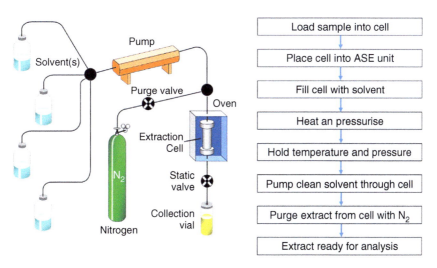

Figure 4.30 Instrument configuration for pressurized liquid extraction ASE™. Source: Image used with permission of Thermo Fisher Scientific Inc.

well as water, or buffered aqueous solutions, are in use for different target analyte groups as listed in Table 4.7.

For the following GC or GC-MS analysis, hexane and DCM/acetone (1:1) are the recommended extraction solvents. However, if LC is the preferred analytical technique, water, methanol, or acetonitrile as LC compatible solvents are in use. For GC analysis, residual water is removed from extracts by adding anh. sodium sulfate into the collection vials. A mixture of hexane/acetone (1:1) is for instance a common solvent for extracting POPs and pesticides. Standard extraction parameters are 100–200 °C cell temperature, five minutes static extraction at a pressure of approx. 100 bar (1450 psi, 10 MPa), followed by 50–60% dynamic cell volume flush. The total extract volume collected depends on the cell volume in use. For a 10 mL cell, it is the cell volume itself plus the 60% flush volume resulting in about 16 mL extract volume, for a 33 mL cell, it is 58 mL respectively. Although this is significantly less than with classical LLE, these volumes still result in a high analyte dilution, not compatible with a direct subsequent online chromatographic analysis. Such extract volumes must be evaporated manually or using an evaporation module on the robotic system. Extracts are frequently subjected to a clean-up process, and finally, aliquots are transferred to autosampler vials for analysis. The large extraction volumes are an obstacle for seamless automation of the regular PLE, in the current format still a semi-automated approach.

4.4.2.2 Miniaturization and Automation

On the background of the high extraction potential and efficiency of PLE, an online realization of PLE is highly desirable. A further development for miniaturization of the PLE process was presented by Ramos et al. [58]. A miniaturised PLE with in-cell purification and subsequent GC analysis using the electron capture detector (ECD) was published for the determination of priority pollutants in a variety of foodstuffs. A small extraction cell of approximately 1.7 mL volume, similar to the dimensions of an LC steel column of 100 mm × 4.6 mm ID × 6.6 mm OD, was used for PLE. The sample size was 100 mg of homogenized meat. Two successive static extraction cycles using n-hexane as solvent, performed at 40 °C and 120 bar (1700 psi, 12MP) at a temperature of 100 °C was used. The extract was collected in autosampler vials for semi-automated analysis. No evaporation was required [58].

In a similar but on-line solution based on the earlier works of Ramos, the PLE cell was further miniaturized with a heated stainless-steel cell of 10 mm × 3 mm ID [59]. A solid sample (50 mg) was packed into the cell for PAH analysis. As solvent 100 µL of toluene was applied at 150 bar (2175 psi, 15 MPa) for 10 minutes in a static and then dynamic mode. Clean-up or filtration of the extracts was not required. The extracts were directly transferred into a GC column by LVI as shown in Figure 4.31. Detection limits for the complete online PLE-LVI-GC-MS procedure were below 9 ng/g for 13 EPA PAHs in real soil samples. The repeatability was better than 15% RSD. A comparison of this PLE approach with Soxhlet or LLE showed that the efficiency of the miniaturized PLE is the same or better for both spiked and non-spiked samples. Other examples demonstrated the applicability for miniaturized PLE for the

Table 4.7 Suggested solvents for PLE of different matrices.

Matrix	Analytes group	Suggested solvents
Fruits and vegetables	Environmental toxins (organic, PAHs, POPs)	DCM/acetone (1:1, v:v), hexane/acetone (1:1, v:v), toluene, ethyl acetate
Fruits and vegetables	Environmental toxins (inorganic, anions, kations)	Water
Fruits and vegetables	Antioxidants	7% Acetic acid in methanol, 0.1% TFA in methanol
Grains and seeds	Oils	Hexane, hexane/isopropanol (3:2, v:v), petroleum ether
Grains and seeds	Pesticides	DCM/acetone (1:1, v:v), hexane/acetone (1:1, v:v), toluene, ethyl acetate
Grains and seeds	Mycotoxins	Methanol/water (1:1, v:v), acetonitrile/water (1:1, v:v), methanol/water/H_3PO_4
Meats (beef, pork and poultry)	Pesticides	Acetone/hexane (1:1, v:v), DCM/acetone (1:1, v:v)
Meats (beef, pork and poultry)	Sterols	Chloroform/methanol (2:1, v:v), hexane/isopropanol (3:2, v:v)
Meats (beef, pork and poultry)	Sulfates and nitrates	Water
Meats (beef, pork and poultry)	Antibiotics	Water
Fish and shellfish	Environmental toxins (organic)	DCM/acetone (1:1, v:v), hexane/acetone (1:1, v:v), toluene
Fish and shellfish	Environmental toxins (inorganic)	Methanol, methanol with 1% acetic acid
Dairy	Environmental toxins (organic)	DCM/acetone (1:1, v:v), hexane/acetone (1:1, v:v)
Dairy	Environmental toxins (inorganic)	Water
Dairy	Fats and lipids	Hexane, hexane/isopropanol (3:2, v:v), petroleum ether
Snack foods	Fats and lipids	Hexane, hexane/isopropanol (3:2, v:v), petroleum ether
Snack foods and potatoes	Acrylamide	Water with 10 mM formic acid or acetonitrile (100%)

Source: Redrawn from Dionex Corporation [57].

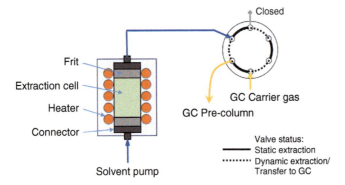

Figure 4.31 Miniaturized online PLE with LVI to GC analysis. Source: Based on Ramos et al. [58].

extraction of VOCs and SVOCs from oak wood samples of wine barrels with GC-MS analysis [60].

These examples demonstrate well the important potential of future developments in miniaturized PLE for automation with online analysis capabilities. "We are not aware of commercially available instruments" [61]. A customization for individual solutions by system integrating value-added resellers would be required.

4.4.2.3 In-Cell Clean-Up

In order to avoid an additional clean-up after an automated extraction, sorbent material can be added to the sample before extraction, called "in-cell clean-up." For many sample types, this approach has proven successful in producing clean extracts that are ready for direct analysis. See the principle of the in-cell clean-up in Figure 4.32.

The sorbents used for clean-up were packed into the cell. The adsorbent can either be mixed with the sample, or as done most often, put first into the extraction cell as a bottom layer, separated by hydromatrix, frits or simply using glass fibre filter material. Suitable adsorbent materials for in-cell clean-up can be aluminium oxide, magnesium silicate (Florisil™), or silica gel, see Table 4.8. The most used aluminium oxide (Al_2O_3, alumina, activated by oven drying at 350 °C for 15 minutes)

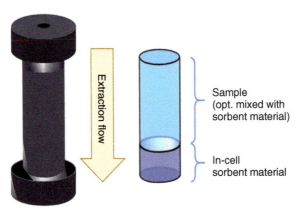

Figure 4.32 PLE extraction cartridge packing for in-cell clean-up. Source: Image used with permission of Thermo Fisher Scientific Inc.

Table 4.8 Sorbent materials used for PLE in-cell clean-up.

Adsorbent	Application
Florisil™ (basic magnesium silicate)	Adsorbs mainly lipids, fats, and other organic compounds of the matrix
Alumina (aluminium oxide, basic, acidic or neutral)	Adsorbs lipids
C18 resin	Lipid organic contaminants
Silica gel	Binds polar and unsaturated compounds
Sulfuric acid saturated silica gel	Oxidizes the hydrocarbon matrix for the isolation of chlorinated POPs

prevents the extraction of unwanted lipids. Hexane should be used with alumina as the extraction solvent to keep the lipids retained on the alumina material. In contrast, if a hexane/acetone (1:1) solvent mix is used, almost no lipid material will be retained on the alumina. Generally, 60–70 mg of lipid material can be retained per gram of alumina. For example, a 10 g sample of fish tissue contains approx. 10% lipid. About 17 g of alumina is required to bind the lipid content. A larger 33 mL extraction cell would be required [57]. Mixing the sample with C18 resin (1:2) retains many unwanted organic contaminants. Typical applications with GC-ECD analysis are shown from egg and sea bream analyses using the miniaturized PLE with in-cell clean-up [58], or for POPs analysis by GC-HRMS (high-resolution mass spectrometry) [62].

4.4.2.4 International Standard Methods

Traditional sample extraction methods were standardized including not only the classical Soxhlet extraction with EPA Method 3540 [63], but also the automated Soxhlet extraction EPA Method 3541 [41]. The accelerated extraction method using PLE was standardized and is included in the SW-846 Methods with US EPA Method 3545a, applicable to the extraction of SVOCs, organophosphorus pesticides, organochlorine pesticides, chlorinated herbicides, PCBs, and polychlorinated dibenzodioxins and -furans (PCDDs/PCDFs) [64, 65].

4.4.3 Liquid/Liquid Extraction

LLE are common sample pre-preparation methods in many analytical applications, if not the most used extraction process at all. LLE was reported as the sixth most common analytical technique in a laboratory, behind weighing, dilution, filtration, centrifugation, and pH adjustments; within those, LLE is the first in row and most important extraction technique [66]. LLE is based on the liquid–liquid partitioning of the target analytes from a usual aqueous sample to a non-aqueous organic solvent. The nature of the solvents used affect the degree of transfer of both analytes and co-extractives. LLE always entails partitioning that leads to equilibrium. Some analytes are going to partially remain in the different phases, leading to <100% recoveries and/or less than complete clean-up. Matrix components can cause changes in the partitioning, leading to more variability. Hence, the use of internal standards (ISTDs) is recommended. The general scope of LLE application is often narrow, which is good for selectivity but often not suitable for multi-residue analysis.

The traditional LLE methods are manually intensive and require a hood and bench space. Large volumes of organic solvent are required, which is not compliant with current efforts for green analytical techniques. Risks to operators can be caused by the manual handling with often hazardous solvents, spillage, and dangers of accidentally dropping the large glass funnels when shaking (Figure 4.33). The basis for many of the LLE methods still in use today was laid with publications by Milton A. Luke et al. from the US Food and Drug Administration in Los Angeles for the sample preparation in pesticide multimethods which coined the well-known name "Luke method" [68–70]. The solvent volumes were lately reduced by the introduction of the "mini-Luke method" [71]. A typical Luke method "E1 Extraction with Acetone, Liquid-Liquid Partitioning with Petroleum Ether /Methylene Chloride" requires 220 mL acetone, 250 mL petroleum ether, and 350 mL methylene chloride [72]. The highly diluted extracts need to be pre-concentrated prior to analysis, an additional potential source of contamination, loss of analytes, and irreproducibility.

The miniaturization and automation of LLE methods was continuously developed and is an accepted alternative in the analytical community as a replacement of the manual procedure [73]. "The move towards miniaturization has resulted in improved techniques that use smaller amounts of organic solvent, provides superior extraction efficiencies, permits the on-line coupling to analytical measurement techniques, and allows easier automation and higher extraction throughput" [74]. Also, new alternative solvents are under investigation to replace toxic and volatile organic solvents for LLE. For instance, ionic liquids (ILs) find increased interest due to their particular handling and analytical benefits of low volatility and flammability, and the inherent potential for customization by the introduction of different cations and anions for the extraction of specific analytes [75].

Figure 4.33 Classical liquid/liquid extraction procedure in a routine water laboratory.

The miniaturized "traditional" LLE can be attained in standard 2–20 mL vials with the aid of a strong vortex mixer. The vortexing unit should be able to intensely disperse even high-density chlorinated extraction solvents like chloroform or tetrachloromethane in a solvent/water system (see Section 4.10.1). A common vortexing unit operation is shown there with Figure 4.118. Some practical limitations with emulsion formation can occur with LLE of matrix samples. Increasing the ionic strength by salt addition, e.g. sodium chloride, is a typical measure. The emulsion can also be broken with additional centrifugation after the vortexing step to achieve a sharp boundary between the aqueous and organic phases. A clear phase separation is required for taking up extracts reliably for further processing steps in the automated workflow, in particular when using low-density hydrocarbon extractants forming the upper layer. A bottom layer can be safely aspirated by using the bottom sensing function of the robotic samplers, if available. Vials with conical bottom support the bottom sensing function, as shown in Figure 4.37.

A very practical solution for automated workflows is the optical detection of the phase boundary during the LLE workflow as developed by the Swiss Brechbühler AG. A camera is positioned within the modules of an x,y,z-robotic sampler as shown in Figure 4.34. Placing a vial in front of the camera, the phase boundary, or a solvent level meniscus in case of a volumetric flask, can be monitored and measured in its height by the control program. Also, for the centrifugation step first, the liquid level in the vial can be checked for safety before moving the vial into the centrifuge rotor. This step assures the correct weight balance of the rotor and avoids potential loading errors. After centrifugation, a second control step verifies the phase separation. The camera monitor screen with phase boundary detection of a 2 mL autosampler vial after centrifugation is shown in Figure 4.35. The meniscus search is done top-down along the shown blue vertical between the previously set-top and bottom limits (yellow lines). The top phase meniscus detected is marked with a green line, the phase separation meniscus with a red line. From the determined meniscus levels, the needle penetration depth into the vial can be calculated by the control program within the workflow. As required, the top or bottom layer can be safely aspirated. This optical control feature allows safe miniaturized LLE

Figure 4.34 Camera module on a x,y,z-robotic sampler for meniscus and phase separation monitoring. Source: Courtesy Brechbühler AG.

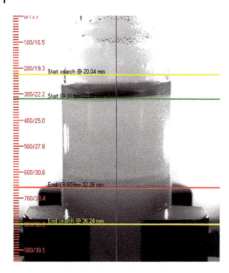

Figure 4.35 Automated optical control of the LLE phase separation level (red line) in a 10 mL vial for needle penetration calculation. Source: Courtesy Brechbühler AG.

extraction steps with low solvent volumes in automated workflows before injection into the chromatographic system.

4.4.4 Dispersive Liquid/Liquid Micro-Extraction

The DLLME is the newest and most promising green analytical chemistry development for automated liquid extractions in the recent decades. In May 2006, the method was first published by Mohammad Rezaee et al. as "a very simple and rapid method for extraction and pre-concentration" of organic compounds such as the demonstrated PAHs extraction from water [76]. In the same year 2006, the publication by Sana Berijani et al. extended the application range to the extraction of organochlorine pesticides (OCPs) from water [77]. DLLME found immediate attention in the analytical community confirmed with the notable increase in published papers during the following years (Figure 4.36). Despite the outstanding scientific interest, the method was mainly manually performed and the lack of automation

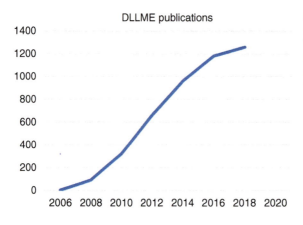

Figure 4.36 Increase of scientific publications using DLLME for sample preparation since the first publication in 2006. Source: Google Scholar search for "DLLME".

noticed early and limited its routine use [78]. Today DLLME is primarily used for extraction, clean-up, and concentration for trace analysis from aqueous media. Concentration factors of more than 100- up to 272-fold were obtained for PAHs and phthalate esters from water for direct analysis by GC or GC-MS [79, 80].

DLLME is based on a three-component solvent/sample system. For applications in aqueous samples, a few microliters of a water-immiscible solvent (extractant) are dissolved in a disperser solvent and rapidly injected into the aqueous sample. The extractant solvent should have a high affinity for the target analytes (partition coefficient). The disperser must be miscible with both, the extractant solvent and the sample, which was water in the first reported studies. The extractant/disperser solvent mixture is rapidly injected into the aqueous sample that microdroplets are formed providing a high surface for analyte extraction. The dispersive solvent ensures the miscibility of the organic phase and the aqueous sample phase and allows the immiscible solvent to form an emulsion with a high surface area for the hydrophobic extractant [81]. Due to this very high surface area, the time to reach a maximum recovery is short. Low polar analytes get efficiently extracted from water into the hydrophobic extractant phase. The extraction process usually is instantaneous, but can be supported by vortexing or sonication of the vial. A pH adjustment and the salting-out effect are additionally applied to increase analyte recoveries. Following this extraction step, the sample vial is centrifuged. The centrifugation step separates the emulsion.

Tapered vials of ca. 10 mL volume with a conical bottom are used, also named "centre drain" vials. Special dual-use DLLME vials are also commercially available with conical bottom and a narrow top neck, most suitable for the aspiration of a low-density extract in the narrow neck, as shown in Figure 4.37 [82]. As extractants typically high density and often, chlorinated solvents are applied

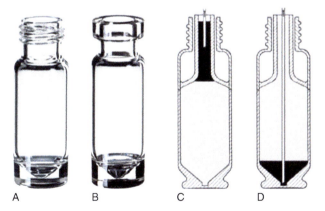

Figure 4.37 Vials with conical bottom or narrow neck for automated DLLME sample preparation. Magnetic caps are used for vial transportation on x,y,z-robotic samplers. (A) Screw cap vial, allowing needle bottom sensing. (B) Crimp cap vial, allowing needle bottom sensing. (C) Dual-use DLLME vial, narrow neck for low-density extractant, for safe needle penetration depth adjustment. (D) Dual-use DLLME vial, conical bottom for high-density extractant and bottom sensing. Source: (C, D) Redrawn from LABC-Labortechnik [82].

that sediment after centrifugation at the vial bottom. The enriched high-density extractant phase can be taken up from the bottom layer with a robotic sampler using the bottom-sensing function with a regular microsyringe for GC injection. The extract can then be directly analyzed by GC-MS without further clean-up or pre-concentration (Figure 4.38). Non-chlorinated solvents as the light hydrocarbons toluene, octanol, or hexane are forming an upper layer above the aqueous sample after centrifugation, as illustrated in Figure 4.39 [83]. Such hydrocarbon extracts are well suited for applications with GC-ECD detection.

The numerous DLLME method advantages including its simplicity, rapidity of operation, low consumption of organic solvents, low cost, and high enrichment factors from low volumes of samples made DLLME an attractive green method for analytical laboratories [84]. Routine laboratories benefit from the increased productivity and ease of automation with x,y,z-robotic samplers. DLLME can be carried out online with GC-MS or LC-MS on the chromatographic timescale in the prep-ahead

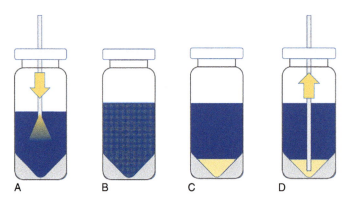

Figure 4.38 DLLME steps with an aqueous sample and *high*-density extraction solvent. (A) Injection of high-density extractant/disperser solvent mixture; (B) Emulsion formation; (C) After centrifugation; (D) Extract take-up with bottom sensing.

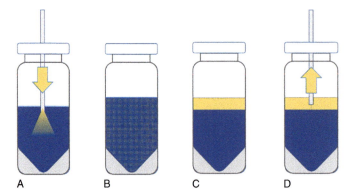

Figure 4.39 DLLME steps with an aqueous sample and *low*-density extraction solvent. (A) Injection of low-density extractant/disperser solvent mixture; (B) Emulsion formation; (C) After centrifugation; (D) Extract take-up from the upper layer.

mode during the analysis of a previous sample (see also Figure 4.74). The flexibility of the three solvents phase system allows the use of the full range of aqueous immiscible organic solvents as extracting solutions customizing DLLME for a wide range of analytes/matrix combinations. Typical disperser solvents include isopropanol, acetone, methanol, or acetonitrile.

4.4.4.1 Automated DLLME Workflows

DLLME applications are straightforward, need only little effort in method development or optimization. Automated workflows use the standard functions of x,y,z-robotic samplers [81]. For the extraction from water or complex samples such as soil or urine solvents like DCM, trichloroethylene (TCE), tetrachloromethane, chloroform, or also ILs are taken from the respectively installed solvent stations. As corresponding dispersive agents often acetone, ethanol, or acetonitrile are chosen. For aqueous samples, a prepared mixture of approx. 1–2 mL of dispersive solvent with about 10–150 µL of extractant is injected rapidly by the robotic sampler into the aqueous sample using a regular liquid syringe of fitting volume. From capped vials, an appropriate air volume can be extracted first before injecting the extraction mixture to compensate for a vial overpressure. The basic DLLME workflow steps for high-density and low-density extractants are illustrated in Figure 4.38 and 4.39.

The DLLME sample preparation workflow is fully automated on robotic x,y,z-robotic samplers [81]. In contrast to a manual application, the injection of the dispersive/extractant solvent mixture can occur with high speed delivering a very fine droplet dispersion. Vial transport, vortexing, and vial centrifugation use standard functions and modules. The take-up of the extract from a bottom layer requires the tool change to a GC injection syringe and is facilitated by the bottom sensing function. For addressing the upper layer of low-density extractants, a precise needle penetration depth can be adjusted. A complete phase separation after centrifugation is required for the small disperser/extractant volumes. With low extractant volumes, the transfer of the complete organic phase into a small or micro-vial can also be used as an intermediate step. From the small diameter vial, a low microliter take-up of extract for injection can safely be accomplished. Alternatively, dedicated narrow neck vials shown in Figure 4.37 are available allowing the safe aspiration of even very low microliter extract volumes.

For LC hyphenation, a solvent evaporation and reconstitution in mobile phase is required. This can be accomplished by tool change to a double-needle device with inert nitrogen flow, shown in Figure 4.84, or by transfer to an evaporation module, as shown in Figure 4.85.

4.4.4.2 DLLME for Soil and Urine

A method for soil and urine analysis was optimized by Mohana Krishna Reddy Mudiam et al. [85]. The 0.5 g of a soil sample was mixed with 1.25 mL of acetone and sonicated for 10 minutes, followed by centrifugation. The supernatant acetone layer was separated for DLLME extraction. TCE (58 µL) was added, and the mixture was rapidly injected into 5 mL of water containing 7% sodium sulfate, then sonicated.

After centrifugation, 1 µL of the sedimented TCE phase was online injected into GC-MS.

For automated urine analysis, 1 mL of the sample was diluted with 4 mL water containing 7% sodium sulfate in a 15 mL conical bottom vial. For DLLME a mixture of 1.25 mL of acetone as the disperser solvent and 58 µL of TCE extraction solvent was rapidly injected by the robotic sampling system. After sonication and centrifugation, the bottom TCE layer was injected into GC-MS for analysis.

4.4.4.3 DLLME for Pesticides in Food

The combination of the QuEChERS acetonitrile extraction with DLLME has the potential for clean-up of the matrix loaded sample extract, and in particular to enrich the analytes from the diluted acetonitrile extract before the GC-MS analysis [86]. In this application, acetonitrile serves in two roles, as the extraction medium with QuEChERS, and as the dispersive solvent in water as an auxiliary DLLME medium. After QuEChERS extraction, 2 mL of the centrifuged acetonitrile extract is transferred into a separate vial and 50 µL of chlorobenzene as an extractant is added. This extract mixture is rapidly injected into a conical bottom vial with 4 mL of water, pH 3 adjusted with formic acid, and centrifuged. The sedimented chlorobenzene phase is used for further analysis, for instance after evaporation and reconstitution also for LC analysis [87].

4.4.4.4 DLLME Hyphenation with LC

A method for the analysis of glyphosate, its primary degradation product aminomethylphosphonic acid (AMPA), and glufosinate in irrigation water was developed by in situ derivatization and DLLME combined with LC-MS/MS [88]. The polar pesticides were first derivatized with fluorenylmethyloxycarbonyl chloride (FMOC-Cl). DLLME was applied as the sample clean-up and concentration method with acetone as dispersant and DCM as the extractant. The derivatized target analytes were concentrated in the DCM phase. After evaporation and reconstitution in mobile phase, the extract is analyzed by LC-MS/MS.

4.4.5 Sorptive Sample Preparation

The sorptive extraction methods are based on the dissolution of the analytes in a liquid polymer material. The analytes are typically extracted into polydimethylsiloxane (PDMS) in pure form, or in combination with polar or porous material coatings, on a rod, film, or bar. PDMS behaves as a stationary liquid. The extraction process with PDMS is controlled by the partition of the analytes between the polymer coating and the sample matrix, as well as by the phase ratio between the polymer coating and the sample volume. For a PDMS coating and aqueous samples, the partition coefficient resembles the octanol–water partition coefficient [89]. As the partition coefficient decreases with higher temperatures, the analytes will release easier from the matrix, but do not absorb with the same proportion into the PDMS coating. Due to this partition effect into PDMS, lower extraction temperatures are in favor of higher recovery rates [90].

4.4.5.1 Solid-Phase Micro-Extraction

A true green analytical extraction technique is available for automation with the SPME, first published by Catherine L. Arthur and Janusz Pawliszyn [91]:

> "Solid phase microextraction (SPME) onto chemically modified fused silica fibres with thermal desorption eliminates the problems associated with SPE while retaining the advantages; solvents are completely eliminated, blanks are greatly reduced, and extraction time can be reduced to a few minutes."

In contrast to many classical extraction techniques, SPME is free from any solvent use. A flexible thin silica fiber or metal rod is coated with sorbent material, a mix of sorbent materials, or applied "overcoated" in different layers, and exposed directly to the sample, either to the sample headspace or by immersion into a liquid sample. Also, passive air sampling [92] or outgassing applications from solid materials are reported [93]. SPME is a non-exhaustive equilibrium method. The extraction is based on the partitioning of the analytes between the SPME stationary phase and the sample matrix. Right after exposure, the sorption rises quickly in a first almost linear phase, then it takes time to approach the equilibrium condition asymptotically as shown in Figure 4.40. For routine analysis, and providing a reasonable sample throughput and quantitative assays, the extraction is stopped in the pre-equilibrium phase upon reaching approx. 60–80% of the equilibrium condition. Also, the penetration depth into the vial headspace should be considered and kept constant for measurements during the non-equilibrium phase. The sorbent to sample distance is reported to be critical for quantitative HS-SPME analyses, being specific to each VOC for the distribution ratio and diffusion constant. Short extraction times command for a close distance to the sample for optimum recovery [94]. The precise time control and optimum reproducibility for quantitative HS-SPME are best achieved with automated systems [95].

The sample temperature during extraction has a strong effect on the analyte partition and equilibrium level. A constantly heated agitating device is required to support a reproducible and fast extraction. Analyte recovery can further be improved by increasing the ionic strength using the "salting-out" effect, and often by the choice of

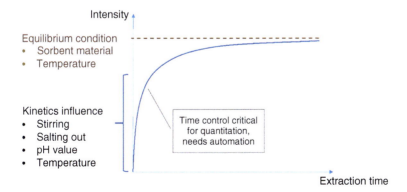

Figure 4.40 SPME extraction – physical factors affecting sample recovery.

a suitable pH-value. Also, the application of vacuum during HS-SPME extraction to facilitate the evaporation of analytes with lower vapor pressures for increased recovery is described [96].

SPME was licensed exclusively to one company for 20 years that led to only few product advancements in that time. New type of products with different form factors, from different vendors and related automated applications increased greatly after the patent expired. The development of diverse coating and sorbent materials of different polarities and thicknesses has expanded continuously the applications for the trace analysis of a wide range of compounds from different sample matrices.

The choice of sorbent materials allows selective or general applications. PDMS is found in most of the available SPME sorbent variations, serving also as the glue of the additionally added sorbent materials to the fiber or rod. Commercially available sorbent materials besides the pure PDMS are the most popular triple-phase divinylbenzene/carboxen/polydimethylsiloxane (DVB/CAR/PDMS), which is used for general applications. Carboxen/polydimethylsiloxane (CAR/PDMS), DVB, PDMS/DVB, polyacrylate (PA) and carbowax/polyethylene glycol (CW/PEG), and other proprietary sorbent mixes can be applied for more selective assays [97]. In the near future, ILs are also expected to play an important role as sorbent materials for further improved ruggedness and thermal stability [98].

For the choice of sorbent materials, the molecular weight, size, and the polarity of the analyte play the major roles [99]. The molecular weight and size of an analyte determine how rapidly it can move in and out of the phase coating and through the sample. A smaller analyte will move faster and is not well retained on thin coatings. Larger analytes migrate through the coating and sample more slowly and take a longer time to reach equilibrium. Thicker sorbent material coating with higher capacity requires longer extraction times to make use of the increased capacity for increased sensitivity, see also Section 4.4.5.6.

Sorbent materials extract via absorption and adsorption mechanisms [100]. In case of absorption, the analytes get dissolved in the highly viscous sorbent material (e.g. PDMS, PA, PEG, ILs). Polar and non-polar interactions drive the extraction process. The sorbents provide a high extraction capacity, especially with high film thicknesses, also fast desorption characteristics. During extraction, a partition of the analytes between the matrix and the phase occurs. Lower extraction temperature favors the analyte collection in the absorbance material.

In contrast, adsorption catches analytes on the surface and in cavities of the sorbent material (e.g. CAR, DVB, CW). Carboxen materials are porous particles with micropore sizes of <20 Å used for trace analysis of small molecules. For instance, Carboxen 1006 is characterized by an average pore size of 12 Å [101]. The capacity is limited, and saturation and matrix competition can occur. Carbon WR material is characterized by a wide range of pore sizes of 30–100 Å. Carboxen with larger pore sizes, with meso pores of 20–500 Å or macro size >500 Å, is available. With porous carbon materials, the molecular size plays the prevailing role in analyte retention besides polarity. The analyte binding is strong. Also, not only higher extraction temperatures can be applied, but also the GC injection requires higher energy to quickly

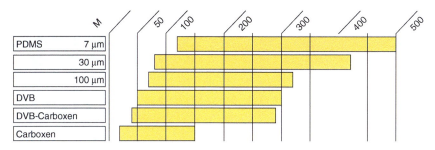

Figure 4.41 Molecular weight extraction range for selected SPME sorbent material. Source: Redrawn from Pawliszyn [99].

release the analytes. An overview of the recommended molecular weight extraction ranges for selected SPME sorbent materials is provided with Figure 4.41.

- PDMS is available in different film thicknesses. The thick 100 μm coating is recommended for low-molecular-weight volatile compounds. Non-polar semi-volatiles or large molecular weight compounds are more effectively extracted with a 30 μm or a 7 μm film. The thin 7 μm film allows higher desorption and conditioning temperatures.
- DVB/CAR/PDMS covers an expanded range of low molecular weight analytes and is recommended for trace compounds in the molecular weight range of C3-C20 (MW 40-275).
- CAR/PDMS is used for gases and trace level volatiles analysis.
- PDMS/DVB is recommended for volatile polar analytes, such as amines and alcohols, and also as a general-purpose SPME sorbent for LC amenable compounds.
- PA is used to extract very polar analytes from polar samples.
- CW/PEG is recommended for the extraction of alcohols and similar polar compounds.

An overview of applications for different SPME sorbent materials is given with Table 4.9. Application-specific material for the use of SPME is provided with a comprehensive SPME Applications Guide [101].

High matrix samples, in particular high in proteins, lipids or sugars with biological samples, blood or serum, are difficult to clean off in a thermal conditioning step. Such matrix can increasingly cover the sorbent material leading to short useful lifetimes. Overcoating of fibers consists of PDMS and protects the sorbent material from fouling. Other workflow solutions are the introduction of wash steps with polar solvents for water removable matrices before the conditioning.

For analysis, the extracted analytes get thermally desorbed into a GC injector, directly into MS [102], or desorbed into compatible solvents for LC or LC-MS analysis [103, 104]. Typical GC desorption temperatures are in the range of 200–300 °C. Thicker coatings require longer desorption times than thin ones. For chromatographic performance, it is important to use a specific SPME inlet liner closely fitting the OD of the SPME device (Table 4.12). After desorption into the GC (and before first use), it is advised to condition the SPME device in a dedicated conditioning station, if possible, and not exceeding the recommended sorbent maximum

Table 4.9 SPME applications by sorbent phase.

Application	Technique	Fibre coating	Film thickness [μm]	Polarity
Volatiles	GC/HPLC	Polydimethylsiloxane (PDMS)	100	Non-polar
Non-polar semivolatiles	GC/HPLC	Polydimethylsiloxane (PDMS)	30	Non-polar
Medium- to non-polar semivolatiles	GC/HPLC	Polydimethylsiloxane (PDMS)	7	Non-polar
General purpose	HPLC	PDMS–divinylbenzene (DVB)	60	Bipolar
Polar volatiles	GC	PDMS–divinylbenzene (DVB)	65	Bipolar
Polar semivolatiles (phenols)	GC/HPLC	Polyacrylate (PA)	85	Polar
Gases and volatiles	GC	Carboxen–PDMS	75	Bipolar
Odours and flavour	GC	Carboxen–DVB–PDMS	50/30	Bipolar
Polar analytes (alcohols)	GC	Carbowax–DVB	65	Polar
Surfactants	HPLC	Carbowax-templated resin (TPR)	50	Polar

Source: Based on Vas and Vékey [97].

temperature, up to 30 °C higher than the later desorption temperature. The maximum operating and conditioning temperatures specified by the manufacturer may not be exceeded. The application of SPME can be performed manual for feasibility studies, but is in routine executed by automated by x,y,z-robotic systems for a reproducible timing of the extraction, quantification, and 24/7 unattended operation.

SPME devices are commercially available in different sizes, geometries, and physical arrangements. Wide application for automated and semi-automated sorptive extraction analyses found the classical SPME fiber, but also the recent developments with the SPME Arrow, thin-film, and the Twister™ stir bar sorptive extraction (SBSE). Another form is available with the rugged HiSorb™ high-capacity sorptive extraction probes for use with the Centri™ multi-mode system developed by Markes International Ltd, shown in Figure 4.42. While these devices share the available variety of sorbent materials, the volume of sorbent material differs significantly and addresses different application areas (Table 4.10).

SPME is very widely applied today in water, food, environmental, drug, clinical analysis, and beyond. Table 4.11 provides an overview of the official methods and applications applying SPME. A practical guideline for selecting the appropriate sorbent material coating for SPME is provided by Janusz Pawliszyn [99]:

- Adsorbent coating is better for analytes at low concentration levels.
- Adsorbent coatings have limited capacity. The linear range for each analyte needs to be determined.

Figure 4.42 HiSorb sorptive extraction probe with the enlarged view of the sorbent phase section. Source: Courtesy Markes International Ltd.

Table 4.10 Sorbent material volumes of different micro-extraction techniques.

SPME technique	Sorbent volume [µL]
Solid phase micro-extraction fiber (SPME fiber)	0.1–0.6
Needle trap	1–4
Solid phase dynamic extraction (SPDE, in-column)	4.5
Solid phase micro-extraction arrow (SPME Arrow™)	3.8–16
HiSorb™ sorptive extraction probes	65
Sorbent Pen™ probes	150–200
In-tube extraction (ITEX DHS™)	160
Micro-extraction by packed sorbent (MEPS™)	100–250
Stir bar sorptive extraction (SBSE, Twister™)	100–300

- Absorbent materials are better for complex samples with varied concentrations.
- DVB/Carboxen/PDMS material is good for complex samples at low concentration levels due to the two adsorbent beds.
- A 30 µm PDMS coating is good for screening samples at high concentration levels over a broad molecular weight range.
- Absorbent coatings are better suited for dirty samples that may contain multiple unknown compounds.
- PEG coatings with polar selectivity are suitable for a wide range of analytes.
- PA coatings are suitable for the extraction of substituted aromatic analytes.

4.4.5.2 SPME Fiber

Making SPME an attractive and easy to use solution for users a syringe-like device was developed [105]. An SPME assembly comprises a holder with a syringe needle of gauge 23 and a flat and blunt needle tip (point style 3). The SPME Fiber devices with 23 gauge needles are recommended for use with autosamplers and are compatible with the Merlin Microseal® system [106], as well as other septum-less seals [107]. Septum-less GC injection is preferred as the blunt tip of the SPME fiber tube is

Table 4.11 Official methods and applications applying SPME.

Method	Application	Title	Detection	References
ASTM D6438, 2005	Paints, coatings	Standard test method for acetone, methyl acetate, and parachlorobenzotrifluoride	GC	www.astm.org/Standards/D6438.htm
ASTM D6520, 2000	Water quality	Standard practice for the solid phase micro extraction (SPME) of water and its headspace for the analysis of volatile and semi-volatile organic compounds	GC-FID, -ECD, -MS	www.astm.org/Standards/D6520.htm
ASTM D6889, 2003	Water quality	Standard practice for fast screening for volatile organic compounds in water using solid phase microextraction (SPME)	GC-FID	www.astm.org/Standards/D6889.htm
ASTM D7363 – 13a:2013	Sediment water	Standard test method for determination of parent and alkyl polycyclic aromatics in sediment pore water using solid-phase microextraction and gas chromatography/mass spectrometry in selected ion monitoring mode	GC-MS	www.astm.org/Standards/D7363.htm
ASTM E2154, 2001	Arson	Standard practice for separation and concentration of ignitable liquid residues from fire debris samples by passive headspace concentration with solid phase microextraction (SPME)	GC	www.astm.org/Standards/E2154.htm
DIN 38407-F34	Water quality	German standard methods for the examination of water, waste water and sludge – Jointly determinable substances (group F) – Part41: Determination of selected easily volatile organic compounds in water	GC-MS	www.wasserchemische-gesellschaft.de/dev/validierungsdokumente?download=30:f34-din-38407-34-2006-05&lang=de
EPA method 8272 (Dec 2007)	Sediment water	Parent and alkyl polycyclic aromatics in sediment pore water by solid-phase microextraction and gas chromatography/mass spectrometry in selected ion monitoring mode	GC-MS	www.epa.gov/sites/production/files/2015-12/documents/8272.pdf

Standard	Field	Title	Method	URL
GB/T 24281-2009 (China)	Textiles	Determination of volatile organic compounds – gas chromatography/mass spectrography	GC-MS	
GB/T 24572.4-2009 (China)	Arson	Standard practice for separation and concentration of ignitable liquid residues from fire debris samples. Part 4: Solid phase microextraction (SPME)	GC or GC-MS	
GB/T 32470-2016 (China)	Water quality	Organic compounds in drinking water. Test methods of geosmin and 2-methylisoborneol	GC-MS	
ISO 17943:2016-04	Water quality	Determination of volatile organic compounds in water – Method using headspace solid-phase micro-extraction (HS-SPME) followed by gas chromatography-mass spectrometry (GC-MS)	GC-MS	www.iso.org/standard/61076.html
ISO 27108:2010-04	Water quality	Determination of selected plant treatment agents and biocide products	GC-MS	www.iso.org/standard/44000.html
OENORM A 1117: 2004-05-01	Cellulose	Determination of volatile compounds in cellulose-based materials by solid phase micro extraction (SPME)	GC-MS	
UNICHIM 2237:09 (Italy)	Workplaces	Determinazione delle aldeidi aerodisperse – Metodo per microestrazione su fase solida (SPME) ed analisi mediante gascromatografia accoppiata a spettrometria di massa (GC-MS) (Determining airborne aldehydes – Method for solid phase micro-extraction (SPME) and analysis by gaschromatography coupled with mass spectrometry (GC-MS))	GC-MS	www.unichim.it/metodi/

prone to septum coring, limiting the number of injections in an automated sequence by a more frequent preventive septum exchange [95]. Inside of the needle, a partially coated fiber of polymer-coated fused silica (80 μm fused silica with 20 μm polymer protection), or a flexible Nitinol metal core is connected to the plunger [108]. The automated use of SPME fibers with large sample series can be limited due to the inherent brittleness of the silica fiber, GC septum coring, and a potential surface poisoning by the matrix with direct immersion applications. PDMS-overcoated fibers are recommended for matrix and life science applications.

The sorbent material-coated fiber section can be exposed and retracted from the needle by plunger movement, as illustrated in Figure 4.43. After penetrating a vial, the extraction process is achieved with the coated fiber part exposed out of the needle sleeve into a sample headspace. Retracting the fiber after exposure time into the needle provides the mechanical protection during the septum penetration and protects additionally from unwanted laboratory background contamination. For thermal desorption, the needle penetrates a regular GC inlet, and the sorbent-coated part is exposed for desorption. The length of the sorbent material coating with 1–2 cm is positioned for efficient desorption in the center of the GC inlet liner with optimum exposure to the set temperature. The used GC inlet liner ID is critical for achieving a sharp GC peak profile. Excess dead volume should be avoided to prevent peak broadening by dilution of the desorbed analytes in the carrier gas. Optimized GC liner for SPME Fiber desorption is commercially available for the most common GC instruments (see Table 4.12).

4.4.5.3 SPME Arrow

The development of the patented SPME Arrow device by CTC Analytics AG [109] led to improved robustness and sensitivity of the SPME technique [110]. In contrast to the SPME Fiber tool with an open fiber tube, the Arrow tip, as the name

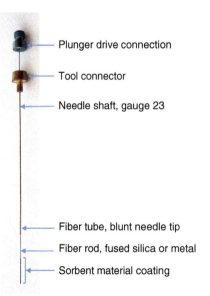

Figure 4.43 SPME fiber assembly for automated applications. Source: Courtesy CTC Analytics AG.

Table 4.12 Recommended GC inlet liner ID for SPME fiber and SPME Arrow desorption.

	ID [mm]	OD [mm]	L [mm]	SPME fiber	SPME Arrow 1.1*	SPME Arrow 1.5*
Injector inlet liner	0.8	as of GC type below		best fit	①	①
	0.8	as of GC type below		best fit	①	①
	1.3	as of GC type below		②	best fit	①
	1.3	as of GC type below		②	best fit	①
	1.7	as of GC type below		②	②	best fit
	1.7	as of GC type below		②	②	best fit
Conditioning station	1.7	6.3	78.5	③	④	④
GC type inlet liner dimensions						
Agilent 6890/7890/8890	s.o.	6.3	78.5			
Shimadzu 2010/2030	s.o.	5.0	95.0			
Thermo TRACE 1300/1310	s.o.	6.3	78.5			
Thermo TRACE Ultra	s.o.	8.0	105.0			

1 ⚠ WARNING: Use of this liner will lead to serious damage of SPME Arrow and injector!
2 This liner ID can be used as well with a slight loss of GC peak performance.
3 Bended fibers can contact the liner surface, replace upon visible contamination or fiber damage.
4 Check monthly. Replace upon baseline problems, at least once a year with continuous operation.
* Denotes the diameter of the Arrow rod in mm
General note: The SPME injector liners cannot be used for splitless liquid injection. Maximum volume is 216 µL with a 1.7 mm ID!

points to, carries a sharp tip. The area of sorbent material is placed right behind the tip in a recess of the metal rod. The arrow-type SPME device is available in outer diameters of 1.1 and 1.5 mm. Although featuring a wider diameter, the arrow tip penetrates GC inlet septa without coring for an increased number of analyses like liquid syringe injections, important for high throughput automated applications. The sharp SPME Arrow tip is shown magnified in Figure 4.44. It could be demonstrated that the force required for septum penetration using the sharp Arrow is less than with the open SPME Fiber rod. The SPME Arrow with 1.1 mm OD required 799 g, the 1.5 mm OD Arrow 908 g, compared to an SPME Fiber rod of 0.63 mm OD that required 1188 g. The data are taken from a balance read when piercing a regular 20 mL vial septum. For use with GC inlets, an adaptation of the septum nut and inlet liner guide for the wider arrow diameter is required, and commercially available for the most common GC instruments. The solid metal rod makes the arrow-type SPME device unbreakable. In particular, for automated applications, the SPME Arrow benefits from higher mechanical stability and improves the method reliability [111]. The larger surface compared to SPME fibers makes the arrow-type device also more suitable for DI-SPME applications (Figure 4.45) [110, 112].

The analytical method parameters for headspace SPME fiber extractions (HS-SPME) can remain unchanged with the SPME Arrow. All published applications can be used as a starting point for further optimization. The increased sorbent volume and surface compared to SPME fibers deliver shorter extraction times at the same sensitivity level, or increased extraction efficiency for more sensitive analyses compared to fibers as shown in Figures 4.46 and 4.47 [111, 113]. The larger surface and robustness of the arrow-type device also allow the extension of the routine

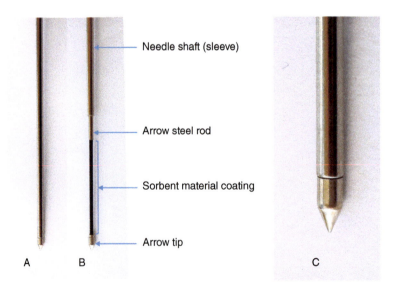

Figure 4.44 SPME Arrow tip closed (A) and with exposed black WR carbon/PDMS sorbent phase (B). Close-up view of the SPME Arrow tip (C) with sleeve closed. Source: Courtesy CTC Analytics AG.

1.5 mm PDMS Smart SPME Arrow   Surface: 63 mm², volume 12 µL

1.1 mm PDMS smart SPME Arrow   Surface: 44 mm², volume 3.8 µL

100 µm PDMS SPME fiber   Surface: 9.4 mm², volume 0.6 µL

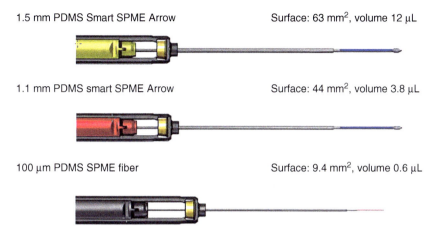

Figure 4.45 SPME Arrow and fiber dimensions. Source: Courtesy CTC Analytics AG.

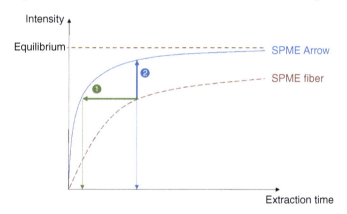

Figure 4.46 SPME Arrow analytical improvements compared to SPME fiber extractions: (1) Shorter extraction times at similar sensitivity; (2) higher sensitivity at the same extraction time.

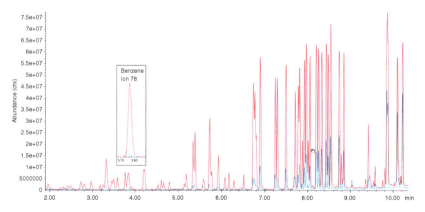

Figure 4.47 Comparison of HS-SPME response using SPME Arrow (red) and SPME fiber (blue) for VOC analysis. Source: Herrington et al. [111]. Licensed under CC BY 4.0.

application range to immersion extractions, mainly applied for aqueous media like drinking water and all kind of beverages for less volatile and polar compounds.

In the DI-SPME analyses, the aqueous boundary layer around the SPME rod affects the rate of diffusion of analytes to the sorbent material coating. A high frequency of shaking is required during the extraction phase in DI-SPME, for instance by usage of cycloidal shakers as described in Section 4.10.5. The optimum SPME immersion depth into the liquid sample needs to be considered. It must be assured that the SPME rod is positioned above the liquid surface and may not immerse itself into the sample. The SPME sorbent material then gets extended from the shaft into the sample. High rotating frequencies above 1000 rpm create a deep central vortex cavity pushing the liquid sample to the wall of the vial preventing sufficient exchange with the extended SPME phase. Depending on the sample volume, maximum vial penetration depths for DI-SPME must be taken into account. Table 4.13 informs about recommended penetration depth values for a 20 mL vial for up to 1000 rpm and typical fill volumes for aqueous samples.

4.4.5.4 Solid-Phase Micro-Extraction with Derivatization

During SPME headspace sampling, but particularly in direct immersion applications, also less volatile and more polar SVOC compounds are collected. Direct thermal desorption into a GC injector will not provide the expected chromatographic performance and insufficient detection limits, as many polar compounds require derivatization in gas chromatography. Several options are available for an automated derivatization step after micro-extraction [114–116]. In this context, only some selected guiding applications as typical examples are referenced.

On-fiber Derivatization A straightforward procedure and most suitable for automated workflows is the derivatization of the analytes after sorption on the fiber or arrow

Table 4.13 Recommended sample volume and penetration depth for DI-SPME liquid immersion methods.

Sample volume [mL][a)	Penetration depth [mm]
10	55
11	55
12	55
13	55
14	50
15	50
16	45
17	40
18	30

a) For a 20 mL vial sample, volumes < 10 mL and > 18 mL are not recommended.

phase as illustrated in Figure 4.54. The so-called "on-fiber" derivatization can be executed as part of a workflow with only a few additional sequence steps:

- Loading the sorbent phase with derivatization agent first, then proceed with the sample extraction.
- Perform the sample extraction first, then expose the collected analytes to the derivatization agent (post-derivatization).
- Combining both – load the derivatization agent first, extract the analytes, and then expose to the derivatization agent again.

The most suitable analytical strategy depends on analytes, derivatization agents, and matrices. Successful implementations are found manifold in the literature. Silylating reagents such as, BSTFA, MSTFA, and MTBSTFA were widely used (Figures 4.48–4.50), see also Section 4.8.2. For instance, the analysis of PAH metabolites in the urine of smokers used the derivatization after extraction. The fiber was placed in the headspace of a vial containing BSTFA before desorption to GC. The derivatization took place at 60 °C for 45 minutes. On-fiber derivatization methods were also applied for carbonyl compounds with O-(2,3,4,5,6-pentafluorobenzyl)hydroxylamine hydrochloride (PFBHA·HCl, Figure 4.51) or with pentafluorophenyl hydrazine (PFPH, Figure 4.52) for the analysis of aldehydes from wine and grape seed oil by GC-MS [117], or with pentafluorobenzyl bromide (Figure 4.53) for carboxylic acids from air [118]. A key SPME method was developed for the analysis of airborne formaldehyde with on-fiber derivatization [119]. The derivatizing agent PFBHA·HCl is first loaded onto the SPME fiber, then exposed to air for sampling. The analysis is finally done by GC with FID detection. An automated DI-SPME method with on-fiber silylation was developed for the analysis of endocrine-disrupting chemicals (EDCs) and steroid

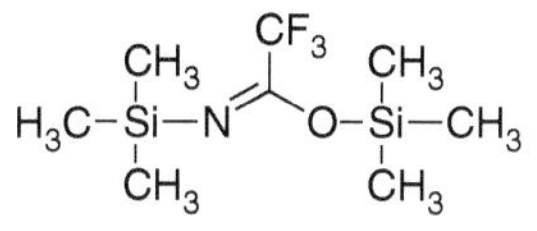

Figure 4.48 BSTFA.

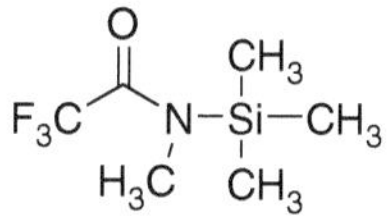

Figure 4.49 MSTFA.

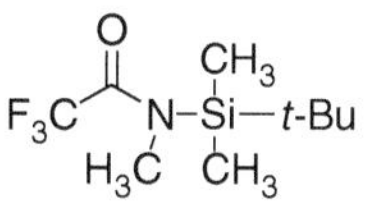

Figure 4.50 MTBSTFA.

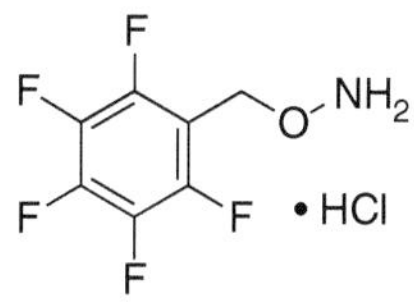

Figure 4.51 PFBHA·HCl.

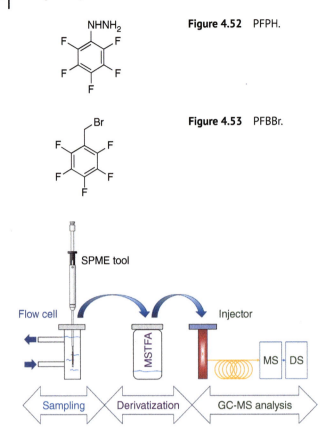

Figure 4.52 PFPH.

Figure 4.53 PFBBr.

Figure 4.54 Automated SPME on-line flow cell sampling and derivatization for GC-MS analysis.

hormones in aqueous and biological samples using the analysis concept shown in Figure 4.54 [120]. A systematic study for the SPME GC-MS analysis of degradation products of chemical warfare agents in water demonstrates the advantages of pre- and post-derivatization [35].

In situ Derivatization In situ derivatization, also labeled as "in-sample", "in-matrix", or "in-solvent" derivatization, is carried out directly in the investigated sample before submitting it to SPME or SBSE extraction. Derivatization performed in aqueous media includes the acylation of phenols with acetic anhydride, or of amines with ethylchloroformate, the esterification of acids, or the oximation of aldehydes and ketones with PFBHA and PFPH (Figures 4.51 and 4.52) [114]. Typical applications are the production relevant analyses of beverages, wine, and beer [121, 122], and the routine analysis of phenols in drinking water [123]. The fully automated analysis of organometallic compounds uses the in situ derivatization with sodium tetraethyl borate in aqueous samples followed by the SPME extraction. Also, high throughput drug analyses from urine samples for amphetamine drugs with the rapid in situ formation of their ethylformate derivatives are reported to provide good sensitivity and quantitative linearity [124]. A cooled tray extends the useful lifespan of the derivatization reagent in these applications [125].

In-port Derivatization With SPME and SBSE applications, the "in-port" derivatization is less used. For SPME in-port derivatization, the reagent is first injected into the injection port, followed by insertion of the loaded SPME phase. Different tools in an automated system have to be used or changed in the programmed workflow making this approach less practical. Silylation is the most versatile derivatization technique introducing a trimethylsilyl (TMS) group into the analyte molecule [126]. Following SPME extraction, the in-port silylation is reported for the determination of phenols and nitrophenols in water [127].

4.4.5.5 Direct Solid-Phase Micro-Extraction Mass Spectrometry

The direct desorption of the SPME device into MS offers a high potential for automation. The time determining and maintenance-intensive chromatography separation is avoided by direct MS analysis. Analytes extracted by SPME can thermally desorbed into a discharge ionization source [102]. This unique setup, developed by the Plasmion GmbH in Augsburg, Germany, allows the automation of SPME-MS analyses using x,y,z-robotic systems for high sample throughput with low maintenance effort.

The desorption takes place in a GC inlet like device, accessible for the x,y,z-robot from the top. The desorbed analytes are completely transferred into the ion source as illustrated in Figure 4.55. In contrast to a conventional electrospray ion source design (ESI), the loss of compounds in front of a cone is avoided with a dielectric barrier discharge ion (DBDI) source (Figure 4.56). The efficiency of ionization and

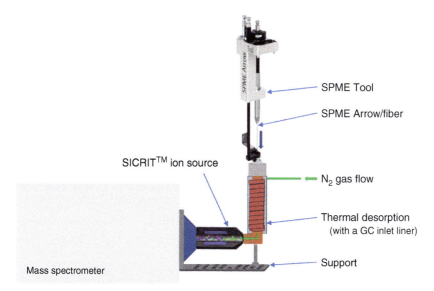

Figure 4.55 Schematic of the instrument configuration for automated direct SPME-MS analysis. Source: Courtesy Plasmion GmbH.

Figure 4.56 Conventional ESI vs. SICRIT™ DBDI ionization technique. Particles shown in green – desorbed analyte molecules; purple – ionized species. Source: Courtesy Plasmion GmbH.

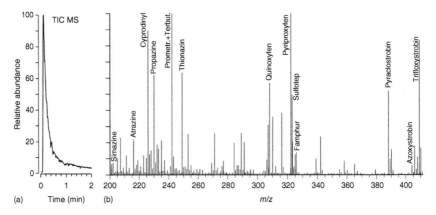

Figure 4.57 TIC-MS (a) and HRMS spectrum (b) of the direct SPME-MS analysis of a mix of pesticides. Source: Mirabelli et al. [128] / with permission of Royal Society of Chemistry.

the inherent sensitivity of the DBDI source are significantly higher compared to ESI type ion sources. Sensitivities in the low ng/L range from SPME water sample analyses are reported [128] (Figure 4.57).

The ion detection in direct MS occurs preferably by high-resolution accurate-mass MS (HR/AM-MS) or MS/MS technique. A comprehensive review of direct ionization methods is provided by Klampfl and Himmelsbach [129], an area of high potential also for automation but the subject would be beyond the scope of this textbook. The DBDI ionization provides high intensity molecular and quasi molecular ions ideally suited for targeted and general unknown analysis [130]. All analytes are detected in parallel within the short SPME desorption phase of approx. two to three minutes. The analyte-specific accurate mass signal is used for calibration and quantification. MS/MS techniques by using the fragmentation pattern from the (quasi-) molecular ion can be used for unknown identification. The selective quantification is based on compound-specific transition signals delivering high signal/noise values even with matrix present. The automated SPME-MS methods show high potential for a fast high throughput screening and trace analysis in such important fields like pesticide analysis, aroma profiling, microbial contamination, process monitoring, origin confirmation, or metabolomics [131]. In the automated setup with x,y,z-robotic

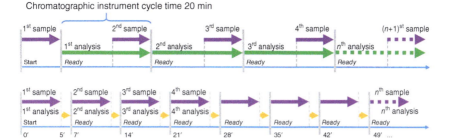

Figure 4.58 Automated direct SPME-MS analyses (bottom, five minutes extraction, two minutes desorption) compared to chromatographic techniques in prep-ahead mode (top) on the analytical time scale. Source: Based on Huba et al. [102].

samplers, short analysis cycles of a few minutes per sample, mainly depending on the SPME extraction time only, can be achieved. The analysis sequence of SPME-MS compared to chromatographic analysis is shown in Figure 4.58 [131].

4.4.5.6 Stir Bar Sorptive Extraction

Stir Bar Sorptive Extraction (SBSE) was developed in 1999 by the Pat Sandra group at the Research Institute for Chromatography (RIC), Kortrijk, Belgium. The large capacity and surface area of the sorbent material combined with active stirring extended the application range of SPME to a highly sensitive detection of very low levels of volatiles and semivolatiles [132]. PDMS-coated magnetic stir bars were used initially for aqueous media with extraction times of 30–60 minutes or beyond, exploiting the high analyte capacity (Figure 4.59). Headspace applications for liquid and solid samples (headspace sorptive extraction, HSSE, Figure 4.60), the passive on-site and in-field sampling, as well as the liquid desorption to LC-MS, followed as further developments [133]. SBSE became popular based on the Twister™, patented and marketed by GERSTEL GmbH & Co KG, Mülheim an der Ruhr, Germany, as a solventless sample preparation technique combined with automated thermal desorption for GC-MS analysis. The SBSE workflow is semi-automated with the stir bar extraction process as an initial separate step before a fully automated thermal desorption GC-MS analysis is performed. Usually, the sample extraction is performed in parallel on multiple position magnetic stirrer plates unattended for several samples simultaneously. SBSE is considered a green analytical alternative to many traditional extraction methods.

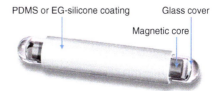

Figure 4.59 Twister™ stir bar. Source: Courtesy GERSTEL GmbH & Co KG.

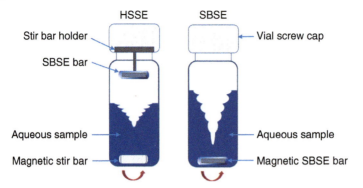

Figure 4.60 Liquid (SBSE) and headspace (HSSE) application of the magnetic SBSE bar.

The extraction with a PDMS phase follows the similar principles and kinetics as the known classical LLEs. PDMS is considered an immobilized extraction "solvent" on the stir bar, albeit reaching equilibrium less quickly. The expected recoveries at equilibrium conditions can be calculated using the Recovery Calculator, downloadable from the internet [134].

The "Twister" is a 10 or 20 mm long magnetic stir bar sealed in glass and coated with sorbent materials (Figure 4.59). Two Twister types are commercially available [135]. The most widely used phase is PDMS with phase volumes available from 24 to 126 µL for high extraction efficiency of non-polar analytes. The ethylene glycol (EG)/silicone-coated stir bars extract polar compounds more efficiently. The mixed phase EG/silicone extends the analyte range for polar and less polar compounds. Typically, stir bars are added to a sample vial and intensely stirred for 60–90 minutes at room temperature for a sample size of 10 mL, or overnight for larger samples. Adjustment of the sample pH and "salting-out" can be used to enhance extraction efficiency. The bar is then manually removed with forceps from the vial, rinsed with distilled water, and quickly dried under a flow of clean nitrogen gas. For automated analysis, the bar is placed into a thermal desorption tube as depicted in Figure 4.61, which is put into the dedicated sample rack of the x,y,z-robotic sampler. Thermal desorption occurs in the thermal desorption unit (TDU) shown in Figure 4.62 in an automated sample sequence. Stir bars can be reused up to 200 times depending on the sample type and desorption conditions. Between analyses, the Twister is conditioned at temperatures up to 300 °C in a nitrogen stream.

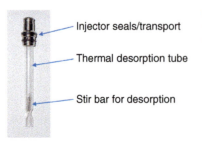

Figure 4.61 SBSE Thermal desorption tube loaded with a stir bar. Source: Courtesy GERSTEL GmbH & Co KG.

Figure 4.62 SBSE Thermal desorption unit for GC-MS. Source: Courtesy GERSTEL GmbH.

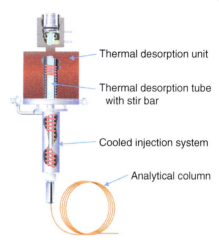

4.4.5.7 Thin-Film Micro-Extraction

The thin-film micro-extraction (TFME) was developed as an alternative to the established SPME and SBSE devices for higher extraction efficiency and sensitivity [136, 137]. The TFME device benefits from both, the larger volume of the sorbent phase and the larger ratio of surface area to phase volume. In contrast to the thick films of SPME fiber and Arrow analytes can be extracted in a much shorter time, an ideal starting point for automated high throughput analyses.

Typical applications are extractions of more polar compounds from aqueous matrixes by immersion [138]. Carbon fiber mesh membranes are applied to carry the sorbent materials glued in a high-density PDMS [139]. The thin-film device can be used for active sampling from vials as illustrated in Figure 4.63, but is also ideally suited for passive and on-site sampling, e.g. from water and air [140, 141]. TFME is semi-automated. After in-vial extraction, or on-site sampling, the thin-film is transferred for analysis to the racks of an x,y,z-robotic system for automated thermal desorption into a GC injector [142], or solvent-desorbed into LC-MS [143]. Due to the high potential of the technique, further solutions of this relatively new SPME format for comprehensive workflow solutions are expected.

Figure 4.63 Thin-film membrane in-vial immersion extraction.

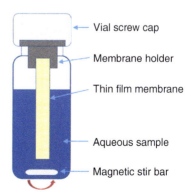

4.5 Clean-Up Procedures

Sample extraction steps, in particular the liquid extraction steps, deliver raw extracts, which carry a large amount of matrix as well. For a subsequent chromatographic analysis, the raw extracts need a mandatory clean-up from co-extractives and particulate matter. The decision of a necessary and the extent of a clean-up depends very much on the extraction method in use and the intended following analysis method. A "just enough" clean-up concept is established in automated methods, "enough" in the view of "fit-for-purpose" and being compatible with the analytical method for analysis. The clean-up is of particular importance in gas chromatography, while analyses via liquid chromatography often tolerate a limited application of matrix extracts. The installation of a pre-column, or pre-column backflush, the application of micro-SPE, or the automated robotic inlet liner exchange are practical solutions for robustness in GC analyses. With modern LC-MS systems of unprecedented high sensitivity, the "dilute-and-shoot" way is often the concept to prevent analyte signal quenching by competitive matrix effects.

Extraction methods that do not require further clean-up steps like static or dynamic headspace (e.g. purge & trap, ITEX DHS), SPME with fiber or Arrow, or SBSE are particularly advantageous in automated operation. Additional clean-up steps typically tend to dilute the extract, require additional evaporation, and pose a risk for analyte loss and additional contamination. Green analytical methods consistently try to reduce superfluous solvent use and evaporation. Finally, the workflows become more efficient and online solutions more productive with an overlap of sample preparation (aka "prep ahead" mode) and the analytical run (see Section 3.3.2).

4.5.1 Filtration

Membrane filtration in automated sample preparation procedures typically makes use of active pressure-driven filtration steps operated with the tools available on a robotic system. Independent of the design of the filtration device, it is the general practice to apply single-use filters to avoid any potential cross-contamination between samples, blanks, or standards. So also the necessity of filter disposal becomes part of the workflow.

Filtration prevents the contamination of sensitive instrumentation by solid impurities, in particular increasing the useful lifetime of chromatographic columns and equipment. A general requirement for the filter materials and housing is for low extractables or surfactants and a low dead volume for an automated micro-volume handling.

Adding filter steps in workflows needs to be considered carefully due to process risks with retaining analytes or added contaminations. Filters can clog, so single-use is mandatory not only with respect to carry-over. Finally, additional time, consumables, and cost are added to the workflows. Centrifugation should be considered as a practical alternative avoiding wasted consumables in the automated method setup.

4.5.1.1 Filter Materials

The choice of filter material follows the intended applications. A wide range of filter materials and pore sizes are commercially available. A major criterion is the hydrophilic or hydrophobic characteristics of the materials for aqueous or organic solvent compatibility. Many filter materials can be autoclaved or are commercially sterilized by gamma radiation so that also sterile filtration of protein solutions, sampling from tissue culture media, or water is possible. For non-aqueous applications, the compatibility of the filter material (and housing) needs to be checked. Leaching of extractables from the filter material itself or the housing need to be considered for trace analyses.

Cellulose Acetate (CA) Cellulose acetate is hydrophilic. The low protein binding characteristics make CA suitable for aqueous protein solutions.

Glass Fiber (GF) Glass fiber is well compatible with organic solvents and also strong acids (apart from hydrofluoric acid) and bases. GF material is often found in composite filter materials for increased mechanical stability and low flow resistance.

Mixed Cellulose Ester (MCE) Mixed cellulose ester is a hydrophilic material with low flow resistance. It allows high filtration speeds with little absorption.

Nitrocellulose (NC) Nitrocellulose is a hydrophilic membrane with a low level of extractables and a rapid flow rate for high throughput applications in life science.

Polyamide (PA) Polyamide is also known by its trade name Nylon™ which is hydrophilic and used for aqueous and organic sample preparation for a wide variety of sample types, often prior to LC or GC, for instance for dissolution sample analysis. Polyamide provides an excellent chemical resistance for filtering aggressive solutions and is chemically well compatible with esters, bases, and alcohols [144].

Polyethersulfone (PES) Polyethersulfone is a hydrophilic material for high filtration speeds, high stability in the pH 3 to 12 range with major usage in life science applications, for instance, filtering tissue cultures and LC applications. PES has low protein binding and low extractables.

Polypropylene (PP) PP is hydrophobic and provides a wide range of chemical compatibility to organic solvents.

Polytetrafluoroethylene (PTFE) PTFE is a hydrophobic material with strong chemical stability and inertness. PTFE is chemically resistant and used for filtering aggressive chemicals that destroy other membrane materials.

H-PTFE is a surface-treated hydrophilized PTFE for aqueous filtrations. Pre-washing with methanol is required to make the membrane more hydrophilic.

Polyvinylidene Difluoride (PVDF) Polyvinylidene difluoride is a hydrophilic inert material with wide applications in life sciences. The main characteristics are the low protein binding, less than Nylon, PTFE, or nitrocellulose, and minimal interaction with sample components, commonly used to clarify and sterilize protein-containing solutions.

Regenerated Cellulose (RC) Regenerated cellulose is hydrophilic, well suited for aqueous and organic media, but not suitable for aggressive acids and bases.

Beyond the standard materials, a wide variety of special materials are commercially available. Ion chromatography (IC) optimized materials are used for analyzing ionic species. Combination material filters with integrated glass fiber pre-filters are manufactured in layers to provide low flow resistance for high particulate matter samples.

Pore sizes can range from 0.1 μm to more than 5 μm. The most common pore sizes in use are 0.20 μm and 0.45 μm. Such filters remove microorganisms, particles, precipitates, and non-dissolved particulate matter larger than 0.20 or 0.45 μm, respectively.

The filter housings are made from different materials like PP, PTFE, or PA. Leachables and extractables from the filter housing need to be considered in applications. Commercially available dimensions are 4, 13, 25, or 30 mm diameter formats requiring customized storage racks in automated systems for proper syringe access. The filter housing burst pressure rates are specified by manufacturers in the range of up to 10 bar (145 psi).

4.5.1.2 Syringe Filter

Filtration steps can be integrated into automated workflows by using syringe filters. The applications cover any kind of filtration of suspended matter from liquid samples or gases. Manual methods can be transferred and automated directly. Typical applications are the removal of particles from a sample before analysis by LC or spectroscopic measurements to protect the column and instrumentation, or the clean-up of extracts from protein precipitation or cell cultures.

The syringe filters are disk filters with a large filter surface area for reduced flow resistance, see Figure 4.64. Due to the low flow resistance, high throughput applications are supported. For the attachment of filters to syringes, Luer tip fittings are used (refer also to Section 4.1.2.5). For the automated use with syringes, the filters are held ready in customized racks, shown as an example in Figure 4.65. The head of the robotic system pushes a Luer tip syringe with a defined force into the Luer connector of the filter. As Luer connectors are sealing by friction, and not safe under pressure, they are often operated by aspirating the sample rather than pushing through the membrane.

In an automated operation, the syringe barrel termination slips into the Luer tip enabling an automated syringe filter exchange, demonstrated in Figure 4.65 [145]. If uncontrolled dispensing happens, the usage of a simple Luer termination must be limited to careful aspiration of liquids. Dispensing at high speeds or against resistance creates the risk of dropping the filter from the syringe tip. With

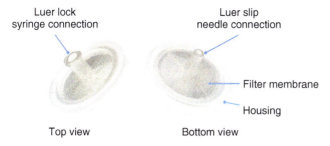

Figure 4.64 Luer tip syringe filter.

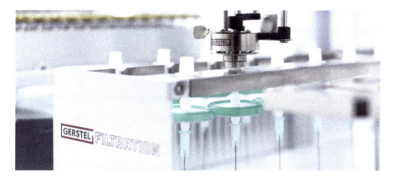

Figure 4.65 Automated use of syringe filters with Luer tip and pre-installed needle. Source: Courtesy GERSTEL GmbH & Co KG.

a programmable tool control and limitation of plunger speed and force with the GERSTEL Filtration Station, the syringe filter can be used safely and more flexible in dispensing direction. Syringe filters and needles are preassembled and kept ready for automated top-down filter applications in a dedicated rack, as it is shown in Figure 4.65. Trays for different sizes of syringe filters, also with a high capacity for overnight runs, provide maximum flexibility, enabling the use of all standard syringe filter sizes in automated workflows. The filtrate can be dispensed into open or sealed vials. During the filtration process, the robot monitors and limits the required force for the plunger operation. This inherent feature of robotic systems secure filter membranes from damage by overpressure, avoids leakages, and provides process safety. The lower needle guide of the syringe tool strips off the filter to waste after use.

4.5.1.3 Filter Vials

Filter vials are a single system which replaces both, autosampler vials and syringe filters, for the filtration of samples. They are consisting of two parts, an outer vial shell, and a plunger tube with a filter at the bottom side and a septum vial cap on top. Samples are filtered by pipetting the sample into the filter vial outer shell, inserting the plunger, and pushing the plunger down into the vial shell, as illustrated in Figure 4.66. The filtrate above the filter can be used for further sample processing. It is protected by the septum cap. The dimensions of the filter vials are compatible

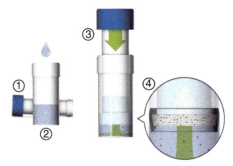

Figure 4.66 Filter vial operation. (1) Plunger with filter material on bottom, top blue-capped with septum; (2) Filter vial, added with sample; (3) Filter plunger pushed into sample; (4) Filtered sample is collected for processing in the plunger part. Source: Courtesy Thomson Instrument Company.

with the standard 2 mL GC or LC autosampler vials (12 × 32 mm) and can be used in x,y,z-robotic autosamplers. Filter vial systems are commercially available from several suppliers with partially patented solutions [146, 147]. Currently available filter materials are PVDF, PTFE, PES, PA, or multilayer membranes, with pore sizes of 0.20 µm or 0.45 µm. The filter vials patented by the Thomson Instrument Company [147] are available in different dimensions with maximum sample loading capacity 450 µL and 120 µL dead volume, and in a smaller version of 250 µL maximum fill volume with 10 µL dead volume. Multi-layered filter material is recommended for viscous and high particulate media. A high capacity filter vial even allows the handling of large volumes up to 630 µL of sample with 420 µL dead volume.

The filter plunger parts carry color-coded septum caps for membrane identification. Vials and filter plungers are held ready on automated systems in separate racks as shown in Figure 4.67 during operation with an x,y,z-robotic

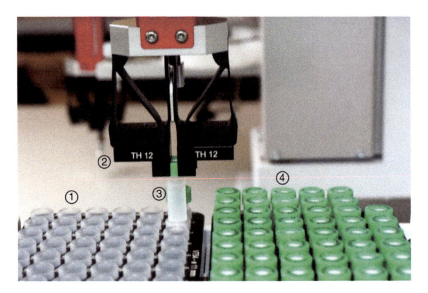

Figure 4.67 Automated sample filtration using Thomson filter vials. (1) Vial rack loaded with sample vials; (2) Universal gripper tool; (3) Vial filter plunger being pushed into sample vial; (4) Rack with Thomson filters hold ready for processing. Source: Courtesy Brechbühler AG.

system. After pipetting, the sample solution into the vial shell a gripper tool is used to take the filter plunger from the storage rack and presses it slowly with controlled speed into the sample chamber. After filtration, the filtered sample is ready for further processing by taking up with a regular syringe from the septum capped plunger reservoir. The sample pipetting step, insertion of the filter plunger and pushing down for filtration is automated for high sample throughput and online analysis. A critical point to be observed in automated system operation is the sealing of the filter plunger with the outer vial shell. A convincing practical application is the analysis of doping drugs by automated urine filtration and online LC-MS analysis, eliminating the time-consuming SPE steps for high sample throughput analyses [148].

4.5.2 Solid-Phase Extraction

Solid phase extraction (SPE) is a broad term used to describe the separation technique in which liquids contact modified solid surfaces and a component of the liquid adheres to the solid. In a separate step, the solid releases the component.

Mike Telepchak [149]

SPE is a well-established sample preparation technique in almost every laboratory today for a wide range of organic analyses. Also, samples often contain particulate matter, which can be filtered out in an SPE sample preparation process. The SPE clean-up and analyte concentration are based on the selective partitioning of compounds, the analytes, and those from the matrix, between a solid and a liquid phase. The separation is effected by the different affinity of the dissolved or suspended analytes and matrix compounds for the solid phase (sorbent material) through which the sample liquid is passed (liquid phase). Either, the desired analytes of interest, or undesired impurities in the sample are retained on the sorbent material [149–151].

One of the first applications was the "chemical filtration" application for the extraction of active compounds in pharmaceutical laboratories. The SPE cartridge works in a so-called "scavenging" mode. The matrix is retarded and the active compound or analytes of interest eluted. It is always preferable, if possible, to retain matrix co-extractives in solution on the SPE sorbent and let the analytes pass through. This avoids additional dilution of the sample extract which is of particular advantage for automated workflows as additional evaporation steps with all incurred risks of possible analyte loss, additional time, and likely contamination become obsolete. The approach of "chemical filtration" is used for the automated clean-up of QuEChERS extracts for pesticides using miniaturized SPE cartridges (see Section 4.5.2.3). In organic trace analysis, SPE started with the analysis of drinking water. Octadecyl-bonded silica sorbent material became the most popular solid phase in the beginning.

Today, SPE is used in many sample processing procedures to concentrate and clean-up samples mainly for chromatographic analysis from a variety of matrices, including food, water, beverages, soil in aqueous dispersion, also from urine, blood, or tissues. In many of these SPE applications, the analytes are retarded on the sorbent material, get concentrated from a diluted sample, and after washing the sorbent material extracted with a stronger solvent. The less affine matrix compounds get washed away. A significant limitation of the traditional SPE is the usage of high amounts of organic solvents, which can reach several 100 mL per sample, e.g. in the standard method for the gas chromatographic determination of chlorophenols in water DIN EN 12673 [152]. The cleaned extracts are collected in high dilution and are subject to evaporation prior to chromatographic analysis.

Following the LC terminology, normal phase, reversed-phase, ion exchange, and adsorption phase types based on modified silica are available [153]: The *normal phase* procedures use a nonpolar liquid phase and polar-modified solid phase for retention by hydrophilic polar-polar interactions, hydrogen bonding, $\pi-\pi$ interactions, dipole–dipole, and dipole–induced interactions. Analytes involved are typically polar within a mid- to nonpolar matrix. Elution solvents are more polar than the sample matrix. *Reversed-phase* separations use a polar liquid, usually aqueous phase, and a nonpolar solid phase. Analytes are typically mid- to non-polar. Interactions of the analyte with the stationary phase are of hydrophobic nature and nonpolar-nonpolar van der Waals or dispersion forces. A nonpolar solvent is used to elute. The *ion exchange* mechanism uses electrostatic attraction of charged compounds to the charged groups on the sorbent surface. Anion and cation exchange materials are in use. *Adsorption* effects occur by interactions of compounds with unmodified materials. Hydrophobic and hydrophilic interactions may apply depending on which solid phase is applied. Ready-to-use SPE materials are available in a variety of form factors of which tubes and cartridges, as well as disks, are the most popular forms. Detailed guides for choosing the most suitable SPE material for specific applications are available from many manufacturers. Table 4.14 shows the most common solvents applied in SPE clean-up according to their polarity, miscibility with water, and their use for reversed or normal phase procedures.

SPE has almost completely replaced today the traditional LLE with the usage of even larger quantities of organic solvents of several hundred milliliters, manual handling of expensive breakable glassware, and leading to significant volumes of chemical solvent waste. SPE with dedicated sorbent materials is faster and more efficient than LLE, and it is available automated with many implementations. While manual SPE procedures usually work vacuum-assisted, automated systems use closed systems allowing a positive pressure control of the sample load and clean-up steps, and for analyte collection. A particular challenge in the design of an automated system is caused when using the very different geometries of SPE cartridges and volumes. Here, flexible solutions with a variety of different rack sizes, tube adapters, and appropriate sample and solvent delivery are required (see Figure 4.68). Such automated SPE instrumentation is commercially available for routine applications [154]. An automated solution using the shown SPE cartridges, processes, and volumes follows and mimics the manual workflow, not yet taking advantage of the potential for further miniaturization, maintaining the analyte concentration, and for a seamless analytical integration with on-line capabilities.

Table 4.14 Characteristics of elution solvents commonly used in SPE.

Polarity	Elution	Strength	Solvent	Miscible in Water
Nonpolar	Strong Reversed Phase	Weak Normal Phase	Hexane	No
			Isooctane	No
			Carbon tetrachloride	No
			Toluene	No
			Benzene	No
			MTBE	Yes
			Chloroform	No
			Dichloromethane	No
			THF	Yes
			Diethyl ether	No
			Ethyl acetate	Poorly
			Acetone	Yes
			Acetonitrile	Yes
			Isopropanol	Yes
	Weak Reversed Phase	Strong Normal Phase	Methanol	Yes
			Water	Yes
Polar			Acetic acid	Yes

Source: Adapted from Smith [153].

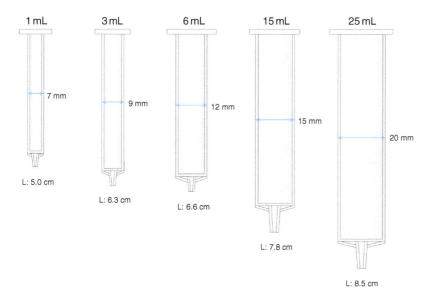

Figure 4.68 Various dimensions of commercial SPE tubes.

Automated solid-phase extraction and clean-up systems are commercially available in different design and capacity since long, and became a standard in many laboratories. In common, they are flexible for use of the different cartridge types shown in Figure 4.68, use vacuum for sample processing, and work mostly in the deciliter solvent range. Such popular systems are designed to work off-line with analytical instruments and not to be installed for online operation. Instrument-top installed automated SPE systems were developed by AiSTI SCIENCE CO., Ltd., Wakayama, Japan [155] and by the GERSTEL GmbH & Co KG, Mülheim, Germany [156], for online GC-MS or LC-MS analysis. The solution by AiSTI SCIENCE is based on a SCARA-type robot with the unique feature of proprietary stackable cartridges [155]. The GERSTEL solution is based on an x,y,z-robotic sampler and designed to work with standard SPE cartridges, uses a precise positive pressure solvent delivery as shown in Figure 4.69. The prepared workflow also offers the welcome potential for further miniaturization using 1 mL cartridges. Laborious manual procedures can be transferred directly to the automated robotic platform. The benefit of the positive pressure solvent delivery is that each cartridge is processed with highly reproducible flow rates and volumes. Besides the SPE process, on a robotic system, the extraction can be combined with additional workflow steps, for instance the addition of standards, evaporation, solvent exchange, or derivatizations. Finally, the online injection to GC-MS or LC-MS for unattended

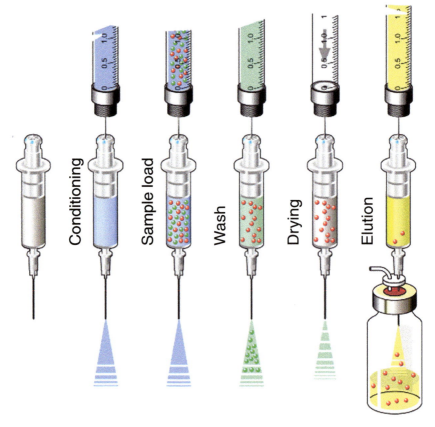

Figure 4.69 Automated SPE workflow steps using a positive pressure system. Source: Courtesy GERSTEL GmbH & Co KG.

analysis is performed after the change of the preparation syringe tool to the liquid injection syringe in the programmed workflow.

4.5.2.1 The General SPE Clean-Up Procedure

The general steps for an SPE sample preparation procedure are outlined in the following [150]. A graphical representation of an automated SPE extraction and clean-up workflow is illustrated in Figure 4.69.

Activation/conditioning: This initial step permeates the material with solvent and activates the functional groups to allow efficient interaction with the analyte.

Equilibration: The function of this step is to create a sorbent chemistry environment similar to that of the sample. If the pH is important for the extraction chemistry, the sample pH should be adjusted properly in both the sample pre-treatment and the sample conditioning step.

Sample loading: During sample loading, the free analyte binds with the extraction sorbent. In this step, the flow rate of the loading step is directly related to the residence time of the analyte on the extraction sorbent. A too high flow could result in a breakthrough of the analytes. Some of the matrix compounds will also be retained on the sorbent. Usually there is enough sorbent material so this does not negatively impact the capacity. A recovery standard can be added to the sample. In the chemical filtration mode (also called scavenging mode), the sorption function is reversed so that unwanted matrix compounds are retained on the sorbent material, here usually the higher molecular and lipid components, and small analyte molecules elute already during the load sequence. The next elution step with a limited solvent volume elutes the analytes for high recovery before matrix components can elute. A breakthrough of the matrix must be avoided.

Washing: With a wash step, most of the retained matrix compounds are flushed to waste. The wash solvent usually has higher elution strength (% organic solvent) than the solvent used for sample loading. Matrix compounds with a weaker interaction with the sorbent compared to the analyte will go to waste, while matrix compounds with stronger interactions will remain on the cartridge. The washing is stopped (limited solvent volume) before some of the analyte starts to elute from the cartridge (breakthrough).

Drying: A drying step of the sorbent material using a gentle nitrogen flow might become necessary for the following solvent exchange for elution of the analytes.

Elution: After the majority of the matrix compounds have been washed to waste, the analyte is eluted from the sorbent. Typically, a strong elution solvent is used to disrupt all the interactions of the analyte with the sorbent. Matrix compounds with stronger interactions are retained on the cartridge.

Evaporation: As several deciliters of solvent are typically applied in the SPE elution step, the evaporation of the solvent is required to concentrate the analyte fraction again for the subsequent mostly chromatographic analysis. Evaporation to dryness is avoided to prevent reduced recovery.

Reconstitution: The analyte is reconstituted in a solvent compatible with the mobile phase in LC, or for GC injection. An internal quantification standard can be added to the sample before injection.

4.5.2.2 On-Line SPE

Integrating the SPE sample preparation from the sample transfer to the cartridge, performing all the above-described cartridge treatment steps until the final analysis is called "on-line SPE". The on-line setup allows the miniaturization of the sample preparation to the microliter scale and facilitates the automation, making the online approach a true green analysis concept. Important advantages are the lower sample volumes, considerably reduced solvent consumption, and significantly higher sensitivity for lower quantitation limits with shorter processing times.

Several of the manual handling steps can be skipped. The cartridge conditioning, sample load, and washing are automatically performed in a prepared workflow. The drying of the cartridge, eluate collection in vials, the solvent evaporation for concentration, and further reconstitution for chromatographic analysis are not required. This does not only save substantial processing time but also reduces sources of potential human error and variability. Critical analytical steps like the use of additional glassware, potential evaporation to dryness, the known sources of losses and contamination are avoided. With on-line SPE, the samples are loaded on the cartridge with a positive pressure and controlled flow rate from regular sample vial racks or are taken in the required volume with sample loops directly from on-line water streams. With the use of positive pressure loading cartridges prepared for high-pressure applications (>4000 psi, 280 bar) can be applied. The high loading pressure allows the use of sorbent materials with lower sorbent particle sizes in the range of 5–15 μm for an improved separation. Finally, the elution of analytes from the cartridge and injection to chromatography becomes a joint workflow step. All of the analytes, initially loaded with the sample on the SPE cartridge, is eluted concentrated to the analytical LC or GC column with no dilution of the extract. The manual LLE reconstitution step even with low take-up volumes leads to the dilution of the extract to for instance 1:10 or 1:100 if just 10 μL, respectively only 1 μL of the reconstituted extract is injected. In contrast, on-line SPE injects without splitting the undiluted and complete analyte fraction. On the practical side, less sample is required for smaller cartridges and a higher sensitivity can be achieved.

These features, usage of low particle sorbent materials and the injection of the entire analyte fraction, are fundamentally different from manual SPE, and the major reason for the significantly improved analytical performance of the automated on-line SPE.

On-Line SPE Elution Profile With a high-pressure sample load and elution, small SPE particles in the range of 5–15 μm can be utilized, in contrast to the traditional SPE sorbents with typical particle sizes of about 40 μm. The smaller particle size increases the sorbent surface and hence analyte interaction, similar to the known LC behavior with smaller packing material sizes. Finally, smaller SPE cartridges can be used with only 50 mg of sorbent material compared with 100–1000 mg in the standard manual cartridges. The result in on-line SPE is a much sharper analyte elution profile providing a significantly improved selectivity from matrix compounds. The principle comparison of an on-line SPE elution profile from 8 μm sorbent material to a good manual cartridge material of 40 μm is shown in Figure 4.70. The analyte fraction

Figure 4.70 On-line SPE analyte elution profile. Source: Courtesy Spark Holland B.V.

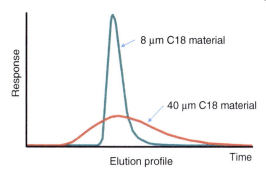

elutes in less solvent volume with higher concentration superseding the manual concentration step by evaporation.

The potential of on-line SPE includes an automated approach for the systematic method development. Variation of the several possible SPE parameters in a manual way is very time-consuming and is probably limited in practice to a few tests only. Starting from a generic SPE method optimum conditions for online-SPE can be developed systematically, and completely unattended. Figure 4.71 illustrates the concept of automated on-line SPE method development covering the evaluation of sorbent materials, optimization of washing and elution steps in different check stages, finally leading to the method validation [157].

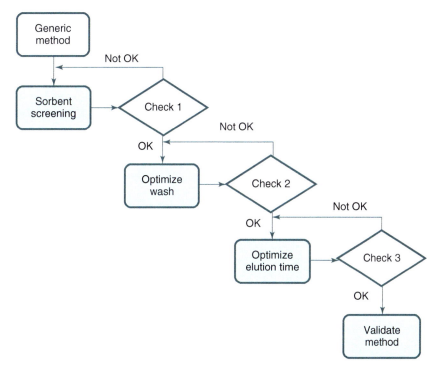

Figure 4.71 On-line SPE concept for automated method development. Source: Courtesy Spark Holland B.V.

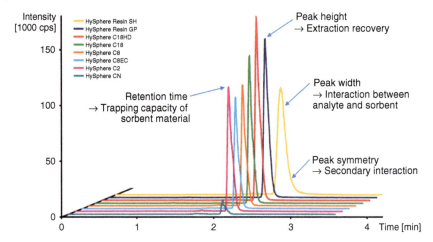

Figure 4.72 Results from the automated method development for the most suitable sorbent material selection with on-line SPE. Source: Courtesy Spark Holland B.V.

The choice of the most suitable sorbent material is key. An example of an automated sorbent material screening is shown in Figure 4.72. The analytes of interest are trapped onto the SPE cartridge utilizing a generic method. A selected number of cartridges with different sorbent materials, chosen according to the expected separation mechanism, are automatically applied using the same method. The comparison of the chromatograms achieved gives a good indication of the most suitable SPE cartridge for a particular application.

With the cartridge selected type, the wash conditions optimization can be performed by using a mix option of the high-pressure dispenser. The wash solvent volume can be altered stepwise in consecutive runs allowing the evaluation of peak width and symmetry, and most important peak height after completing the programmed runs. The effect of wash volume on trapping analytes on the cartridge is less critical than the organic modifier ratio. Besides wash volumes, also the wash solvent mixtures should be subject to optimization.

Typical for online SPE is the multiple use of the sorbent cartridge. Workflows must be designed and evaluated that all retained matrix must be removed, for instance after elution of the analyte. Final steps removing the retained matrix compounds extend the LC column lifetime, minimizes the re-equilibration time for the next sample, and improves the overall sample clean-up with high repeatability. Modern on-line SPE systems also allow the frequent and automated change of the clean-up cartridge to maintain optimum conditions [158].

Applications are related but not limited to water analysis, a typical area demanding for high-sensitive multi-compound methods with low limits of detection. For instance, the analysis of personal care products (PCPs) in environmental samples demands very low detection limits in the ppt range (ng/L) [159, 160]. A sensitivity enhancement of up to 90 times was achieved for some of the PCP compounds due to the high enrichment volume compared to the manual method [161]. Other examples are the widely discussed analysis of polar pesticides, e.g. glyphosate/AMPA [162], or PCBs and OCPs in human blood [163].

SPE has a proven success record in the past decades for a multitude of analytical applications. Current new developments for the miniaturization of SPE were very well recognized with high interest and quickly found application in the analytical community. In particular, solutions suitable for routine methods are addressing the increasing demand for automated and green analytical workflows. Further developments particularly in automated micro-SPE applications are expected.

4.5.2.3 Micro-SPE Clean-Up

The terms "micro-SPE" or "mini-SPE" represent the move toward miniaturization of the SPE method. This trend is following many other micro-methods, foremost SPME, for establishing green analytical methods (see also Section 2.1.1) [164]. The design and operation of the micro-SPE cartridges was initially patented by Kimberley R. Gamble and Werner S. Martin [165], commercialized today by ITSP Solutions Inc. (Instrument-Top Sample Preparation). Micro-SPE (µSPE) employs small autosampler vial compatible cartridges with a bed mass of up to 45 mg of sorbent material providing a solution for the miniaturization and full automation of the SPE process [166]. The pre-conditioning, loading of sample, matrix washing, and analyte elution from the cartridges only requires microliter extract and elution volumes and is performed fully automated using x,y,z-robotic systems. The particular design of the cartridge supports the operation of the solvent and sample delivery by a regular syringe using a positive pressure see Figure 4.73. In contrast to the manual SPE method that is using a vacuum manifold, the solvent and sample volumes and flow rates are controlled in an automated workflow by a syringe. The required solvent volumes are significantly reduced to the microliter range compared to the classical SPE making solvent evaporation obsolete. The analyte concentration from the sample is largely maintained or even enriched. Also, the

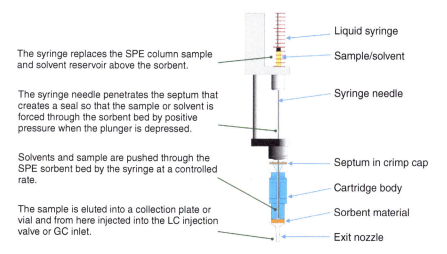

Figure 4.73 µSPE cartridge design and operation. Source: Courtesy ITSP Solutions Inc.

crimped septum of the current μSPE cartridges allows the needle transport from a conditioning station to the collecting vial for elution. A beneficial aspect different from other micro-methods is the one-time use of such cartridges with a disposal to waste after analyte elution for the elimination of any potential carryover or performance degradation by retained matrix.

Similar to the above-described online-SPE, also the μSPE operation can be compared well to the elution from an LC column. Typically packing materials are primary/secondary amine (PSA), octadecylsilane (C18), anhydrous (anh.) $MgSO_4$, zirconium-modified silica (Z-Sep), and graphitized carbon black with particle diameters between 10 and 50 μm. With these small particle sizes, a high separation efficiency is achieved using low flow rates of 2–10 μL/s during load and elution, which is most precisely delivered only by a preparative syringe in a programmable and automated system, as illustrated in Figure 4.73. Extract load and elution volumes are in the range of a few 100 μL only avoiding the concentration of the cleaned extract by evaporation as it is common in the traditional SPE. A detailed comparison of micro-SPE (μSPE) with the traditional SPE is given with Table 4.15.

A well-described application of μSPE is the clean-up procedure of the raw extracts from the QuEChERS extraction method for pesticide analysis by GC-MS and LC-MS [167]. Bruce Morris and Richard Schriner described the optimization of

Table 4.15 Comparison of Micro-SPE with classical SPE.

	Classical SPE	Micro SPE
Cartridge volume	1 mL to > 50 mL	<100 μL
Bed weight	100 mg to > 10 g	<50 mg
Dead volume	high	<20 μL
Sorbent particle size	50 μm to > 100 μm	<50 μm
Selectivity	Limited	High
Sample volume	Several mL	Few 100 μL
Drying	Before elution	No drying
Elution volume	Several mL	Few 100 μL
Elution profile	Wide	Sharp
Concentration	Evaporation with N_2	No concentration
Operation	Vacuum operated	Positive pressure
Workload	Manual operation	Walk away automation
Time exposure	Time consuming	Fast with < 10 min
Productivity	Low sample throughput	High throughput
Integration	Batch processing	Online GC-MS, LC-MS
Automation	Potential	Fully automated
QA/QC	No	Traceable

Table 4.16 Micro-SPE sorbent material mix for pesticide clean-up. Source: ITSP Solutions Inc.

Sorbent material	Bedmass [mg]	Comment	Application
Supelco Z-Sep	8	C18 and zirconia bonded silica, removal of fat, lipids, pigments	
UCT C18EC	21	C18 modified silica, endcapped, removal of lipids and non-polar components	Used for **LC-MS** pesticides clean-up
United Science CarbonX	1	Carbon coated porous substrate, removal of planar molecules, pigments, carotenoids, sterols, chlorophyll	
TOTAL	30		
UCT PSA	12	Primary secondary amine, retains acidic interferences such as fatty acids, polar pigments, sugars,	
UCT C18EC	12	C18 endcapped, removal of lipids and non-polar components	Used for **GC-MS** pesticides clean-up
United Science CarbonX	1	Carbon coated porous substrate, removal of planar molecules, pigments, carotenoids, sterols, chlorophyll	
UCT MgSO$_4$	20	Magnesium sulphate, drying agent	
TOTAL	45		

Source: Based on Morris et al. [168].

the cartridge sorbent material concerning clean-up efficiency and analyte recovery [168]. For routine applications in the contract laboratory, two types of cartridge sorbent mix were found to be most suitable, one for GC-MS and another mix for LC-MS pesticide compounds. The chosen sorbent mix ratios are listed in Table 4.16. The difference of both sorbent material mixes is the Z-Sep material for LC-MS [169], which is replaced by PSA material for GC-MS applications. The cartridge for GC-MS also comprises anh. MgSO$_4$ material for extract drying to retain water by hydration, which improves clean-up by PSA in MeCN and reduces the amount of water introduced into the GC [170]. Excess water in the extracts would expand too much upon vaporization in the GC inlet, worsens the chromatography, and can damage the column and MS filament [171]. The clean-up process works here in the "scavenging" mode as a chemical filtration. The unwanted matrix is retained on the cartridge, while analytes of interest are eluted first. This allows the elution of a concentrated analyte fraction avoiding additional evaporation steps. The high-throughput mega-method "QuEChERSER" (more than QuEChERS), developed by Steven J. Lehotay's group, uses the robotic μSPE clean-up for a wide range of contaminants like veterinary drugs, pesticides and their metabolites, and polychlorinated biphenyls (PCBs) in different even high fat-containing matrices [172, 173]. A successful clean-up for the high matrix-containing spice extracts is described by Goon et al. [174].

The special feature of many x,y,z-robotic systems is the prep-ahead mode (*aka* look-ahead) for automated sample preparation integrated with chromatographic analysis by GC-MS or LC-MS, see also Section 3.3.2. The prep-ahead function measures in the first sample run the time required for sample preparation and analysis until receiving a *Ready* signal again. Based on this time axis, the next sample preparation starts scheduled to be completed when the next *Ready* signal is expected. This way the duty cycle of the analysis instrument is exploited to 100%, reducing idle time and cost of analysis. Also, from the analytical aspect, all samples in the automated sequence are prepared identically without wait time before analysis. In Figure 4.74, the automated processing cycle is illustrated on a timeline as applied for the routine analysis of QuEChERS extracts from food commodities [175].

The μSPE clean-up is fast and can be easily integrated with the prep-ahead function into the chromatographic sequence of LC-MS and GC-MS analyses. A high duty cycle and return is achieved with optimal sample throughput. The complete workflow, for instance of the clean-up of QuEChERS extracts for pesticides, with cartridge conditioning, sample load, and elution, can be completed within 7 minutes [168]. This short processing time allows a parallel clean-up during the analysis of the previous sample. It was shown by Steve Lehotay and team that the clean-up can be even shortened without compromising performance making the μSPE concept compatible with fast GC-MS analysis [175] as outlined in Figure 4.74 with the reported time scale of 7 minutes clean-up time and a 10 minutes GC run for a total analysis cycle time of only 13 minutes per sample.

Besides pesticides, micro-SPE clean-up emerges into much wider fields of automated applications such as alkaloids from herbal products [176], psychoactive substances in organ tissues [177], PCBs in serum [178], 1,4-dioxane in drinking water [179], estrogens in serum [180], PAHs in vegetable oil [181], or as described in Section 6.18 the residue analysis of chemical warfare agents [182].

Figure 4.74 Micro-SPE prep-ahead clean-up on the time axis parallel to the GC-MS analysis: 7 minutes clean-up time parallel to 10 minutes GC run and 3 minutes equilibration time. Source: Based on Lehotay et al. [175].

4.5.2.4 Syringe-Based Micro-SPE

Micro-Extraction in Packed Sorbent The "micro-extraction in packed sorbent" (MEPS), also called "micro-extraction in a packed syringe," is a syringe-based device suitable for robotic autosampler to facilitate the SPE process and perform rapid method development exploring the number of cycles for loading, washing and elution, and control the flow rate with syringe precision. MEPS is a particularly useful extraction method for sampling small volumes typically in the range of only 10 μL, or taking multiple aliquots for repeated analysis from low microliter samples. As with regular SPE, larger sample volumes are handled as well. The outstanding feature of the MEPS device is its small size built into the needle of a syringe with only 7 μL sorbent volume (see Figure 4.75). The packing is approx. 1–2 mg (100–250 μL) of regular SPE silica-based sorbent material with 40–50 μm particle size available in different surface modifications. Also restricted access sorbent media are in common use [183]. The difference to other SPE techniques is the fact that the sorbent packing is directly connected to a liquid syringe, no additional cartridge is required. As MEPS works with regular liquid syringes the workflow is compatible with x,y,z-robotic systems for automated analyses workflows. The MEPS device is used for multiple samples in sequence with appropriate wash cycles between the sample runs to eliminate carryover. The MEPS device is reported to be robust for more than 400 analyses of for instance environmental water samples, or in case of the direct extraction from biological matrix samples for more than 100 plasma or urine analysis [184, 185]. The exchange of the MEPS needle device is manual.

As with classical SPE, the MEPS workflow comprises the same steps of conditioning, sample load, washing, drying, and elution. A liquid syringe of typical 100 μL to 250 μL volume with an exchangeable needle is used. The regular needle is replaced by the MEPS device to aspirate solvent and sample and enables offline work, or a productive automated online elution into GC-MS or LC-MS systems. The sample can be aspirated once or pumped multiple times through the sorbent bed to absorb the analytes with high recovery. This is a unique feature especially beneficial for the enrichment of trace components. The sorbent material is then washed with a suitable solvent, for instance using water to remove proteins from biological samples, then in a last step, eluted with a GC- or LC-compatible solvent directly into the GC injector, respectively, the LC injection valve. MEPS allows sharp, concentrated analyte bands that can be delivered directly to ESI-MS [186]. If polar compounds get extracted and need to be analyzed by GC-MS, MEPS is compatible with in-port derivatization by for instance silylation or acetylation [187].

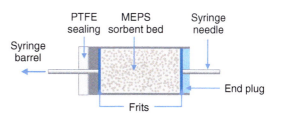

Figure 4.75 MEPS device detail. Source: Courtesy Trajan Scientific and Medical.

MEPS is fast with approx. only 1 minute for a sample, hence reduces significantly the sample preparation time and solvent consumption. It works fully automated on x,y,z-robotic systems and allows the unattended analysis of large sample series. With these features, MEPS is a true green extraction method and can replace as micro-method the regular SPE and LLE in many cases.

Typical applications range from environmental water samples to biological matrices with the extraction of drugs [188] or metabolic profiling [189] from saliva or whole blood, or active agents from saliva, plasma, and urine [190, 191]. MEPS has been directly hyphenated with ESI-MS for the rapid screening of opiates and codeine metabolites in urine [186]. Also, the highly sensitive haloanisole analysis from cork and wine off-flavors is reported [192]. Due to the high sorbent surface, the flexible ability to absorb and wash away the matrix, and with a concentrated elution profile, MEPS reaches high recoveries for high sensitive analyses [193].

Micro-SPE with Miniature Check-Valve Cartridges Another development within the miniaturization of SPE for the design of syringe integrated μSPE methods can be seen as an advancement of the above-described MEPS device. While with MEPS aspiration and dispensing of samples and solvents goes through the sorbent material in both directions, the further improved solution features a two-way operation facilitated by a miniature check-valve in this little still syringe-based device (Figure 4.76). The particle size of the used sorbent material of the "μSPEed" called cartridges developed by Peter Dawes is <3 μm providing for high analyte/matrix separation [67]. Pulling the syringe plunger up makes the sample and solvents bypassing the μSPE cartridge filling the syringe barrel. Pushing the plunger down now directs the solvents and sample passing the sorbent bed [66]. The analytical syringe provides the high pressure required for the particular small particle size material. The small and constant sample flow can be well controlled in automated systems focusing the analytes onto the top of the sorbent bed. The stepwise operation of the μSPEed cartridges is illustrated in Figure 4.77. Washing and elution of analytes can be achieved with a few hundred microliters of solvent only resulting in a significant saving of organic solvents making this μSPE device a true green analytical solution.

Working with such μSPE devices allows a simple and fast extraction and clean-up procedure, amenable to single sample manual use, but most productive in programmable automated systems. Applications range with MEPS and μSPEed cartridges from environmental samples, beverages [195], to biological matrices like saliva, urine, serum, or blood [196]. For integration with GC or GC-MS analyses, also the on-line derivatization can become a step in an automated workflow [197].

Figure 4.76 Syringe base μSPEed device. Source: Courtesy Trajan Scientific and Medical, after Dawes et al. [194].

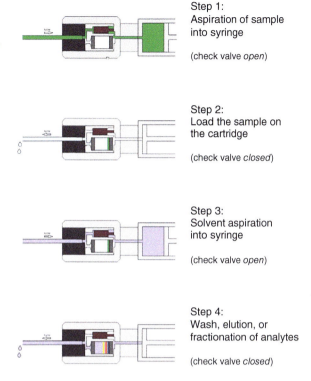

Figure 4.77 Principle µSPEed operation steps. Source: Courtesy Trajan Scientific and Medical, after Raynie [66].

Step 1:
Aspiration of sample into syringe

(check valve *open*)

Step 2:
Load the sample on the cartridge

(check valve *closed*)

Step 3:
Solvent aspiration into syringe

(check valve *open*)

Step 4:
Wash, elution, or fractionation of analytes

(check valve *closed*)

4.5.3 Gel Permeation Chromatography

GPC is based on size exclusion chromatography (SEC) initially developed for the separation of high molecular weight polymers [198]. Several names were given to different types of SEC applications since its discovery in 1955 by G.H. Lathe and C.R. Ruthven [199], but all of them are based on the same separation principle. Lathe and Ruthven were working on swollen starch granules and incorrectly called their process "gel filtration," which implies that the high molecular weight components are physically retained on the column [200]. Historically, the porous medium was made of an agarose gel, and therefore, the term "gel permeation" was coined. In life sciences, the method is also known as gel filtration chromatography (GFC) for the desalting of protein solutions.

GPC separates molecules based on their hydrodynamic size and shape, also called the hydrodynamic volume (radius of gyration) of the analytes. This differs significantly from other chromatographic techniques that depend on chemical or physical interactions [198]. For this particular separation principle, the GPC column material is porous. The solvent present in the pores of the sorbent beads can be considered as a stationary phase. Small molecules enter the cavities and are detained as illustrated in Figure 4.78. Larger molecules are increasingly excluded from the cavities making the largest compounds not retained and eluted first. The smaller molecules have a longer path length because they enter more of the porous beads, whereas larger molecules remain more often in interstitial regions between beads leading to

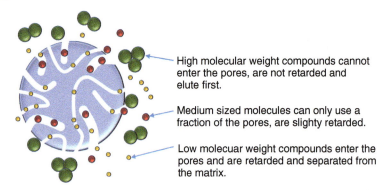

High molecular weight compounds cannot enter the pores, are not retarded and elute first.

Medium sized molecules can only use a fraction of the pores, are slighty retarded.

Low molecuar weight compounds enter the pores and are retarded and separated from the matrix.

Figure 4.78 Schematic of GPC separation – Bed material pores vs. analyte size.

a separation by size. In the clean-up use of GPC, these large molecules are typically bigger biomolecules mostly lipids, in particular triglycerides. For such kind of GPC sample clean-up, sorbent bead pore sizes of 50–100 Å with particle sizes of 5–10 µm are in use. The required separation quality is also depending on the length of the column. Column dimensions of 20 mm ID with 300 mm length are considered standard [201].

The GPC separation by molecular size brands the method as an ideal choice for automated sample preparation and clean-up in trace analysis. The clean-up by GPC is particularly beneficial for chromatographic methods as the analytes of interest are of small molecular weight compared to the co-extracted matrix. This prepares raw sample extracts well for chromatography by removing the unwanted and not compatible higher molecular weight matrix. The GPC clean-up eliminates matrix compounds like lipids, pigments, natural polymers as proteins or humic acids from the raw sample extracts prior to GC-MS or LC-MS analysis. The high molecular matrix elutes from the GPC column first. The small analyte compounds are retarded and can be collected from a calibrated elution window for further chromatographic analysis. Samples with high amounts of low-molecular-weight triglycerides like butterfat and palm kernel oil can cause interferences [202]. Butter can contribute up to 10% of residual fat in the pesticide fraction [203].

A typical GPC elution profile for matrix compounds and pesticides is illustrated in Figure 4.79 [204]. The analyte fractions are of only small volume and can be either collected or directly transferred for chromatography. During the chromatographic runtime, the GPC column can be regenerated for the repeated use in large sample series. Applications range from the clean-up of pesticide residues, semi-volatiles, PAHs, PCBs, mycotoxins to antibiotic residues from animal tissue, fatty foods, plant tissue, soil, sediment, sludge, or wastewater. In the current context, the focus is on the GPC usage for matrix clean-up for automated sample preparations.

The column capacity for matrix samples need to be considered for the clean-up of high fat carrying samples, in particular vegetable oils. The sample capacity is primarily limited by the amount of fatty matrix injected. With the given standard column, dimensions up to 40 mg of dissolved sample can be injected [206]. With samples of a high fat content, the tailing of the large triglyceride peak must be taken into account.

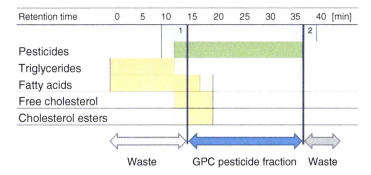

Figure 4.79 GPC Elution profile for the clean-up of pesticides extracts. (1) Elution of Avermectin, Ivermectin; (2) Elution of Aminopyralid, Clopyralid, Quinmerac, Topramezanone. Source: Based on Hildmann et al. [205].

The matrix peak tailing caused by high sample loads can partially overlap the analyte fraction. In practice, for the setup of the method, the maximum sample load on the GPC column is determined by the tolerated amount of triglyceride matrix transferred to GC or LC with the cleaned extract. Also, for the mostly applied GC analysis, the chosen oven program needs to make sure that the injected high molecular matrix is completely eluted from the column. An appropriate oven temperature ramp and maximum temperature isothermal must be chosen as needed for the particular samples [207]. Also, the GC injection using a pre-column with backflush of the high boiling matrix should be considered [205]. Due to the inherent limitations, GPC as a clean-up method for pesticides is increasingly replaced by the QuEChERS extraction combined with dispersive SPE or automated µSPE clean-up also for fat-containing samples [175, 208], as described in Section 4.5.2.3.

4.5.3.1 Standardized Methods

Due to the favorable long year experience and wide application range, the GPC clean-up became standard in many official methods worldwide. The Japan Ministry of Health, Labor, and Welfare notification requires GPC as the extract clean-up method for meat, fat, seafood, milk, and egg [209]. For similar pesticide analysis applications in food, the China National Standards Method GB 23200.113-2018 was established [210]. Beyond that, China has adopted many standards, including GB/T 19650-20065, SN/T 0123-20106, SN/T 2149-20087, SN/T 2915-20118, on GPC as a purification method for the analysis of pesticide residues in various types of food [211, 212]. In the United States, the GPC clean-up method is recommended by the Environmental Protection Agency (EPA) with the official method 3640A for more than 170 SVOC compounds [213].

4.5.3.2 Workflow and Instrument Configuration

The instrumental setup for GPC clean-up compares well to LC systems requiring a high-pressure solvent pump capable of the typical flow rates up to 10 mL/min at pressures up to approx. 4000 psi (276 bar), and the injection module accessible for automated x,y,z-robots including a sample loop. The GPC column is typically connected to an ultraviolet (UV) or photodiode-array LC detector for the calibration of the analyte elution windows based on the retention time. Detection wavelengths of 254 or 280 nm are used and meet the EPA method guidelines [212].

The workflow is straightforward. The sample is first dissolved in a GPC compatible normal phase solvent or solvent mix and injected to the sample loop. Typical mobile phases for pesticides clean-up are acetone/cyclohexane (20:80, v/v) at flow rates of 4.0 mL/min with, e.g. a Shodex column CLNpak EV-G AC, 20 mm ID × 100 mm. Gradient elution methods are not used. Separations are run at room temperature. A thermostatically controlled column oven is recommended for improved reproducibility and in case of more viscous solvent systems. Various GPC column types are commercially available from former low-pressure glass columns to current high-pressure stainless steel columns. For positive pressure with higher flow rates, the robust cross-linked polystyrene packings with particle sizes 5–7 µm are in use, providing sharp peaks with low fraction volumes and fast separations. A representative elution profile of a large number of pesticides is provided for instance by the Japanese GPC column manufacturer Showa Denko [214]. The GPC system must be software controlled and able to be calibrated on the retention time axis to facilitate the fraction collection. A retention time calibration with standards is performed. Spiked vegetable oil is used to determine the triglyceride elution, and early and late eluting pesticides. For collection of the cleaned extract fractions can be collected in separate vials for off-line or the online analysis by the x,y,z-robot. The direct hyphenation to GC for online GC-MS analysis is established by a time-controlled LC-GC valve as outlined below.

4.5.3.3 GPC-GC Online Coupling

For the hyphenation of the sample clean-up via GPC and online chromatographic analysis, the coupling of GPC with GC and GC-MS is established. The traditional off-line setup employing a fraction collector for collecting high volume pesticide eluates from a GPC run implicates additional manual steps, which involve the concentration of the extract, a transfer to AS vials, and load them on a GC-MS autosampler. Potential introduction of additional contamination, losses, and handling errors can be avoided by the direct transfer of pesticide fraction into the inlet of a GC or GC-MS analysis system. The on-line coupling with GC is consistent as the mobile phase of GPC is normal-phase LC and very well compatible for GC injection [215].

Already in 1991, the concept of on-line GPC-GC was described as SEC-LC-GC by Koni Grob and Iren Käelin [216]. Referring to the cost of sample extraction and clean-up they stated: "It is no surprise that there is a corresponding interest in an automated sample workup process integrated with the final GC analysis." The important step ahead was the miniaturization of the GPC setup. For the on-line coupling to GC, small ID columns of 2 mm were introduced for a reduced elution volume of the analyte fraction, in this case, pesticides. A total elution volume of 200–300 µL was completely transferred into a GC pre-column. A loop type interface with concurrent evaporation from the pre-column during the transfer phase was used. Limitations reported were related to the tailing of the triglyceride peak of oil or high-fat samples into the approximately 3 minutes later starting elution of the pesticide fraction. It was found that significant tailing was caused by the used injection valve and connectors, which later could be avoided by suitable measures. The automated on-line GPC-GC method simplified significantly the pesticide analysis in food. Finally, it only required the extraction of the sample with DCM, and after filtration, the raw extract was injected without further treatment into the GPC-LC-GC system. A significant reduction in solvent use was achieved.

The on-line GPC-LC-GC method was used already at that time for the determination of 22 organophosphorus pesticides in fruits with recoveries of 88–100%. Two pesticide compounds were used in ppm concentration for identification of the start and end of the eluting pesticide fraction by UV detection at 225 nm. Ethion marked the beginning of the pesticide elution, azinphos-methyl the end of the fraction in this application. A 3 m long uncoated but deactivated fused silica pre-column with 0.53 mm ID, and a short 20 m × 0.32 mm ID analytical column with SE-54 stationary phase of 0.15 µm film thickness was used for GC analysis. Today, rugged metal-coated pre-columns are preferred for the LC-GC coupling due to the inherent brittleness of 0.53 mm ID fused silica columns.

Typical applications of on-line GPC-GC are the determination of GC amenable pesticides [204, 217], in particular the chlorinated pesticides, but also PAHs [218], PCBs and PBDEs [219], and plasticisers [220] in high lipid-containing foods like edible oil, egg yolk, meat, fish and similar matrices.

4.5.3.4 Micro-GPC-GC Online Coupling

Another step ahead to green analytical chemistry is the further solvent reduction with the usage of micro-methods also in GPC. An on-line micro-GPC method (µGPC) coupled to GC-MS by using a robotic x,y,z-sampler is described and complies with the requirements of the official Chinese methods for the analysis of pesticides in foods [221, 222]. The mobile phase used for the described clean-up application of avocado or vegetable oil extracts is acetone/cyclohexane (3:7, v/v) at a flow rate of only 0.1 mL/min. The GPC column used here was a Shodex CLNpak EV-200 column with dimensions 2 mm ID and 150 mm length held constant at a temperature of 40 °C. A significant reduction of the lipid matrix background could be achieved with recoveries of OCP pesticides in the range of 73–115%.

The µGPC unit is integrated on the rail of an x,y,z-robotic sampler system and can be served with samples using appropriate liquid tools for extract injection. The schematics of the integrated µGPC solution are shown in Figure 4.80. Figure 4.81 shows the instrument setup. Vials with extracts, for instance from the QuEChERS extraction for pesticides, are placed into a dedicated rack of the robotic sampler. The matrix samples are automatically applied in a sample sequence of the GC-MS system to the µGPC purification. The pesticide fraction in this configuration is collected in another dedicated vial rack. From here, a large volume injection (LVI) is made into the GC-MS system for compound analysis and quantification. The workflow is designed to process the sample extract purification and analyses in prep-ahead-mode so that an individual sample clean-up is started right in time to be completed with the collected fraction so that the GC-MS *Ready* signal can trigger the immediate injection (see also Figure 4.74).

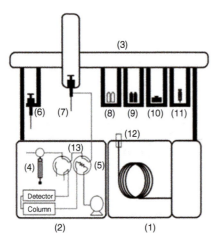

Figure 4.80 On-line µGPC configuration for GC-MS, combined with optional µSPE. Source: Courtesy LabTech Instruments Ltd.

1 GC-MS unit
2 µGPC module
3 x,y,z-Robot
4 Syringe pump
5 µGPC sampling valve
6 Liquid syringe tool
7 Collection/large volume injection tool
8 Collection vial tray (cleaned extract)
9 Sample vial tray (QuEChERS raw extract)
10 µSPE solvents tray
11 µSPE cartridge tray
12 Large volume GC injector
13 µGPC collection line

Figure 4.81 Automated µGPC clean-up system for on-line GC-MS analysis. Source: Courtesy LabTech Instruments Ltd.

4.6 Centrifugation

A centrifugation step is required and often applied to separate suspended particulate matter from a solution or solvent mix for sedimentation or phase separation in LLEs. The separated liquid layers are further processed by using pipets or syringes. The phase separation is based on the gravitational force generated by a high-speed rotor with a vial holder. The magnitude of the applied gravitational force, the centrifugal acceleration, is expressed as the "relative centrifugal force" (RCF) measured in multiples of the standard earth acceleration "g". The "*g force*" of a centrifuge is positively and linearly related to the radius, and quadratic to the angular speed of the rotor as shown in (4.3).

$$g \text{ force (RCF)} = r \cdot \omega^2 \cdot g^{-1} \tag{4.3}$$

with

g = the standard earth acceleration (m·s^{-2})
r = the radius of the rotor (cm)
ω = the angular velocity (radians/unit time)

The rotating speed and radius of a rotor converts into the gravitational force. Hence, a centrifuge specification in rpm without information on the radius is meaningless. The radius of the rotor is measured from the center axis to the bottom of a vial when deflected in its rotating position. For a known centrifuge from the radius in cm and speed in rpm, the g force conversion is derived from (4.3) as follows:

$$g \text{ force (RCF)} = 1.118 \times 10^{-5} \cdot r \cdot \text{rpm}^2 \tag{4.4}$$

with

RCF = Relative Centrifuge Force, expressed in multiples of g
r = the rotor radius (cm)
rpm = the rotation speed in rotations per minute (min^{-1})

Centrifuges develop heat during operation, which may be of negative impact to biological or low boiling samples. Operational heat is not only developed by the motor itself but can also be contributed by internal friction with air due to persistent operation. Centrifuges for automated systems require cooling. Most used are air-cooled rotors in which the rotor itself provides the ventilation with ambient air. Rotor temperatures in the range of 10–15° above laboratory room temperatures are typically maintained. A temperature sensor and display should be available. Temperature tracking allows traceability.

Centrifuge safety requires preventive maintenance and attention, in particular for the rotor. Follow the manufacturer's guidelines. Centrifuge septum closed vials only to prevent spill and exposure with hazardous materials. The centrifugation of flammable liquids like the often used hydrocarbon solvents, alcohols, or ethers requires specifically certified equipment. For operation always ensure that rotor loads are evenly balanced. For use of centrifuges in automated configurations, the programmed workflow needs to take care of the rotor balancing with appropriate vial fill levels and positioning. A symmetrically loaded centrifuge rotor for the combined use of 2 mL and 20 mL vials is shown in Figure 4.82. Monitor the rotor for fatigue and corrosion in a preventive maintenance plan regularly.

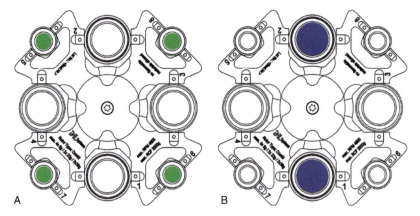

Figure 4.82 Centrifuge rotors symmetrically loaded with (A) 4 × 2 mL and (B) 2 × 20 mL vials. Source: Courtesy CTC Analytics AG.

4.7 Evaporation

Evaporation steps are common in sample preparation for extract concentration, derivatization, or solvent exchange. A wide range of solvent evaporation apparatuses are used for manual operation in every laboratory necessary to concentrate diluted sample extracts for instance from column chromatography, fractionation, or SPE, to achieve lower detection limits, or for solvent exchange. Solvent evaporation comes at the risk of introducing new contamination from glassware and loss of analytes due to absorption or volatility. It is usually a manual step, hard to control the extent and endpoint of evaporation. With automated systems running miniaturization of sample preparation steps, additional evaporation can often be avoided, see Section 4.5.

Also in miniaturized workflows, evaporation steps may become necessary, on a small scale, yet the solvent volumes for evaporation are significantly less. Typically, 2 mL vials with small solvent volumes are subject to evaporation. In some cases, vials up to 20 mL are in use as well. An important step here, and first to consider, is the safe removal of solvent vapors via an exhaust line. For that reason, double needles or similar solutions for gas in- and outlet are in use, working with septum sealed vials. This way the exposure of the laboratory staff with potentially hazardous solvent vapors is significantly reduced. Nitrogen as an inert gas for blow-down is the recommended purge gas minimizing the risk of oxidation reactions. Evaporation devices allowing programmed temperature control and vial agitation are supporting the evaporation process and commercially available. Usually, with current devices, there is no endpoint detection available. The automated evaporation steps are calibrated by evaporation time. Practically, a time calibration is performed once by weighing vials filled with the solvent in use at increasing evaporation times. At a given incubation temperature and a set gas flow linear, calibrations are achieved, which allow a precise timing in automated workflows. Solvents with different evaporation rates need to be

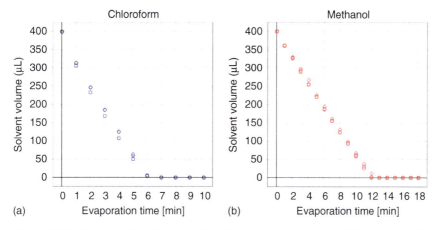

Figure 4.83 Time-based gravimetric calibration for the evaporation of chloroform and methanol from a 2 mL vial by a gentle N_2 flow at 35 °C ($n = 3$). The weight of the vials is converted to volume based on solvent density. Source: Redrawn from Hopfgartner et al. [223].

considered. Figure 4.83 illustrates the evaporation of chloroform and methanol from a 2 mL vial for an automated metabolomics derivatization application. A linear and reproducible solvent volume decrease over time can be achieved [223]. The temperature setting should also consider the often observed azeotropic reduction of boiling points (or in some rare exceptions an increase). A typical example is the evaporation of toluene with water, which carries 80% of toluene with 20% of water already at 84 °C. See a table of common azeotropic solvent mixtures, which can be exploited for solvent evaporation in Table 4.17 [224]. A compatible "keeper" solvent with a higher boiling point can be applied in a small volume before evaporation to a defined endpoint volume, and to reduce analyte losses during solvent blow-down by avoiding a completely dried residue [225].

Technical solutions commercially available for automated robotic systems include double-needle techniques for nitrogen blow-down, or vacuum-assisted procedures. Vials of 2–20 mL size can be penetrated while located in a thermostated incubator (agitator) with a pre-heated nitrogen flow through the side port of a headspace syringe (Figure 4.84). A vacuum-assisted multi-vial evaporation station is offered by GERSTEL with the "mVAP" [226]. The vacuum-assisted evaporation is achieved batch-wise for up to six sample vials in parallel and can be operated in the "prep-ahead" mode. Vials up to 10 mL size can be used in a temperature-controlled agitator shown in Figure 4.85. The vial evaporation adapters are moved by the head of the robot from their park position to the vials and penetrate the septum with a needle. A programmable frequency for vial agitation facilitates the evaporation process. The mVAP is typically used in fully automated SPE workflows for rapid eluate evaporation [156, 227], or metabolomics sample preparation [228]. A practical automated workflow provided with the mVAP module facilitates a program-controlled solvent exchange.

Table 4.17 Azeotropic mixture compositions and boiling points (after [224]).

Azeotrop mixture	BP solvent 1 [°C]	BP solvent 2 [°C]	Composition S1 : S2 [%]	BP Azeotrop [°C]
Toluene–acetic acid	110.6	118.5	72 : 28	105.4
Water–CCl_4	100	76.7	4 : 96	66
Water–benzene	100	80.6	9 : 91	69.2
Water–ethylacetate	100	77.1	9 : 91	70
Water–ethanol	100	78.3	4 : 96	78.2
Water–toluene	100	110.6	20 : 80	84.1
Water–dioxane	100	101.3	20 : 80	87
Water–acetic acid	100	100.7	12 : 77	107.3
Ethanol–chloroform	78.3	61.2	7 : 93	59.4
Ethanol–CCl_4	78.3	76.7	16 : 84	64.9
Ethanol–benzene–water	78.3	80.6/100	19 : 74 : 7	64.9
Ethanol–ethylacetate	78.3	77.1	30 : 70	72
Ethanol–benzene	78.3	80.6	32 : 68	66.2
Ethylacetate–CCl_4	77.1	76.7	43 : 57	75
Methanol–benzene	64.7	80.6	39:61	48.3
Methanol–CCl_4	64.7	76.7	21 : 79	55.7
Chloroform–acetone	61.2	56.4	80 : 20	64.7

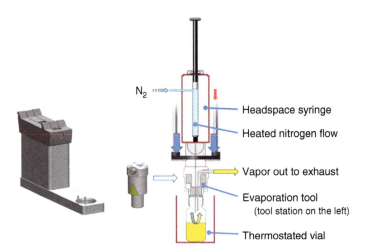

Figure 4.84 Workflow evaporation with a tool-based double-needle device. Source: Courtesy CTC Analytics AG.

1 Robot head
2 Evaporation unit with agitator
3 Transport mechanism for vials and adapters
4 Vial positions (6×)
5 Vacuum tubes
6 Needle adapters

Figure 4.85 Vacuum-assisted evaporation module. Source: Courtesy GERSTEL GmbH & Co KG.

4.8 Derivatization

Derivatization procedures are required for many chromatographic methods. The manual handling is usually an additional and unwanted, laborious, and time-consuming process. Also, hazardous reagents are handled, and often critical organic solvents are involved. Within the Significance Principles of Green Analytical, the "*a*" stands for "*a*void derivatization," see Section 2.1.1.3. This goal of avoiding potentially hazardous reagents became increasingly possible with new chromatographic phases, and the move of polar analytes from GC to LC analysis today, with equally high detection capabilities. Also, alternative derivatization strategies are discussed [229]. But still, for many current chromatographic standard methods, derivatization is inevitable. The complete transfer of derivatization steps to robotic systems, embedded in the workflow from sample preparation to analysis, is a requirement in many analytical laboratories. Proven automated solutions are available. Here, derivatization can be operated from closed vials unattended, with minimal operator contact, and benefit from a significant reduction of expensive reagent volumes. Automated derivations as part of comprehensive workflows should be fast on the chromatographic timescale, at best, working in the prep-ahead mode during the analysis run of a previous sample. A large number of general and specific derivatization reactions are available for GC and LC applications and published with useful guidelines [230, 231]. In the context of this textbook, the focus is on the most common derivatization agents and their integration in automated workflows.

In LC analysis, a known limitation is the detection of analytes with classical detectors, if it is not mass spectrometry. Many compounds, although well separated on the column, do not carry a chromophore for sensitive UV detection. Light scattering detection may become a potential solution here, but lacks sensitivity in trace analysis. Derivatization agents offer the potential to introduce a chromophore to the target analytes for sensitive UV detection, or, a fluorophore for the even more selective fluorescence detection. Post-column derivatizations are typical for the analysis of amino acids, carbamates, mycotoxins, antibiotics, and many other applications [232]. Also

LC-MS/MS benefits by using derivatization for small molecules to avoid low mass chemical noise and gain more and better ion transitions in the higher mass range.

4.8.1 For LC and LC-MS

4.8.1.1 Aromatic Acid Chlorides

For LC detection as chromophores, usually conjugated aromatic compounds with high UV absorption are introduced [229]. Common reagents are aromatic acid chlorides such as benzoyl chloride, p-nitrobenzoyl chloride, p-methoxybenzoyl chloride, m-toluoyl chloride, or 3,5-dinitrobenzoyl chloride (Figures 4.86–4.91). They react

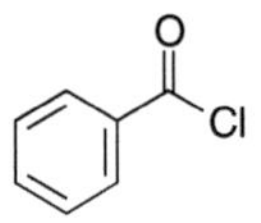

Figure 4.86 Benzoyl chloride.

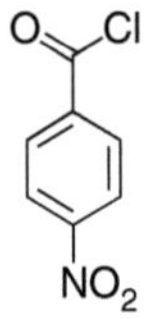

Figure 4.87 p-Nitrobenzoyl chloride.

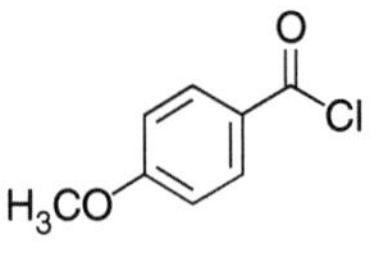

Figure 4.88 p-Methoxybenzoyl chloride.

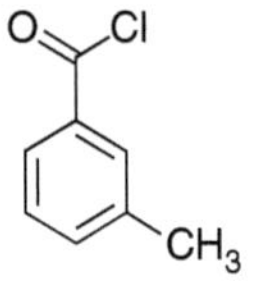

Figure 4.89 m-Toluoyl chloride.

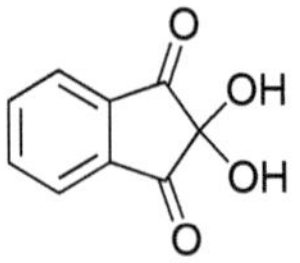

Figure 4.90 Ninhydrin.

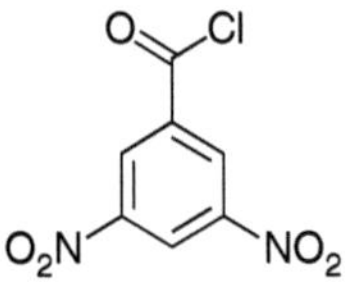

Figure 4.91 3,5-Dinitrobenzoyl chloride (DNBC).

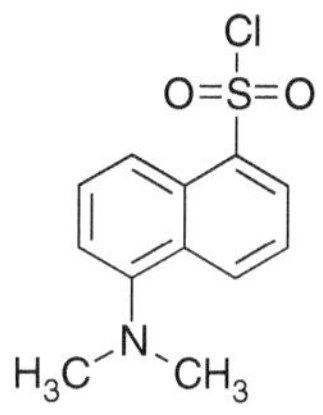

Figure 4.92 Dansylchloride.

with amines, amino acids, and hydroxyl compounds to yield amides and esters. The reaction is fast for primary amines or alcohols with only gentle heating.

4.8.1.2 Dansylchloride

A very common reagent for the introduction of a fluorophore is dansylchloride (DNS-Cl, 5-dimethylaminonaphthalene-1-sulfonyl chloride, Figure 4.92) [233]. DNS-Cl reacts with amines at a slightly alkaline pH in an acetone/water mixture at 38°C within 90 minutes. The excitation wavelength is 350–370 nm; the fluorescence emission is measured at 490–530 nm.

4.8.1.3 Ninhydrin Reaction

The ninhydrin reaction (2,2-dihydroxyindane-1,3-dione, Figure 4.90) is widely used with automatic amino acid analysers for post-column derivatization, known for the dark purple color of the reaction products with the absorbance maximum at 570 nm [234, 235]. Except for proline and hydroxy-proline, their alpha-amino group is part of a five-membered ring, all the alpha-amino acids react readily with ninhydrin.

4.8.1.4 FMOC Derivatization

For LC pre-column derivatization, fluorenylmethyloxycarbonyl chloride (FMOC-Cl) (Figure 4.93) is not only recommended for fluorimetric and UV detections, but also well used for trace analysis with LC-MS detection. FMOC-Cl reacts rapidly at room temperature with primary and secondary amines, amino acids, and alcohols. The reaction products yield a highly fluorescent derivative for subsequent determination by LC with fluorescence detection. The FMOC derivatization increases the chromatographic retention of polar compounds in a reversed-phase column and facilitates, for instance, the routine LC-MS/MS analyses for the pesticide residues of glyphosate, AMPA, and glufosinate in water and beverages [236, 237]. The reaction takes place "in situ" in buffered aqueous media within minutes and is applied with subsequent extraction and concentration by LLE or SPE. The complete sample preparation workflow is automated including online SPE and LC-MS/MS analysis, see also Section 6.15 [162].

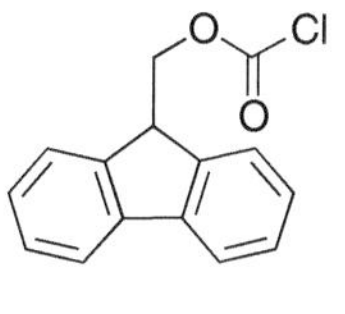

Figure 4.93 Fluorenylmethyloxycarbonyl chloride (FMOC-Cl)

4.8.2 For GC and GC-MS

In GC applications, derivatization reactions are required to decrease the analyte polarity, thereby enhancing the substance volatility (reducing the boiling point). Many analytes are amenable to GC separation only or with better quality of analysis after derivatization. Once derivatized, small molecular weight analytes get detected in GC-MS in a higher mass range, less prone to small fragment noise with higher S/N ratio detection, which is of particular importance for single quadrupole instruments in the selected ion monitoring mode (SIM). Also, the detection with triple quadrupole MS instruments in the targeted MS/MS mode (MRM, multiple reaction monitoring) benefits from parent ions in the high mass range, and a reduced compound fragmentation. Although polar analytes are typically subject to LC-MS analysis, the GC-MS detection in trace analysis is often preferred due to quantification issues caused by the interfering matrix effects occurring in LC atmospheric pressure ionization (API), which are avoided with GC-MS analysis.

4.8.2.1 Silylation

The formation of trimethylsilyl-esters by silylation (TMS-esters) is most probably the most common and most versatile derivatization method for GC analysis [238]. The popular reagents are BSTFA and MSTFA (Figure 4.48, Figure 4.49), which react with a wide range of compounds including alcohols, phenols, carboxylic and amino-acids, amines, amides, saccharides, thiols, indoles, or nucleosides. Labile polar compounds can be derivatized using the more gentle HDMS (Figure 4.94) [229]. For moderately hindered or slowly reacting analytes, small amounts of trimethylchlorosilane (TMCS) (1–10%, Figure 4.95) are added as a catalyst reducing the reaction time. Ketones, when derivatized using MSTFA or BSTFA, form partially enol TMS esters. These esters can be eliminated by forming a methoxime as a protection group first (see Section 4.8.2.4).

Silyl derivatives are of high volatility and good thermal stability. The silylation reaction is fast and hence well suited for the derivatization reaction also in the GC inlet liner *aka* "in-port" derivatization. The reactivity of the functional groups for silylation follows the order of alcohol > phenol > carboxyl > amine > amide/hydroxyl [239]. In similar order is the general hydrolytic stability of the TMS-derivatives with alcohols > phenols > carboxylic acids > amines > amides. However, variations can occur among these groups [240].

For the automated processing, typically 2 mL vials sealed with septum are used. The TMS derivatives are sensitive to moisture. It is important to keep the reaction vials closed and use as anhydrous as possible solvents. The silylating reagent is used

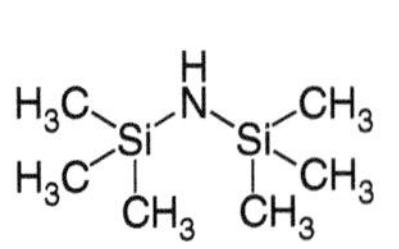

Figure 4.94 HDMS.

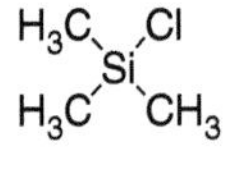

Figure 4.95 TMCS.

in large excess. No additional solvent is required as long as the analytes dissolve in the reagent. Reactions take place at room temperature or gentle heating (50–100°C), shaking in an agitator facilitates the quantitative reaction. Derivatization reactions in micro-vials or micro-inserts require vortexing for proper mixing of the reagent. The reaction times can vary between a few minutes and hours depending on the analyte reactivity. Microwave heating can accelerate the reaction. The TMS products are not recommended for storage, decompose with humidity. On a practical note: For unknown analysis with GC-MS, a huge number of library spectra of silylated compounds facilitates compound identification by library search.

4.8.2.2 Acetylation

Water samples for phenolic compounds and personal care products (PCPs) can be treated for derivatization by the addition of acetic anhydride and a soluble base (Figure 4.96). The acetylated analogues, now as VOCs, can then be analyzed by GC. Instead of pyridine as a base for the reaction, potassium carbonate is also used. One approach is using the in situ water acetylation with ensuing dynamic headspace extraction of the derivatized phenols for GC-MS analysis [241]. Another application extracts the derivatized compounds with SPME, a straightforward solution for full method automation [242].

4.8.2.3 Methylation

Probably, the most used GC analysis with derivatization in the food industry is the determination of fatty acid methyl esters (FAMEs) for fatty acid composition (FAC), nutritional value analysis with saturated-, mono-, and polyunsaturated fatty acids contents (SFA, MUFA, PUFA), cis/trans fatty acid determination and more. In petrochemical applications, the characterization of biodiesel quality and alternative plant sources via FAMEs is an important application. The intact triglycerides (triacylglycerols, TAGs) are analysed by GC only for special tasks, e.g. the determination of cocoa butter equivalents (CBEs) in milk chocolate [243]. TAG chromatograms reveal the large variety of fatty acid substitution. The fatty acid profile of an oil or fat as a sum parameter of all occurring fatty acids is determined by conversion of TAGs and other acyl-substituted lipids to the FAMEs before GC injection. Two major methods are established with automated workflows in food and life science applications.

Free fatty acids (carboxylic acids) can be converted to methyl esters directly by alkylation with methanol using BF_3 as a catalyst. Triglycerides must be extracted first from a food sample, if not available as oil or fat, followed by saponification [244, 245]. The alternative method uses sodium methylate (H_3CO^- Na^+, sodium methoxide) for transesterification [246]. The analytical benefit for automation of the methylate method is that there is no prior fat extraction from the food and no saponification required, and the reaction is fast. The transesterification even works with water containing samples like milk and cheese. Fat containing food is dissolved in dioxane

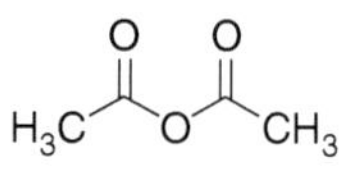

Figure 4.96 Acetic acid anhydride.

or methyl tert-butyl ether (MTBE) before the reaction. Both methods comply with several international standard methods for diverse matrices.

4.8.2.4 Methoxyamination
In the lipidomics and metabolomics applications, the lipid extraction is the initial step of the standard workflow. The traditional Bligh and Dyer total lipid extraction for life science applications was shown to be well automated on x,y,z-robotic samplers: "The wet tissue is homogenized with a mixture of chloroform and methanol … the chloroform layer containing all the lipids and the methanolic layer containing all the non-lipids" [223]. In GC-MS-based metabolomics, a two-step derivatization is applied by using methoxyamine hydrochloride (CH_3-O-NH_2 · HCl) and MSTFA. In the first step, methoxyamine hydrochloride is used, the second is the silylation of polar functional groups. The primary methoxyamination protects ketones and aldehydes by conversion to methoxyamino groups [247, 248].

$$R_1, R_2\text{-C=O} + CH_3\text{-O-}NH_2 \rightarrow R_1, R_2\text{-C=N-O-}CH_3 + H_2O$$

The methoxyamination is a precautionary step for quantitative metabolite profiling. Without protection, alpha-keto acids tend to undergo chemical loss of carboxyl groups as carbon dioxide. More importantly, many carbohydrates are present in cyclic and open-chain (linear) form. Each of the various forms would lead to different peaks in gas chromatography. Cyclization is inhibited when the carbonyl groups are methoximated [247]. Both steps, methoxyamination with following trimethylsilylation, are performed automatically in sequence before GC-MS analysis by robotic autosampler [249]. It could be shown that the repeatability of detected metabolites, as well as the quantification of lipids in serum samples, improved significantly using the automated workflow. The dried organic phase from the Bligh & Dyer extraction step [244, 245] is evaporated and dissolved in methoxamine hydrochloride in pyridine, incubated at 30°C for 90 minutes with continuous shaking, and then added with MSTFA for silylation at 37°C for 30 minutes, see Section 6.2.3 [250].

4.8.2.5 Fluorinating Reagents
The introduction of fluorine by derivatization is a common measure to improve detection limits and selectivity with GC-ECD and GC-NCI-MS detection. While BSTFA and MSTFA introduce fluorine as well, the goal is to use derivatization agents with high fluorine content like perfluorinated acids and anhydrides as acylation reagents like trifluoroacetic anhydride (TFAA), pentafluoropropionic anhydride (PFPA), heptafluorobutyric acid (HFBA), pentafluorobenzaldehyde (PFBAY) (Figures 4.97–4.100), or the fluorinated aryl bromides tetrafluoro-4-(trifluoromethyl)benzyl bromide (TTBB), pentafluorobenzyl bromide (PFBB) for alkylation (Figures 4.101 and 4.102). PFBAY is well known for the derivatization of primary amines, in particular amphetamines [251, 252]. PFBB is used for the in situ derivatization of carbonyl and carboxyl compounds in water samples, followed by HS-SPME of the generated volatile products [118]. The NCI-MS spectra typically show low fragmentation and are dominated by high ion intensities in the less noisy upper mass range, facilitating the selective and sensitive detection e.g. for drugs of abuse in SIM mode [253].

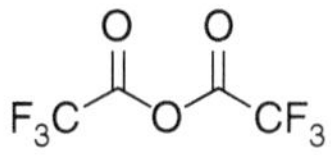

Figure 4.97 TFAA.

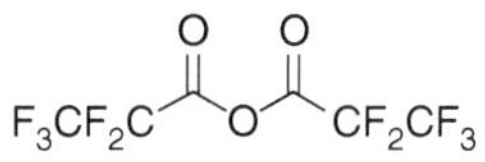

Figure 4.98 PFPA.

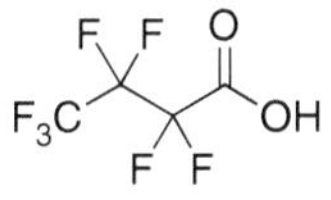

Figure 4.99 HFBA.

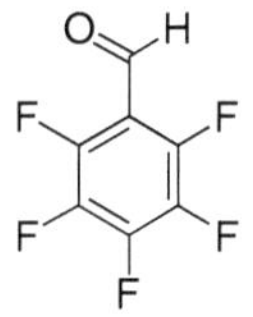

Figure 4.100 PFBAY.

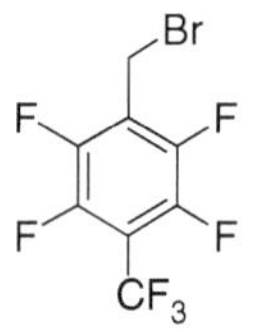

Figure 4.101 TTBB.

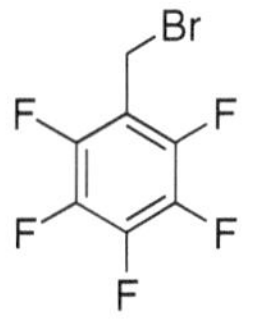

Figure 4.102 PFBB.

4.8.3 For GC and GC-MS In-Port Derivatization

The derivatization in the GC injector, also known as "injection-port," or in short "in-port" derivatization, is a classical derivatization procedure very suitable for automation purposes with liquid injections. Instead of preparing the derivatization steps separately in 2 mL vials by mixing the sample with derivatization agent and heating for defined times, the derivatization process can be achieved directly in the GC injector for fast-reacting analytes [254, 255]. In general, the most common and widely applied derivatization reactions are all available for in-port derivatization as well [187]. Reagents like MSTFA and BSTFA are used successfully for in-port silylation.

The in-port derivatization reaction is carried out by injecting the sample together with the derivatization reagent directly into the GC inlet [256]. The in-port derivatization works well on both, the hot split/splitless and the temperature programmable

(PTV type) injectors. Regular injection temperatures above 250 °C are required. The derivatization is easily achieved with automated samplers by using the sandwich injection technique (see Section 5.2.1). Using the sandwich technique allows the injection of sample extract and derivatization reagent at once at the same workflow step. The derivatization occurs in the gas phase in the hot injector.

Alkylation reactions are mainly used for compound methylation. A very versatile derivatization method is described with the use of tetraalkylammonium salts due to their high reaction rate, high derivatization efficiency, and ease of application in sandwich injections. Reagents such as phenyltrimethylammonium hydroxide (TMAH), trimethylsulfonium hydroxide (TMSH), tetrabutylammonium hydrogensulfate (TBAHS), tert-butyl acetate (TBAC), tert-butyl hydroperoxide (TBH), or tetrabutylammonium hydroxide (TBAH) are in use for in-port derivatization of acidic compounds (Figures 4.103–4.108) [254, 255].

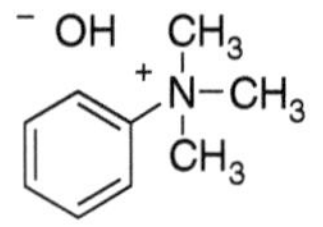

Figure 4.103 TMAH.

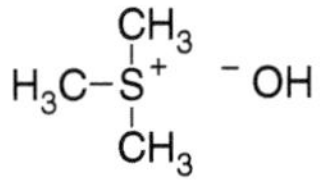

Figure 4.104 TMSH.

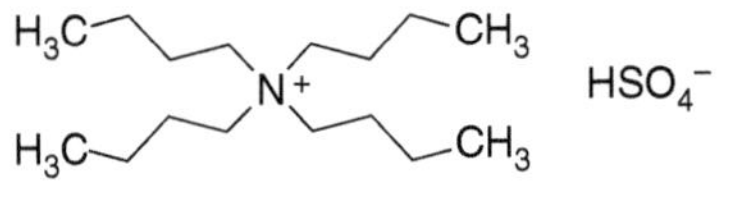

Figure 4.105 TBAHS.

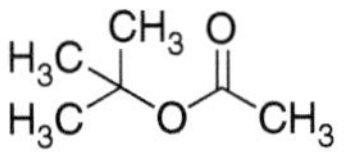

Figure 4.106 TBAC.

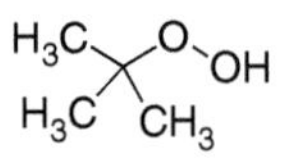

Figure 4.107 TBH.

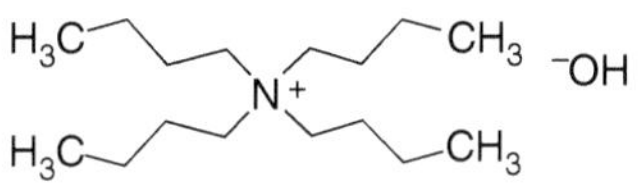

Figure 4.108 TBAH.

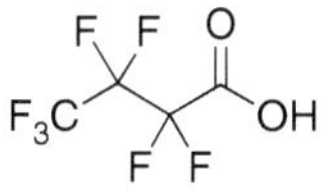

Figure 4.109 HFBA.

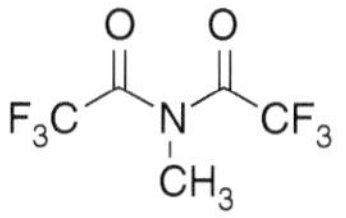

Figure 4.110 MBTFA.

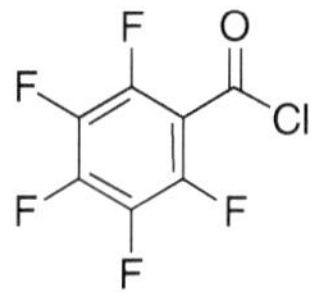

Figure 4.111 PFBCl.

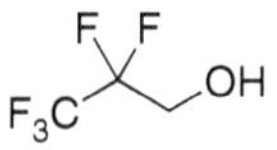

Figure 4.112 PFPOH.

Acylating reagents readily derivatize highly polar, multi-functional compounds with active hydrogens such as -OH, -SH, and -NH into esters, thioesters, and amines, such as carbohydrates and amino acids. GC separations of polar analytes such as sugars or amphetamines are facilitated. The introduction of perfluoracyl groups is used for the selective compound detection by ECD or GC-MS with negative chemical ionization. Common perfluorinated reagents are HFBA, N-methyl-bis(trifluoroacetamide) (MBTFA), pentafluorobenzoyl chloride (PFBCl), or pentafluoro-1-propanol (PFPOH) (Figures 4.109–4.112).

A special injector for LVI and in particular for in-port derivatization was developed in Japan by AiSTI SCIENCE CO., Ltd. with the patented "Lavistoma" LVI system using a GC inlet liner in the "shape of a stomach". A cutaway of the injector is shown in Figure 4.113. The siphon-shaped inlet liner has about 600 µL total volume, up to 200 µL of a liquid sample can be injected. With the injector split open, a diluted sample extract can be concentrated in the solvent split mode before GC analysis. The special inlet liner is built into an air heated housing with programmable temperature control. Access for maintenance is fast and easy when removing the septum holder. Two O-rings seal the top and bottom of the Lavistoma inlet.

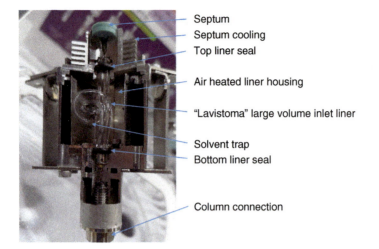

Figure 4.113 "Lavistoma" large volume injector cutaway. Source: Hans-Joachim Hübschmann.

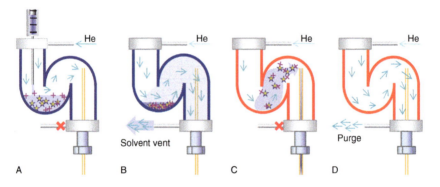

Figure 4.114 In-port analyte derivatization using the "Lavistoma" injector. (A) The sample and reagents are injected at low inlet temperature below the solvent BP. (B) The solvent vapor is vented via the split exit. The sample gets concentrated. (C) The analytes are transferred into the analytical column in splitless mode by raising the inlet temperature. (D) The split is opened to remove remaining contaminants. Source: Courtesy Leco/AiSTI SCIENCE CO. Ltd. [257].

The use of the injector in automated workflows is compatible with robotic systems using regular micro-syringes, also with sandwich injection. Large volume injection provides a significantly enhanced sensitivity (up to >100 times) over the regular few microliter GC injections. For derivatization, a liquid sample and a small amount of reagent are injected into the solvent trap of the inlet liner (Figure 4.114). The derivatization reaction occurs at a temperature below the solvent boiling point in the solvent. After the reaction is completed, the excess solvent can be vented via the split exit. The injection into the analytical column is achieved by rapidly raising the injector temperature [258]. Limitations of the large volume solvent split injection are with the risk of volatile losses. Other applications use the Lavistoma injector for direct online injection of the extracts from SPE or LLE. At moderate temperature and open split exit, the excess solvent is evaporated. In the following workflow phase, for instance, the two-step derivatization with methoximation and silylation for metabolomics applications is achieved in the same liner after adding the reagents [259]. It is demonstrated that the injector can be used for direct micro-SPE elution into the large volume inlet liner with online derivatization.

4.9 Temperature Control

Precise heating or cooling of samples and reagents to a constant and reproducible temperature is a general requirement for extractions, reactions, or storage within many sample preparation workflows. Different technical solutions depending on vial sizes and required temperatures are available using integrated modules on robotic systems or connected to external thermostatting devices.

4.9.1 Heating

Many workflow steps require sample heating over a certain time maintaining a constant pre-set temperature. Most important, a constant incubation time at a constant temperature with agitation is required for equilibrium methods like the static headspace or multiple headspace (MHE) extractions (see Section 5.1). Other applications, foremost derivatization reactions, evaporation or simply the melting of samples or dissolution, require program-controlled heating.

The classical widespread metal "dry bath" or "heating block" is realized for robotic sample preparation as well. Also, there is often an eccentric agitating function integrated that allows gentle shaking of a liquid sample during the incubation time. The frequency is controlled by the workflow and needs to be adapted to the sample viscosity. Vials can be transported from the storage rack by a gripper or magnetic transport into the agitator. Vial sizes of 20 mL volume are standard. Vial sizes from 2 mL to 10 mL can be inserted by using vial adapters inside of such an agitator.

4.9.1.1 Incubation Overlapping
The six agitator positions shown in Figure 4.115 can be used in a pre-programmed overlapping mode for sample incubation. If necessary, one position can be permanently reserved for a vial with reagent or standard used repeatedly in the workflow. A scheduler, integral part of the system firmware, monitors in a first run the processing and incubation times. Once the GC cycle time is known, the incubation for the next sample can be started during the GC run time. The timing is scheduled that all samples are incubated identical, with the same timing till the injection to GC for best possible reproducibility and maximized sample throughput, see Section 3.3.2. The vial of the injected sample is moved back to the vial rack and replaced in time with the next vial as shown in Figure 4.116.

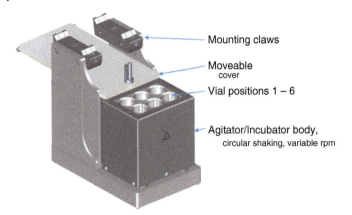

Figure 4.115 Incubator/Agitator device for vial heating and shaking. Source: Courtesy CTC Analytics AG.

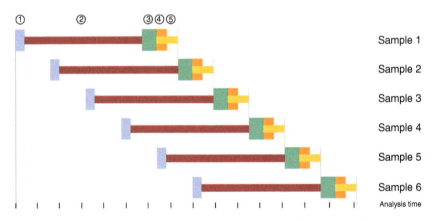

Figure 4.116 Headspace overlapping incubation principle. (1) Set/check the agitator temperature, transfer the vial into the agitator, switch agitation on; (2) Sample incubation time; (3) Upon GC *Ready*, move the heated syringe to a vial, aspirate the headspace, inject to GC injector; (4) Transfer vial back to rack; (5) GC cycle time. Source: Courtesy CTC Analytics AG.

4.9.2 Cooling

Samples and reagents in life science applications often require vial cooling for the reduction of a potential reactivity and eliminate variability during processing. Also, in case of the storage of stock solutions for calibrations dilutions, keeping the vials at a lower temperature than ambient is recommended to prevent solvent evaporation. Sample racks are placed typically into cooled drawers or cooled stacks for an increased sample capacity. The drawers are opened and closed using the tools hooked up to the x,y,z-robotic systems.

An additional challenge with cooling is to avoid the condensation of ambient humidity and the formation of ice at the cooling device. The dew point of humid air at the cooling device must be considered. Practically, cooling temperature ranges start from +5°C to prevent water condensation and freezing at the cooler fins. Freezing samples below zero is not a typical analytical workflow requirement, but special devices are available for deep freezing required for instance for the analysis of 3-monochloropropane-1,2-diol (3-MCPD) as a process contamination in edible oil raffination (refer to Section 6.12). Measures to prevent condensation are the gentle flooding of the cooled space with dry air or nitrogen to avert ambient humid air to enter, and suitable workflow programming measures, which keep the cooled stack drawers closed when not accessed.

Peltier devices are often used for heating and cooling making additional external thermostatting needless. The "Peltier effect" allows the development of a thermoelectric heat pump between two different types of metals if an electrical current is applied (named after the inventor J.C.A. Peltier, a French physicist, 1725-1845). It is the inverse "Seebeck effect" that generates a voltage between different metals, used as thermocouple temperature sensors (J.T. Seebeck, a Baltic-German physicist, 1770-1831). The generated heat flux creates a cold and hot side of the element. The cold side of the Peltier element is used as the cooler for the sample compartment, the hot side is placed outside of the cooled stack, usually ventilated by ambient air for heat dissipation. The benefit of Peltier devices for automated systems is the lack of moving parts or circulating cooling liquid, being safe from leaks, maintenance-free, and its long lifetime. A limitation is seen in the available capacity and cooling power. Maintaining constant low temperatures with automated robotic systems for standard vial sizes of 2 or 10 mL, the Peltier devices proved to be the most economical and practical solution despite the low efficiency.

The challenge of avoiding water condensation within the vial racks can be solved by separating the Peltier cooling fins from the sample vial compartment. A recently patented development by CTC Analytics AG uses cooled and dried air for the cooling of sample vial racks. A fan-driven airflow passes the separate cooling fins first, condenses, and collects humidity before the dried cool air is directed to the sample vial racks, as illustrated in Figure 4.117. The racks with sample vials are flooded horizontally from side exits pushing humid ambient air out of the rack compartment. This solution allows the controlled condensation and discharge of condensed water and the precise temperature control of a dry cooling airflow. On the mechanical side, this cooling device does not require a closed compartment. The open concept allows the free access by the robot head to all vial positions from the top at any time of a workflow without additional time-consuming tasks as it is required by opening and closing drawers. Different vial sizes can be used with standard or customized racks.

The classical alternative to Peltier cooled stacks uses external circulating cooling baths filled with a 1:1 mixture of water and ethylene glycol or silicon oil. The temperature ranges and cooling capacity vary widely and are determined by the specifications of the external cooler. Due to the situation with robotic sample preparation that only low heat capacities from sample vials need to be handled, and the focus is kept on maintaining a constant low temperature, the use of external circulating baths recedes into the background.

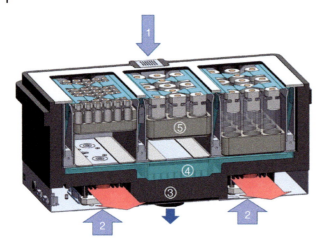

Figure 4.117 Operation principle of a vial cooling device using dried air. (1) Air inlet for vial cooling; (2) Air inlet for Peltier element cooling; (3) Condensed water drain (tube not shown); (4) Cooled air distribution to vials; (5) Vial racks for different vial sizes. Source: Courtesy CTC Analytics AG.

4.10 Mixing

4.10.1 Vortexing

A vortex mixer, or "Vortexer," is a small and simple device commonly found in all analytical laboratories to mix small vials of liquid [260]. The name stems from the description of a whirling mass of water like a whirlpool [261]. The vortex mixer was invented by Jack A. and Harold D. Kraft. A patent was filed by the Kraft brothers on April 6, 1959 and granted on October 30, 1962 [262]. The basic principle of the Vortex mixer is precisely claimed in the patent application [261]: "… provide a high-speed fluent material mixing device having an eccentrically mounted dynamic mounting for receiving one lower end of a laboratory test tube and a relatively static mounting substantially universally pivotally supporting the opposite upper end of the test tube …"

Vortex mixers are usually running at high speed with up to several 1000 rpm and a small orbital diameter of few millimeters only. An excenter drive delivers a strong shaking force to the vial in all three spatial axes creating a turbulent mixing effect. Typically, the bottom of a vial is agitated by the excenter drive, while the top of the vial is kept in its horizontal position as shown in Figure 4.118. This allows the high-frequency shaking in the vertical axis as well. The strong forces mix even immiscible solvents e.g. for LLEs, disperse solid-liquid suspensions e.g. protein/water solutions, in general, facilitate any kind of mixing purposes quickly and reliable.

In automated systems, the vortexing units need to be dimensioned to accept different vial sizes from the classical 2 mL autosampler vial or micro-vial dimensions up to the often used 10 or 20 mL vials as well. Custom size sample vials can be made

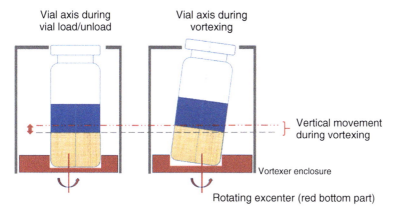

Figure 4.118 Principle of the vortex mixer with shaking in three spatial axes.

possible easily on existing vortex platforms using tailored vial holders. High volume vortexing of the heavy immiscible chlorinated solvents with aqueous media is the benchmark of a well-performing vortexer, e.g. for LLEs. The vortexing frequency in rpm must be selectable according to the intended vial size, volume, and solvent viscosity.

4.10.2 Agitation

The common understanding of sample agitation is the more gentle mixing mechanism compared to vortexing. An agitator is a general laboratory device used for shaking (orbital shaker) or stirring. With automated sample preparation systems, mechanisms can be found that use a vial spinning around the axis or are equipped with magnetic stir bars. More efficient sample agitation can be achieved by a circular eccentric motion as it is found in many x,y,z-robotic systems. The rotation frequency can be optimized according to the sample viscosity. In many cases, heating and/or cooling is integrated to offer a constant and reproducible sample temperature setting.

Heated agitators are mostly in use for static or dynamic headspace, or the SPME analysis from capped 10 or 20 mL headspace vials. The standard 2 mL autosampler vials are used with the corresponding adaptors shown in Figure 4.119. Also, derivatization reactions are carried out in heated agitators at elevated temperatures. In case of solid or viscous samples at room temperature, the heated agitators are versatile for melting purposes before further processing. Temperatures from close to room temperature up to typically 200 °C can be used. Cooled agitators from 10 °C up serve specific applications for thermolabile compounds [263]. Headspace analyses of aqueous samples are performed typically at 60–80 °C. Agitation of the sample at constant temperature is necessary to establish a reproducible sample/headspace equilibrium in a short time, and, for dynamic extraction methods, replenishing the headspace with analytes during the extraction process. In these applications, a moveable and

Figure 4.119 Agitator inserts for 2 mL vials, left the insert with 2 mL vial, right the handling with the agitator.

perforated lid is used keeping the temperature-controlled compartment closed for temperature accuracy also during the sampling phase.

The equilibration time for headspace analyses depends on the compounds and matrix to be analyzed. For the analysis of aqueous samples analytes with low partition coefficient k value, see Equation (1), e.g. hexane, benzene, toluene etc., only short equilibration times are required. They quickly equilibrate with a high concentration in the gas phase [264]. Special care for these compounds is required to fill the sample and cap the vials quickly for quantitative calibrations [265]. More polar analytes with high k values require longer equilibration time with ongoing agitation to establish a constant analyte equilibrium, e.g. ketones, alcohols, etc. See also the Handbook of GC-MS for a detailed discussion on static and dynamic headspace analysis [266].

$$\text{Partition Coefficient } k = C_s \cdot C_g^{-1} \tag{4.5}$$

with

C_s = analyte concentration in the sample
C_g = analyte concentration in the gas phase

The required agitation speeds can be very different depending on the sample viscosity and vial size. Commercial agitators provide an agitation speed range from 250 to 750 rpm in parallel for up to six vial positions. For solid samples, the agitation can be shut off. A suitable wait time in the workflow needs to be reserved until equilibration is achieved. Small vials e.g. with the use of 2 mL standard autosampler vials require higher agitation speeds for instance during the typical derivatization reaction times. Mixing of small volumes in small vials, in particular micro-vials or with micro-vial inserts, requires the vortexing module for sufficient mixing.

4.10.3 Spinning

The mere spinning of a vial for solubilization, adding standards, dilution, or mixing layers of solvents is depreciated in automated systems. The mixing is limited to a spinning of the immiscible solvent layers only and does not allow the vortex typical force in the three spatial dimensions. Upper and lower levels of solvents with different densities remain separate in the vial during spinning. A possible measure would be a rapid stop of spinning and proceed again in the reverse direction with several repetitions, but this type of mixing is not known from current commercial systems.

4.10.4 Mixing with Syringes

A very feasible way of mixing small solvent volumes e.g. in 2 mL vials is the use of syringes or pipettes with appropriate capacity, e.g. during liquid handling. This is the most productive and efficient solution in workflows with many dispensing steps. Adding small amounts of standard solutions, reagents, or solvents can be efficiently mixed by using the syringe of the last dispensing step. Using bottom sensing (or a low needle penetration depth), a larger volume of the prepared mix is pulled up slowly into the syringe and dispensed again quickly. The syringe and needle is kept in place for several plunger/mixing strokes until leaving the vial for a following washing step.

4.10.5 Cycloidal Mixing

A highly powerful shaking of vials is achieved by using a cycloid shaped mixing pattern. This mode keeps the head of a vial in place while the bottom of the vial follows a flower-like eight-turn circular cycloid pattern (Figure 4.120). This multiplies the shaking frequency of the vial 8-fold with the set rpm value by repeat inversion on the liquid acceleration inside the vial in opposite directions. The mixing power reaches the maximum with a maximum deflection of the vial bottom from the center.

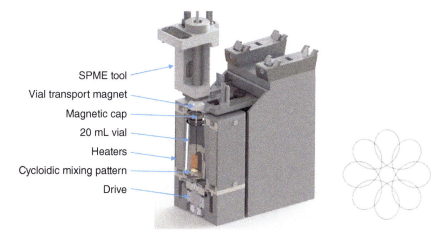

Figure 4.120 Cycloid mixing module with an 8-fold cycloidal mixing pattern. Source: Courtesy CTC Analytics AG.

Long vials with high deflection at the vial bottom like the 20 mL headspace vials are preferred.

With the high shaking frequency and inversed liquid acceleration, the solvent viscosity also needs to be considered. Very high frequencies may not resonate with the sample/solvent level and leaves the liquid unaffected. For unknown samples, a course optimization of the cycloidal mixing from low to high speed is recommended.

The cycloidal mixing turned out to be beneficial for the dynamic extraction methods with SPME in particular for DI-SPME sampling and the ITEX DHS extraction methods. The special design ensures that the delicate SPME fiber is kept in a vertical place, solidly kept by the position locked vial septum, and not damaged, while the vial bottom rapidly turns around the SPME device following the cycloidal pattern. The septum of the vial is kept in position by design and does not pose a force on the SPME rod, as it would happen using a circular shaking agitator (see Section 4.10.2).

The power of cycloidal mixing exceeds the usage of stir bars at the same rotating speed. This advantage is of special importance for DI-SPME experiments. The liquid layer around the SPME fiber or arrow gets depleted of the analyte during the extraction process and must be replenished with analyte molecules from the usually aqueous sample. Strong shaking is required to break the depleted layer for achieving a high extraction efficiency. An example of the extraction of PAHs from drinking water is shown in Figure 4.121. The extraction efficiency is demonstrated in comparison with stir bar mixing at the different speeds of 750 and 1250 rpm. With 1600 rpm speed on a cycloidal mixer, the extraction process is faster than a medium-fast stir bar rotation, and even more efficient than the highest stir bar mixing speed. As it can be seen with a vortex mixer, also the cycloidal mixing creates a central cavity with high rotating speeds. Due to different sample viscosities, the optimization of the rotating speed might become necessary. Viscous liquids can behave like an inert mass and do not follow rapidly a very fast movement of the vial with only low mixing effect. For setting the SPME fiber and arrow penetration depth into the vial, the minimum and maximum filling volumes need to be considered, see Table 4.13. Beyond DI-SPME applications, the cycloidal mixing is a viable alternative to vortexing for LLEs for intense exchange between organic and aqueous layer when compared to a magnetic stirrer.

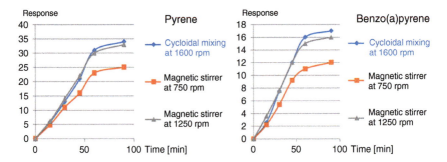

Figure 4.121 Comparison of cycloidal with stir bar mixing for PAH analysis from drinking water by DI-SPME. Source: Courtesy CTC Analytics AG.

References

1. Malissa, H. (1969). Analytical chemistry and automation. *Pure and Applied Chemistry* 1 (18): 17–34. https://doi.org/10.1351/pac196918010017. Licensed under CC BY 4.0.
2. Wikipedia (2019). Drop (liquid). https://en.wikipedia.org/wiki/Drop_(liquid) (accessed 16 July 2019).
3. Hamilton (2017). *Syringe Selection Guide*, Lit. No. L20082 Rev. B-05/2017. Hamilton Company.
4. CTC Analytics (2017). PAL System Specifications. Part No. PAL RTC and RSI Specifications, LC & GC, Rev. 7 – October 2017. Zwingen, Switzerland: CTC Analytics AG.
5. Pöll, J.S. (1999). The story of the gauge. *Anaesthesia* 54 (6): 575–581. https://doi.org/10.1046/j.1365-2044.1999.00895.x.
6. Wikipedia (2018). Birmingham gauge definition. Source: https://en.wikipedia.org/wiki/Birmingham_gauge (accessed 25 Mai 2018).
7. Sauter, A.D. (2007). The Nanoliter Syringes. *American Laboratory*. www.americanlaboratory.com/913-Technical-Articles/1328-The-Nanoliter-Syringes/.
8. Wikipedia (2018). Luer-Lock taper. Source: https://en.wikipedia.org/wiki/Luer_taper (accessed 08 August 2018).
9. UCT (2020). Push-Thru Format, Product Information. https://www.unitedchem.com/product-category/spe/push-thru/.
10. Wikipedia (2018). Pipette – history. Source: https://en.wikipedia.org/wiki/Pipette (accessed 14 August 2018).
11. Ewald, K. (2015). Impact of pipetting techniques on precision and accuracy. Eppendorf User Guide No. 20, May 2015.
12. Wiest, L. (2020). Tips for Preparing Calibration Curve Standards and Avoiding Sources of Error, Lit. Cat.# GNAR3169-UNV. Bellefonte, PA: Restek Corporation.
13. Pipette.com (2018). Guide to pipetting. Source: www.pipette.com/public/staticpages/guidetopipetting.aspx? (accessed 04 October 2018).
14. Sander, L.C. (2019). Volumetric Transfer of Liquids. National Institute of Standards and Technology, Chemical Sciences Division, Material Measurement Laboratory. Tutorial. www.nist.gov/programs-projects/tutorials-analytical-chemistry.
15. AccuTek Laboratories (2007). Guide to pipetting. See lifeserv.bgu.ac.il/wp/zarivach/wp-content/uploads/2017/11/GuideToPipetting.pdf (accessed 15 September 2018).
16. ANSI – Society for Laboratory Automation and Screening (2011). Footprint Dimensions for Microplates. Washington, DC: ANSI American National Standards Institute. https://www.slas.org/SLAS/assets/File/ANSI_SLAS_1-2004_FootprintDimensions.pdf.
17. Shukla, A. and Majors, R.E. (2005). Micropipette tip – based sample preparation for bioanalysis. *LCGC North America* 23 (7): 646–660. http://www.chromatographyonline.com/print/235132?page=full&id=&sk=&date=&=&pageID=2.

18 Brewer, W.E. (2003). Disposable pipette extraction. US Patent 6, 566, 145 B2. https://doi.org/10.1038/incomms1464.

19 DPX Technologies (2019). INTip™ Solutions for Your Lab. https://dpxtechnologies.com/overview/ (accessed 19 December 2019).

20 Horne, M., Mastrianni, K.R., Amick, G. et al. (2020). Fast discrimination of marijuana using automated high-throughput cannabis sample preparation and analysis by gas chromatography–mass spectrometry. *Journal of Forensic Sciences* 65 (5): 1709–1715. https://doi.org/10.1111/1556-4029.14525.

21 Ellison, S.T., Brewer, W.E., and Morgan, S.L. (2009). Comprehensive analysis of drugs of abuse in urine using disposable pipette extraction. *Journal of Analytical Toxicology* 33 (7): 356–365. https://doi.org/10.1093/JAT/33.7.356.

22 Lehotay, S.J. and Lightfield, A.R. (2018). Simultaneous analysis of aminoglycosides with many other classes of drug residues in bovine tissues by ultrahigh-performance liquid chromatography–tandem mass spectrometry using an ion-pairing reagent added to final extracts. *Analytical and Bioanalytical Chemistry* 410 (3): 1095–1109. https://doi.org/10.1007/s00216-017-0688-9.

23 Bremer, R. (2009). Drugs of Abuse: Extraction in Seconds. *GERSTEL Solutions Worldwide*, March 2009. Mülheim: GERSTEL GmbH & Co KG.

24 Evosep (2019). Evotips. www.evosep.com/evotip/ (accessed 19 December 2019).

25 Bache, N., Geyer, P.E., Bekker-Jensen, D.E. et al. (2018). A novel LC system embeds analytes in pre-formed gradients for rapid, ultra-robust proteomics. *Molecular and Cellular Proteomics* 17 (11): 2284–2296. https://doi.org/10.1074/mcp.TIR118.000853.

26 Geyer, P., Bache, N., S. Doll et al. (2017). Gradient off-set focusing HPLC instrument for robust and high throughput clinical proteomics. Application Note, Evosep 2017.

27 Watson, J., Greenough, E.B., Leet, J.E. et al. (2009). Extraction, identification, and functional characterization of a bioactive substance from automated compound-handling plastic tips. *Journal of Biomolecular Screening* 14 (5): 566–572. https://doi.org/10.1177/1087057109336594.

28 Huebschmann, H.-J. (2015). *Handbook of GC-MS*, 3rd Ed., Chapter 3.4, Frequently Occuring Impurities.

29 Xu, Z., Gartia, M.R., Choi, C.J. et al. (2011). Quick detection of contaminants leaching from polypropylene centrifuge tubes with surface-enhanced raman spectroscopy and ultraviolet absorption spectroscopy. *Journal of Raman Spectroscopy* 42 (11): 1939–1944. https://doi.org/10.1002/jrs.2950.

30 McDonald, G.R., Hudson, A.L., Dunn, S.M.J. et al. (2008). Bioactive contaminants leach from disposable laboratory plasticware. *Science* 322 (5903): 917. https://doi.org/10.1126/science.1162395.

31 Starlab (2017). Why RPT? Product Brochure.

32 Tittmas, R. (2015). All About Low Retention Tips! www.linkedin.com/pulse/all-low-retention-tips-ryan-titmas, published on 27 October 2015.

33 CTC Analytics (2018). *PAL System Specifications*. Part No. PAL RTC and RSI Specifications (LC & GC), Rev 8, July 2018.

34 Huntscha, S., Singer, H.P., McArdell, C.S. et al. (2012). Multiresidue analysis of 88 polar organic micropollutants in ground, surface and wastewater using

35 Sng, M.T. and Ng, W.F. (1999). In-situ derivatisation of degradation products of chemical warfare agents in water by solid-phase microextraction and gas chromatographic-mass spectrometric analysis. *Journal of Chromatography A* 832 (1–2): 173–182. https://doi.org/10.1016/S0021-9673(98)00990-X.

online mixed-bed multilayer solid-phase extraction coupled to high performance liquid chromatography–tandem mass spectrometry. *Journal of Chromatography A* 1268: 74–83. https://doi.org/10.1016/J.CHROMA.2012.10.032.

36 Choi, J., Ahn, S.-Y., Kim, Y. et al. (2016). Development of real time monitoring of off-flavour compounds using multi functionalized auto sampler with SPME-GC-MS/MS. *Poster and presentation at the Korea Mass Spectrometry Conference 2016*.

37 The Engineering Toolbox (2020). Volumetric or Cubical Expansion Coefficients of Liquids. https://www.engineeringtoolbox.com/cubical-expansion-coefficients-d_1262.html (assessed 30 June 2020).

38 Rothmeier, S., Nestola, M., A. Hentschel et al. (2018). Fully automated dilution workstation for pesticides working standards mixtures. Berlin, Germany: Institut Kirchhoff GmbH. Presented at the European Pesticides Workshop 2018 in Munich, Germany.

39 Chemspeed Technologies AG (2019). Swing Powderdose, Product Information.

40 Wikipedia (2019). Soxhlet extractor. en.wikipedia.org/wiki/Soxhlet_extractor (accessed 14 September 2019).

41 US Environmental Protection Agency (1994). Method 3541 – Automated Soxhlet Extraction. Rev. 1.0, September 1994. www.epa.gov/sites/production/files/2015-12/documents/3541.pdf (accessed 5 November 2020).

42 Ramos, L. (2020). Greening sample preparation: new solvents, new sorbents. In: *Challenges in Green Analytical Chemistry*, 2nd Ed. (eds. M. de la Guardia and S. Garrigues), 114–153. Royal Society of Chemistry. https://doi.org/10.1039/9781788016148-00114.

43 Ambrus, A., Dobrik, H.S., Majzik, E.S. et al. (2016). Practical Experience Gained with Testing the Effect of Sample Processing on the Stability of Residues and Uncertainty of the Results. Miskolc, Hungary: National Food Chain Safety Office, NFCSO Pesticide Analytical Laboratory.

44 SANCO (2011). Guidance Document on Analytical Quality Control and Validation Procedures for Pesticide Residues Analysis in Food and Feed. European Commission Health & Consumer Protection Directorate General. www.eurl-pesticides.eu/library/docs/allcrl/AqcGuidance_Sanco_2013_12571.pdf.

45 Jain, R. (2017). Microextraction techniques for forensic drug analysis in saliva. *Forensic Research & Criminology International Journal* 5 (4): 164. https://doi.org/10.15406/frcij.2017.05.00164.

46 Costas-Rodriguez, M. and Pena-Pereira, F. (2014). Method development with miniaturized sample preparation techniques. In: *Miniaturization in Sample Preparation* (ed. F. Pena-Pereira), 276–307. Warsaw/Berlin: De Gruyter Open Ltd.

47 Bandelin (2019). Ultraschall – Eine kurze Einführung. bandelin.com/ultraschall/ (accessed 12 December 2019).

48 US Environmental Protection Agency (2018). Method 8270E – Semivolatile Organic Compounds by Gas Chromatography/Mass Spectrometry. SW-846 Update VI, Revision 6, June 2018.

49 US Environmental Protection Agency (2007). Method 3550C – Ultrasonic Extraction. Revision 3 February 2007.

50 US Environmental Protection Agency (1999). USEPA Contract Laboratory Program - Statement of Work For Organics Analysis. OLM04.2 May 1999.

51 Mudiam, M.K.R., Jain, R., and Singh, R. (2014). Application of ultrasound-assisted dispersive liquid–liquid microextraction and automated in-port silylation for the simultaneous determination of phenolic endocrine disruptor chemicals in water samples by gas chromatography-triple quadrupole mass spectrometry. *Analytical Methods* https://doi.org/10.1039/c3ay41658e.

52 Sander, L.C. (2017). Pressurized fluid extraction. *Journal of Research of the National Institute of Standards and Technology* 122 https://doi.org/10.1016/B0-12-369397-7/00692-0.

53 Richter, B.E., Jones, B.A., Ezzell, J.L. et al. (1996). Accelerated solvent extraction: a technique for sample preparation. *Analytical Chemistry* 68 (6): 1033–1039. https://doi.org/10.1021/ac9508199.

54 Büchi (2009). Total Petroleum Hydrocarbons in Soil, Sediment, and Waste Samples. *Application Note No. 19*. Flawil, Switzerland: BÜCHI Labortechnik AG. www.buchi@buchi.com (accessed 5 November 2020).

55 FMS (2017). Pressurized Liquid Extraction. *Applications Notebook*. Watertown, MA, USA: Fluid Management Systems.

56 US Environmental Protection Agency (1995). Method 3545 – Pressurised Fluid Extraction, Test Methods for Evaluating Solid Waste, 3rd Ed., Update III, EPA SW-846, US GPO. Washington, DC, USA.

57 Dionex Corporation (2006). Accelerated Solvent Extraction (ASE®) Sample Preparation Techniques for Food and Animal Feed Samples. *Technical Note 209*, Dionex Corporation, Sunnyvale, CA, USA.

58 Ramos, J.J., Dietz, C., Gonz, M.J. et al. (2007). Miniaturised selective pressurised liquid extraction of polychlorinated biphenyls from foodstuffs. *Journal of Chromatography A* 1152: 254–261. https://doi.org/10.1016/j.chroma.2006.11.097.

59 Ramos, L., Vreuls, R.J.J., and Brinkman, U.A.Th. (2015). Pressurised liquid extraction of polycyclic aromatic hydrocarbons from contaminated soils. *Application Note No. 043*. Eindhoven, The Netherlands: GL Sciences B.V. https://doi.org/10.1016/S0021-9673(00)00419-2.

60 Alañón, M.E., Ramos, L., Díaz-Maroto, M. et al. (2009). Extraction of volatile and semi-volatile components from oak wood used for aging wine by miniaturised pressurised liquid technique. *International Journal of Food Science and Technology* 44: 1825–1835. https://doi.org/10.1111/j.1365-2621.2009.02006.x.

61 Ramos, L. and Richter, B. (2016). Extraction of micropollutants from size limited solid samples. *LCGC Europe*, October, pp. 558–568.

62 Bernsmann, T. and Fürst, P. (2004). Comparison of accelerated solvent extraction (ASE) with integrated sulphuric acid clean-up and soxhlet extraction for determination of PCDD/PCDF, dioxin-like PCB and indicator PCB in feeding stuffs. *Organohalogen Compounds* 66: 157–161.

63 US Environmental Protection Agency (1996). Method 3540C – Soxhlet Extraction. Rev. 3, December 1996.

64 US Environmental Protection Agency (1998). Method 3454A – Pressurized Fluid Extraction (PFE). January 1998.

65 US Environmental Protection Agency (1997). SW-846, Update III: Test Methods for Evaluating Solid Waste, Method 3545. Fed. Reg. Vol. 62, 114, 32451 US GPO, Washington, DC, June 13, 1997.

66 Raynie, D.E. (2018). New sample preparation products and accessories. *LCGC Europe*, May, pp. 278–279.

67 Dawes, P. and Minett, A. (2014). High efficiency SPE/fractionation cartridges for direct mass spectrometer sample preparation. *Poster at the ASMS 2014*.

68 Luke, M.A., Froberg, J.E., and Masumoto, H.T. (1975). Extraction and cleanup of organochlorine, organophosphate, organonitrogen, and hydrocarbon pesticides in produce for determination by gas-liquid chromatography. *Journal – Association of Official Analytical Chemists* 58 (5): 1020–1026. www.ncbi.nlm.nih.gov/pubmed/1158821.

69 Luke, M.A., Froberg, J.E., Doose, G.M. et al. (1981). Improved multiresidue gas chromatographic determination of organophosphorus, organonitrogen, and organohalogen pesticides in produce, using flame photometric and electrolytic conductivity detectors. *Journal – Association of Official Analytical Chemists* 64 (5): 1187–1195. www.ncbi.nlm.nih.gov/pubmed/7287614.

70 Luke, M.A. and Doose, G.M. (1983). A modification of the Luke multiresidue procedure for low moisture, nonfatty products. *Bulletin of Environmental Contamination and Toxicology* 30 (1): 110–116. https://doi.org/10.1007/BF01610107.

71 NVWA (2014). Dutch mini-Luke ('NL-') extraction method followed by LC and GC-MS/MS for multiresidue analysis of pesticides in fruits and vegetables. NVWA, Netherlands Food and Consumer Product Safety Authority, NRL for Pesticide Residues in Food and Feed in collaboration with EURL-FV. www.eurl-pesticides.eu/userfiles/file/NL-miniLuke-extraction-method.pdf.

72 Makovi, C. and McMahon, B.M. (eds.) (1999). *Pesticide Analytical Manual: Multiresidue Methods*, Vol. I, 3rd rev. ed. US Department of Health and Human Services, Public Health Service, Food and Drug Administration.

73 Kabir, A. (2017). Recent trends in microextraction techniques employed in analytical and bioanalytical sample preparation. *Separations* 4 (36): 1–15. https://doi.org/10.3390/separations4040036.

74 Majors, R.E. (2006). Miniaturized approaches to conventional liquid–liquid extraction. *LCGC North America* 24 (2): 118–130.

75 De Boeck, M., Dehaen, W., Tytgat, J. et al. (2019). Microextractions in forensic toxicology: the potential role of ionic liquids. *TrAC – Trends in Analytical Chemistry* https://doi.org/10.1016/j.trac.2018.11.036.

76 Rezaee, M., Assadi, Y., Hosseini, M.-R.M. et al. (2006). Determination of organic compounds in water using dispersive liquid–liquid microextraction. *Journal of Chromatography A* 1116 (1–2): 1–9. https://doi.org/10.1016/J.CHROMA.2006.03.007.

77 Berijani, S., Assadi, Y., Anbia, M. et al. (2006). Dispersive liquid–liquid microextraction combined with gas chromatography-flame photometric detection: very simple, rapid and sensitive method for the determination of organophosphorus pesticides in water. *Journal of Chromatography A* 1123 (1): 1–9. https://doi.org/10.1016/J.CHROMA.2006.05.010.

78 Maya, F., Horstkotte, B., Estela, J.M. et al. (2014). Automated in-syringe dispersive liquid–liquid microextraction. *TrAC – Trends in Analytical Chemistry* 59: 1–8. https://doi.org/10.1016/j.trac.2014.03.009.

79 Shi, Z.-G. and Lee, H.K. (2010). Dispersive liquid–liquid microextraction coupled with dispersive μ-solid-phase extraction for the fast determination of polycyclic aromatic hydrocarbons in environmental water samples. *Analytical Chemistry* 82 (4): 1540–1545. https://doi.org/10.1021/ac9023632.

80 Guo, L. and Lee, H.K. (2014). Automated dispersive liquid–liquid microextraction–gas chromatography–mass spectrometry. *Analytical Chemistry* 86 (8): 3743–3749. https://doi.org/10.1021/ac404088c.

81 Hutchinson, A. and Carrier, D. (2017). Fully Automated Method Using Dispersive Liquid Liquid Micro Extraction (DLLME) for Extractable and Leachable Studies. *Technical Note AS182*. Cambridge, UK: Anatune Ltd.

82 LABC-Labortechnik (2017). Dispersive Liquid–Liquid-Microextraction (DLLME) mit Bilimex. Hennef, Germany: LABC-Labortechnik Zillger KG. www.LABC.de.

83 Mudiam, K.M.R., Jain, R., Maury, S.K. et al. (2012). Low density solvent based dispersive liquid–liquid microextraction with gas chromatography–electron capture detection for the determination of cypermethrin in tissues and blood of cypermethrin treated rats. *Journal of Chromatography B* 895–896: 65–70. https://doi.org/10.1016/j.jchromb.2012.03.015.

84 Assadi, Y., Farajzadeh, M.A., and Bidari, A. (2012). Dispersive liquid–liquid microextraction. *Comprehensive Sampling and Sample Preparation* 2: 181–212. https://doi.org/10.1016/B978-0-12-381373-2.00051-X.

85 Mudiam, M.K.R., Ch, R., Chauhan, A. et al. (2012). Optimization of UA-DLLME by experimental design methodologies for the simultaneous determination of endosulfan and its metabolites in soil and urine samples by GC–MS. *Analytical Methods* 4 (11): 3855–3863. https://doi.org/10.1039/c2ay25432h.

86 Rai, S., Singh, A.K., Srivastava, A. et al. (2016). Comparative evaluation of QuEChERS method coupled to DLLME extraction for the analysis of multiresidue pesticides in vegetables and fruits by gas chromatography-mass spectrometry. *Food Analytical Methods* https://doi.org/10.1007/s12161-016-0445-2.

87 Song, N., Guo, M., Liu, Y. et al. (2016). Rapid residue analysis of sulfonylurea herbicides in surface water: methodology and residue findings in eastern Tiaoxi River of China. *Journal of Materials Science and Chemical Engineering* 4: 41–50.

88 Pinto, E., Soares, A.G., Ferreira, I.M.P.L.V.O. (2018). Quantitative analysis of glyphosate, glufosinate and AMPA in irrigation water by in situ derivatization–dispersive liquid–liquid microextraction combined with

UPLC-MS/MS. *Analytical Methods* 10 (5): 554–561. https://doi.org/10.1039/C7AY02722B.

89 Baltussen, E., Cramers, C.A., and Sandra, P.J.F. (2002). Sorptive sample preparation – a review. *Analytical and Bioanalytical Chemistry* 373 (1–2): 3–22. https://doi.org/10.1007/s00216-002-1266-2.

90 Zhang, Z. and Pawliszyn, J. (1995). Quantitative extraction using an internally cooled solid phase microextraction device. *Analytical Chemistry* 67 (1): 34–43. https://doi.org/10.1021/ac00097a007.

91 Arthur, C.L. and Pawliszyn, J. (1990). Solid phase microextraction with thermal desorption using fused silica optical fibers. *Analytical Chemistry* 62 (19): 2145–2148.

92 Ouyang, G. and Jiang, R. (eds.) (2016). *Solid Phase Microextraction – Recent Developments and Applications*. Berlin Heidelberg: Springer-Verlag https://doi.org/10.1007/978-1-4615-1247-9_6.

93 Kusch, P. (2018). Headspace solid-phase microextraction coupled with gas chromatography – mass spectrometry for the characterization of polymeric materials. *LCGC North America* 1: 1–25.

94 Nzekoue, F.K., Angeloni, S., Caprioli, G. et al. (2020). Fiber-sample distance, an important parameter to be considered in headspace solid-phase microextraction applications. *Analytical Chemistry* 92: 7478–7484.

95 O'Reilly, J., Wang, Q., Setkova, L. et al. (2005). Automation of solid-phase microextraction. *Journal of Separation Science* 28 (15): 2010–2022. https://doi.org/10.1002/jssc.200500244.

96 Mascrez, S., Psillakis, E., and Purcaro, G. (2019). A Multifaceted investigation on the effect of vacuum on the headspace solid-phase microextraction of extra-virgin olive oil. *Analytica Chimica Acta* 1103: 106–114. https://doi.org/10.1016/j.aca.2019.12.053.

97 Vas, G. and Vékey, K. (2004). Solid-phase microextraction: a powerful sample preparation tool prior to mass spectrometric analysis. *Journal of Mass Spectrometry* 39 (3): 233–254. https://doi.org/10.1002/jms.606.

98 Ho, T.D., Yu, H., Cole, W.T.S. et al. (2013). Ionic liquids and their applications in sample preparation. *LCGC Europe* 26 (2): 101–109.

99 Pawliszyn, J. (2009). *Handbook of Solid Phase Microextraction* (ed. J. Pawliszyn). Beijing: Chemical Industry Press.

100 Michel, F. and Buckendahl, K. (2011). Basics in solid phase micro extraction (SPME) and applications in food and environmental analysis. *Presentation at the Euroanalysis 2011 Belgrade, Lunch Symposium* (13 September 2011). Sigma-Aldrich Co. Bellefonte, PA, USA: Sigma-Aldrich Co.

101 Sigma-Aldrich (2009). SPME Applications Guide – Bulletin 925F. Sigma-Aldrich Application Publication T199925F CJQ, 225 pp. Bellefonte, PA, USA.

102 Huba, A.K., Mirabelli, M.F., and Zenobi, R. (2018). High-throughput screening of PAHs and polar trace contaminants in water matrices by direct solid-phase microextraction coupled to a dielectric barrier discharge ionization source. *Analytica Chimica Acta* 1030: 125–132. https://doi.org/10.1016/j.aca.2018.05.050.

103 Chen, J. and Pawliszyn, J. (1995). Solid phase microextraction coupled to high-performance liquid chromatography. *Analytical Chemistry* 67 (15): 2530–2533. https://doi.org/10.1021/ac00111a006.

104 Lord, H.L. and Pawliszyn, J. (2000). Microextraction of drugs. *Journal of Chromatography A* 902 (1): 17–63. www.ncbi.nlm.nih.gov/pubmed/21637206.

105 Shirey, R.E. (2009). SPME commercial devices and fibre coatings. In: *Handbook of Solid Phase Microextraction* (ed. J. Pawliszyn), 86–115. Beijing: Chemical Industry Press.

106 Merlin MicroSeal (2019). High performance alternative to traditional rubber disc septa. www.merlinic.com/products/merlin-microseal (accessed 13 September 2019).

107 Millipore Sigma (2019). SPME fibers and holders. www.sigmaaldrich.com (accessed 07 September 2019)

108 Wikipedia (2020). Nitinol is a non-ferrous 1:1 alloy of nickel and titanium. see https://en.wikipedia.org/wiki/Nickel_titanium (accessed 05 May 2020).

109 Schueler, K.H., Schillig, C., and Schilling, B. (late registered) (2018). Extraction device. US Patent US10,001,431 B2 of June 19, 2018.

110 Kremser, A., Jochmann, M.A., and Schmidt, T.C. (2016). PAL SPME Arrow – evaluation of a novel solid-phase microextraction device for freely dissolved PAHs in water. *Analytical and Bioanalytical Chemistry* 408 (3): 943–952. https://doi.org/10.1007/s00216-015-9187-z.

111 Herrington, J.S., German, A.G., Myers, C. et al. (2020). Hunting molecules in complex matrices with SPME arrows: a review. *Separations* 7 (1–20) https://doi.org/10.3390/separations7010012.

112 Kremser, A. (2016). Advances in automated sample preparation for gas chromatography: solid-phase microextraction, headspace-analysis, solid-phase extraction. Ph.D. thesis. Faculty of Chemistry, University of Duisburg-Essen, Germany.

113 Lee, J.Y., Kim, W.S., Lee, Y.-.Y. et al. (2019). Solid-phase microextraction arrow for the volatile organic compounds in soy sauce. *Journal of Separation Science* 42 (18): 2942–2948. https://doi.org/10.1002/jssc.201900388.

114 De Koning, S., Janssen, H.-G., and Brinkman, U.A.Th. (2009). Modern methods of sample preparation for GC analysis. *Chromatographia* 69: 33–78. https://doi.org/10.1365/s10337-008-0937-3.

115 Yang, L. and Luan, T. (2016). Application of solid-phase microextraction combined with derivatization for polar compound sampling in environmental analysis. In: *Solid Phase Microextraction – Recent Developments and Applications* (eds. G. Ouyang and R. Jiang). Berlin Heidelberg: Springer-Verlag https://doi.org/10.1007/978-1-4615-1247-9_6.

116 Merkle, S., Kleeberg, K., and Fritsche, J. (2015). Recent developments and applications of solid phase microextraction (SPME) in food and environmental analysis – a review. *Chromatography* 2 (3): 293–381. https://doi.org/10.3390/chromatography2030293.

117 Schmarr, H.G., Sang, W., Ganß, S. et al. (2008). Analysis of aldehydes via headspace SPME with on-fiber derivatization to their *O*-(2,3,4,5,

6-pentafluorobenzyl)oxime derivatives and comprehensive 2D-GC-MS. *Journal of Separation Science* 31 (19): 3458–3465. https://doi.org/10.1002/jssc.200800294.

118 Stashenko, E.E., Mora, A.L., Cervantes, M. et al. (2006). HS-SPME determination of volatile carbonyl and carboxylic compounds in different matrices. *Journal of Chromatographic Science* 44 (6): 347–353. https://doi.org/10.1093/chromsci/44.6.347.

119 Martos, P.A. and Pawliszyn, J. (1998). Sampling and determination of formaldehyde using solid-phase microextraction with on-fiber derivatization. *Analytical Chemistry* 70: 2311–2320.

120 Yang, L.H., Lan, C.Y., Liu, H. et al. (2006). Full automation of solid-phase microextraction/on-fiber derivatization for simultaneous determination of endocrine-disrupting chemicals and steroid hormones by gas chromatography-mass spectrometry. *Analytical and Bioanalytical Chemistry* 386: 391–397.

121 Moreira, N., Araújo, A.M., Rogerson, F. et al. (2019). Development and optimization of a HS-SPME-GC-MS methodology to quantify volatile carbonyl compounds in port wines. *Food Chemistry* 270: 518–526.

122 Moreira, N., Meireles, S., Brandão, T. et al. (2013). Optimization of the HS-SPME-GC-IT/MS method using a central composite design for volatile carbonyl compounds determination in beers. *Talanta* 117: 523–531. https://doi.org/10.1016/j.talanta.2013.09.027.

123 Kawaguchi, M., Ishii, Y., Sakui, N. et al. (2005). Stir bar sorptive extraction with in situ derivatization and thermal desorption–gas chromatography–mass spectrometry for determination of chlorophenols in water and body fluid samples. *Analytica Chimica Acta* 533 (1): 57–65. https://doi.org/10.1016/J.ACA.2004.10.080.

124 Namera, A., Yashiki, M., Kojima, T. et al. (2002). Automated headspace solid-phase microextraction and in-matrix derivatization for the determination of amphetamine-related drugs in human urine by gas chromatography–mass spectrometry. *Journal of Chromatographic Science* 40: 19–25.

125 Parkinson, D.R., Bruheim, I., Christ, I. et al. (2004). Full automation of derivatization-solid-phase microextraction-gas chromatography–mass spectrometry with a dual-arm system for the determination of organometallic compounds in aqueous samples. *Journal of Chromatography A* 1025: 77–84.

126 Fluka Chemie AG (2005). *Silylating Agents*.

127 Jain, R., Kumar, A., and Shukla, Y. (2015). Dispersive liquid–liquid microextraction-injector port silylation: a viable option for the analysis of polar analytes using gas chromatography-mass spectrometry. *Austin Journal of Analytical and Pharmaceutical Chemistry* 2 (3): 10–122.

128 Mirabelli, M.F., Gionfriddo, E., Pawliszyn, J. et al. (2018). A quantitative approach for pesticide analysis in grape juice by direct interfacing of a matrix compatible SPME phase to dielectric barrier discharge ionization-mass spectrometry. *Analyst* 143 (4): 891–899. https://doi.org/10.1039/c7an01663h.

129 Klampfl, C.W. and Himmelsbach, M. (2015). Direct ionization methods in mass spectrometry: an overview. *Analytica Chimica Acta* 890: 44–59. https://doi.org/10.1016/j.aca.2015.07.012.

130 Wolf, J.C., Gyr, L., Mirabelli, M.F. et al. (2016). A radical-mediated pathway for the formation of [M+H]$^+$ in dielectric barrier discharge ionization. *Journal of the American Society for Mass Spectrometry* 27 (9): 1468–1475. https://doi.org/10.1007/s13361-016-1420-2.

131 Huba, A.K. (2017). Simple high-throughput screening of trace organic contaminants in food matrices by HS-SPME ambient mass spectrometry. *Presentation at the RAFA Conference*, Prague, November 9th 2017.

132 Baltussen, E., Sandra, P., David, F. et al. (1999). Stir bar sorptive extraction (SBSE), a novel extraction technique for aqueous samples: theory and principles. *Journal of Microcolumn Separations* 11 (10): 737–747. https://doi.org/10.1002/(SICI)1520-667X(1999)11:10<737::AID-MCS7>3.0.CO;2-4.

133 Sandra, P. (2015). Stir Bar Sorbtive Extraction (SBSE) Established, Useful and Quite Often Simply the Extraction Technique of Choice. *GERSTEL Solutions Worldwide*, Mülheim an der Ruhr, Germany: GERSTEL GmbH & Co. KG.

134 Sandra, P., Tienpont, B., and David, F. (2003). Stir bar sorptive extraction (SBSE) recovery calculator: easy calculation of extraction recoveries for SBSE. *Application Note 2003/2*. Kortrijk, Belgium: Research Institute for Chromatography.

135 Nie, Y. and Kleine-Benne, E. (2011). Using three types of twister phases for stir bar sorptive extraction of whisky, wine and fruit juice. *Application Note 3/2011*. Mülheim an der Ruhr, Germany: GERSTEL GmbH & Co. KG.

136 Bruheim, I., Liu, X., and Pawliszyn, J. (2003, 2003). Thin-film microextraction. *Analytical Chemistry* 75 (4): 1002–1010. https://doi.org/10.1021/ac026162q.

137 Jiang, R. and Pawliszyn, J. (2012). Thin-film microextraction offers another geometry for solid-phase microextraction. *TrAC – Trends in Analytical Chemistry* 39: 245–253. https://doi.org/10.1016/j.trac.2012.07.005.

138 Pawliszyn Research Group (2019). Thin Film Microextraction (TFME). uwaterloo.ca/pawliszyn-group/research/thin-film (accessed 12 September 2019).

139 Grandy, J., Boyaci, E., and Pawliszyn, J. (2016). Development of a carbon mesh supported thin film microextraction membrane as a means to lower the detection limits of benchtop and portable GC/MS instrumentation. *Analytical Chemistry* 88 (3): 1760–1767. https://doi.org/10.1021/acs.analchem.5b04008.

140 Risticevic, S., Niri, V.H., Vuckovic, D. et al. (2009). Recent developments in solid-phase microextraction. *Analytical and Bioanalytical Chemistry* 393: 781–795. https://doi.org/10.1007/s00216-008-2375-3.

141 Emmons, R.V., Tajali, R., and Gionfriddo, E. (2019). Development, optimization and applications of thin film solid phase microextraction (TF-SPME) devices for thermal desorption: a comprehensive review. *Separations* 6 (3): 39. https://doi.org/10.3390/separations6030039.

142 Stuff, J.R., Whitecavage, J.A., J.J. Grandy et al. (2018). Analysis of Beverage Samples Using Thin Film Solid Phase Microextraction (TF-SPME) and Thermal Desorption GC/MS. *Application Note 200*. Linthicum, MD, USA: GERSTEL Inc.

143 Boyaci, E., Goryński, K., Viteri, C.R. et al. (2016). A study of thin film solid phase microextraction methods for analysis of fluorinated benzoic acids in seawater. *Journal of Chromatography A* 1436: 51–58. https://doi.org/10.1016/j.chroma.2016.01.071.

144 Omnexus (2020) Polyamide (PA) or Nylon: Complete Guide (PA6, PA66, PA11, PA12…). Product Information, SpecialChem SA, Paris, France. https://omnexus.specialchem.com/selection-guide/polyamide-pa-nylon (accessed 21 December 2020) .

145 GERSTEL (2019). Automated Filtration Using the MPS, Product Information. Mülheim an der Ruhr: GERSTEL GmbH & Co. KG.

146 Merck (2020). Whatman Syringe & Syringeless Filters, Product Information. Source: https://www.sigmaaldrich.com/analytical-chromatography/analytical-products.html?TablePage=9655376.

147 Thomson Instrument Company (2019). What is a Filter Vial? Product Information. www.htslabs.com/fv/ (accessed 03 August 2019).

148 Goebel, C., Wanders, L., and Ellis, S. (2014). Improved Sample Preparation Methods for Athlete Doping Analysis of Common Compounds in Urine by LCMS. Australian Sports Drug Testing Laboratory, Sydney, Australia: Thomson Instrument Company, *Application Note*.

149 Telepchak, M.J., August, T.F., and Chaney, G. (2004). *Introduction to Solid Phase Extraction. Forensic and Clinical Applications of Solid Phase Extraction.* Totowa, NJ: Humana Press. https://doi.org/10.1007/978-1-59259-292-0_1.

150 Spark Holland B.V (2006). *SPE Workbook*. Emmen, The Netherlands. www.sparkholland.com.

151 Poole, C.F. (ed.) (2020). *Solid-Phase Extraction*. Amsterdam, The Netherlands: Elsevier Inc. https://doi.org/10.1016/C2018-0-00617-9.

152 DIN EN 12673:1999-05 (1999). *Water quality – gas chromatographic determination of some selected chlorophenols in water*. Berlin, Germany: Beuth Verlag.

153 Smith, S. (2015). Solid Phase Extraction (SPE), An Introduction to Basic Theory, Method Development, and Applications. Sigma-Aldrich Co. LLC.

154 LCTech (2017). Automated Sample Preparation – Intelligent System Solution, Product Information p/n 14303. Obertaufkirchen: LCTech GmbH.

155 Aisti Science (2015). SGI-P100 for Gas Chromatography, Product Information. Wakayama: Aisti Science Co Ltd.

156 Lerch, O., Temme, O., and Daldrup, T. (2014). Comprehensive automation of the solid phase extraction gas chromatographic mass spectrometric analysis (SPE-GC/MS) of opioids, cocaine, and metabolites from serum and other matrices. *Analytical and Bioanalytical Chemistry* 406 (18): 4443–4451. https://doi.org/10.1007/s00216-014-7815-7.

157 Spark Holland (2015). 25 Questions about Symbiosis, *Application Guide*. Emmen: Spark Holland B.V.

158 GERSTEL (2015). Online SPE with Replaceable Cartridges, Product Information. Mülheim an der Ruhr: GERSTEL Gmbh & Co KG.

159 Filippe, T.C., Goulart, F.D.A.B., Mizukawa, A. et al. (2018). Validation of analytical methodology for determination of personal care products in environmental matrix by GC-MS/MS. *Ecletica Quimica* 43 (3): 30–36. https://doi.org/10.26850/1678-4618eqj.v43.3.30-36.

160 Naccarato, A. and Tagarelli, A. (2019). Recent applications and newly developed strategies of solid-phase microextraction in contaminant analysis: through the environment to humans. *Separations* 6 (4): 2–43. https://doi.org/10.3390/separations6040054.

161 Lebertz, S. and Schuhn, B. (2018). Rapid Quantification of Polar and Semipolar Pesticide Metabolites with Combined Online SPE and Direct Injection. Agilent *Application Note 5994-0151en*.

162 Helle, N. and Chmelka, F. (2012). Glyphosate/AMPA a global presence. *GERSTEL Newsletter*, EPRW 2012. Mülheim an der Ruhr, Germany: GERSTEL Gmbh & Co KG.

163 Wittsiepe, J., Nestola, M., Kohne, M. et al. (2014). Determination of polychlorinated biphenyls and organochlorine pesticides in small volumes of human blood by high-throughput on-line SPE-LVI-GC-HRMS. *Journal of Chromatography B* 945–946: 217–224. https://doi.org/10.1016/j.jchromb.2013.11.059.

164 Dugheri, S., Marrubini, G., Mucci, N. et al. (2020). A review of micro-solid-phase extraction techniques and devices applied in sample pretreatment coupled with chromatographic analysis. *Acta Chromatographica* https://doi.org/10.1556/1326.2020.00790.

165 Gamble, K.R. and Martin, S.W. (2000). Sample collection and processing device. US Patent 7, 001, 774 B1, issued 21 February 2006. https://patentimages.storage.googleapis.com/94/c0/86/3b3b529e89d525/US7001774.pdf (accessed 5 November 2020).

166 Hayward, M., Ho, J., Youngblood, R. et al. (2016). Automated Chromatographic Solid-Phase Extraction Using an Autosampler. *American Laboratory (Online)*. www.americanlaboratory.com/914-Application-Notes/190631-Automated-Chromatographic-Solid-Phase-Extraction-Using-an-Autosampler/.

167 Anastassiades, M., Lehotay, S.J., Stajnbaher, D. et al. (2003). Fast and easy multiresidue method employing acetonitrile extraction/partitioning and 'dispersive solid-phase extraction' for the determination of pesticide residues in produce. *Journal of AOAC International* 86 (2): 412–431.

168 Morris, B.D. and Schriner, R.B. (2015). Development of an automated column solid-phase extraction cleanup of QuEChERS extracts, using a zirconia-based sorbent, for pesticide residue analyses by LC-MS/MS. *Journal of Agricultural and Food Chemistry* 63: 5107–5119. https://doi.org/10.1021/jf505539e.

169 Herrmann, S.S. and Poulsen, M.E. (2016). Clean-up experiments of oat extracts for pesticide residues analysis by PSA, C18, Z-Sep or EMR-lipid, individually and combinations. *Poster session presented at 11th European Pesticide Residue Workshop*, Limassol, Cyprus.

170 Schenck, F.J., Callery, P., Gannett, P.M. et al. (2002). Comparison of magnesium sulfate and sodium sulfate for removal of water from pesticide extracts of foods. *Journal of AOAC International* 85 (5): 1177–1180.

171 Huebschmann, H.-J. (2015). *Handbook of GC-MS: Fundamentals and Applications*, 3rd Ed. Weinheim: Wiley-VCH.

172 Monteiro, S.H., Lehotay, S.J., Sapozhnikova, Y. et al. (2020). High-throughput mega-method for the analysis of pesticides, veterinary drugs, and environmental contaminants by ultra-high-performance liquid chromatography-tandem mass spectrometry and robotic mini-solid-phase extraction cleanup + low-pressure gas chromatography-tandem mass spectrometry, Part 1: Beef. *Journal of Agricultural and Food Chemistry* https://doi.org/10.1021/acs.jafc.0c00710.

173 Ninga, E., Sapozhnikova, Y., Lehotay, S.J. et al. (2020). High-throughput mega-method for the analysis of pesticides, veterinary drugs, and environmental contaminants by ultra-high-performance liquid chromatography-tandem mass spectrometry and robotic mini-solid-phase extraction cleanup + low-pressure gas chromatography-tandem mass spectrometry, Part 2: Catfish. *Journal of Agricultural and Food Chemistry* https://doi.org/10.1021/acs.jafc.0c00995.

174 Goon, A., Shinde, R., Ghosh, B. et al. (2019). Application of automated mini–solid-phase extraction cleanup for the analysis of pesticides in complex spice matrixes by GC-MS/MS. *Journal of AOAC International* 10: 1–6. https://doi.org/10.5740/jaoacint.19-0202.

175 Lehotay, S.J., Han, L., and Sapozhnikova, Y. (2016). Automated mini-column solid-phase extraction cleanup for high throughput analysis of chemical contaminants in foods by low pressure gas chromatography-tandem mass spectrometry. *Chromatographia* 79: 1113–1130. https://doi.org/10.1007/s10337-016-3116-y.

176 Chang, Y., Zou, S., Ge, Y. et al. (2019). Simultaneous determination of five alkaloids by HPLC-MS/MS combined with micro-SPE in rat plasma and its application to pharmacokinetics after oral administration of lotus leaf extract. *Frontiers in Pharmacology* 10: 1252. https://doi.org/10.3389/FPHAR.2019.01252.

177 Lehmann, S., Schulze, B., Thomas, A. et al. (2018). Organ distribution of 4-MEC, MDPV, methoxetamine and α-PVP: comparison of QuEChERS and SPE. *Forensic Toxicology* No. 0123456789, 1–14. https://doi.org/10.1007/s11419-018-0408-y.

178 Eguchi, A., Enomoto, T., S. Mikami et al. (2016). Investigation of analytical method for PCBs in serum sample using mini SPE cartridge with intelligent autosampler. Center for Preventive Medical Sciences, Chiba University, Japan, *Poster at the Symposium on Halogenated Persistent Organic Pollutants*, Dioxin 2018, Kraków, Poland. Center for Preventive Medical Sciences, Chiba University, Japan.

179 Korenková, E. and Cepeda-Leucea, M. (2017). The determination of 1,4-dioxane in water by automated SPE gas chromatography-high resolution mass spectrometry (GC-HRMS). Ministry of the Environment and Climate

Change – Laboratory Services Branch E3534 Rev.1, Ministry of the Environment and Climate Change, Ontario, Canada.

180 Jacob, C.C., Martins, C.P.B., K. Van Natta et al. (2019). Automating the analysis of estrogens in serum or plasma using a multi-purpose auto sampler coupled to liquid chromatography triple quadrupole mass spectrometry. *Poster at the ASMS Conference 2019 Atlanta, USA*. San Jose, CA, USA: Thermo Fisher Scientific.

181 Eyring, P., Preiswerk, T., Frandsen, H.L., and Duedahl-Olesen, L. (2020). Automated micro-solid-phase extraction clean-up of polycyclic aromatic hydrocarbons in food oils for analysis by gas chromatography-orbital ion trap mass spectrometry. *Journal of Separation Science* https://doi.org/10.1002/jssc.202000720.

182 Althoff, M.A., Bertsch, A., and Metzulat, M. (2019). Automation of micro-SPE (smart-SPE) and liquid–liquid extraction applied for the analysis of chemical warfare agents. *Separations* 6 (4): 1–15. https://doi.org/10.3390/separations6040049.

183 Leijotode Oliveira, H., Suleimara Teixeira, L., Aparecida Fonseca Dinali, L. et al. (2019). Microextraction by packed sorbent using a new restricted molecularly imprinted polymer for the determination of estrogens from human urine samples. *Microchemical Journal* 150: 104162. https://doi.org/10.1016/J.MICROC.2019.104162.

184 Abdel-Rehim, M. (2004). New trend in sample preparation: on-line microextraction in packed syringe for liquid and gas chromatography applications I. Determination of local anaesthetics in human plasma samples using gas chromatography–mass spectrometry. *Journal of Chromatography B* 801: 317–321.

185 Pereira, J., Câmara, J.S., Colmsjö, A., and Abdel-Rehim, M. (2014). Microextraction by packed sorbent: an emerging, selective and high-throughput extraction technique in bioanalysis. *Biomedical Chromatography* 28 (6): 839–847. https://doi.org/10.1002/bmc.3156.

186 Candish, E., Gooley, A., Wirth, H.-J. et al. (2012). A simplified approach to direct SPE-MS. *Journal of Separation Science* 35 (18): 2399–2406. https://doi.org/10.1002/jssc.201200466.

187 Bizkarguenaga, E., Iparragirre, A., Navarro, P. et al. (2013). In-port derivatization after sorptive extractions. *Journal of Chromatography A* 1296: 36–46. https://doi.org/10.1016/j.chroma.2013.03.058.

188 Dawes, P., Hibbert, R., H.-J. Wirth et al. (2010). The extraction of saliva for the analysis of basic drug residues using MEPS-GCMS. *Poster at the 36th ISCC Conference Riva del Garda*, Italy.

189 Silva, C., Cavaco, C., Perestrelo, R. et al. (2014). Microextraction by packed sorbent (MEPS) and solid-phase microextraction (SPME) as sample preparation procedures for the metabolomic profiling of urine. *Metabolites* 4 (1): 71–97. https://doi.org/10.3390/metabo4010071.

190 Abdel-Rehim, M. (2009). Current advances in microextraction by packed sorbent (MEPS) for bioanalysis applications. *LCGC Europe*, January, pp. 8–19.

191 Ahmadi, M., Moein, M.M., Madrakian, T. et al. (2018). Determination of local anesthetics in human plasma and saliva samples utilizing liquid chromatography-tandem mass spectrometry. *Journal of Chromatography B* 1095: 177–182. https://doi.org/10.1016/J.JCHROMB.2018.07.036.

192 Jönsson, S., Hagberg, J., and Van Bavel, B. (2008). Determination of 2,4, 6-trichloroanisole and 2,4,6-tribromoanisole in wine using microextraction in packed syringe and gas chromatography-mass spectrometry. *Journal of Agricultural and Food Chemistry* 56 (13): 4962–4967. https://doi.org/10.1021/jf800230y.

193 Abdel-Rehim, M. (2011). Microextraction by packed sorbent (MEPS), a tutorial. *Analytica Chimica Acta* 701 (2): 119–128. https://doi.org/10.1016/j.aca.2011.05.037.

194 Dawes, P., Dawes, E., D. Difeo et al. (2007). Online and Offline Applications of Micro-SPE (MEPS). *Application Note*. SGE Analytical Science.

195 Porto-Figueira, P., Figueira, J.A., Pereira, J.A.M. et al. (2015). A fast and innovative microextraction technique, μSPEed, followed by ultrahigh performance liquid chromatography for the analysis of phenolic compounds in teas. *Journal of Chromatography A* 1424: 1–9. https://doi.org/10.1016/J.CHROMA.2015.10.063.

196 Minett, A., Diplock, M., Doble, P. et al. (2019). Flexible automation for all steps in sample preparation, Product Information. ePrep Pty Ltd. www.eprep-analytical.com/copy-of-applications.

197 Pandohee, J. and Jones, O.A.H. (2016). Evaluation of new micro solid-phase extraction cartridges for on-column derivatisation reactions. *Analytical Methods* 8 (8): 1765–1769. https://doi.org/10.1039/C5AY02618K.

198 Wikipedia (2019). Gel permeation chromatography. en.m.wikipedia.org/wiki/Gel_permeation_chromatography (accessed 18 August 2019).

199 Lathe, G.H. and Ruthven, C.R. (1956). The separation of substances and estimation of their relative molecular sizes by the use of colums of starch in water. *Biochemical Journal* 62 (4): 665–674. https://doi.org/10.1042/bj0620665.

200 Barth, H.G. (2013). The early development of size-exclusion chromatography: a historical perspective. *LCGC North America* 31 (7): 550–558.

201 Agilent Technologies (2015). *An Introduction to Gel Permeation Chromatography and Size Exclusion Chromatography*. Primer, doc no. 5990-6969EN. Agilent Technologies Inc.

202 David, F., Devos, C., Dumont, E. et al. (2017). Determination of pesticides in fatty matrices using gel permeation clean-up followed by GC-MS/MS and LC-MS/MS analysis: a comparison of low- and high-pressure gel permeation columns. *Talanta* 165: 201–210. https://doi.org/10.1016/j.talanta.2016.12.032.

203 Pekar, H., Blomkvist, G., S. Ekroth et al. (2009). Analysis of pesticides in butter with GC-MS/MS and LC-MS/MS using EtOAc/CyH extraction and GPC clean-up. *Presentation at the 8th Nordic Pesticide Residue workshop*, Oscarsborg, Norway.

204 DePaoli, M., Barbina, M.T., Mondini, R. et al. (1992). Determination of organophosphorus pesticides in fruits by on-line size-exclusion chromatography-liquid chromatography-gas chromatography-flame photometric detection. *Journal of Chromatography* 626: 145–150.

205 Hildmann, F., Speer, K., and Kempe, G. (2013). Application of the precolumn back-flush technology in pesticide residue analysis: a practical view. *Journal of Separation Science* 36 (13): 2128–2135. https://doi.org/10.1002/jssc.201300007.

206 Majors, R.E. (2014). *Sample Preparation Fundamentals for Chromatography* (eds. T. Robarge, N. Simpson, C. Deckers, et al.). Wilmington, DE: Agilent Technologies, Inc. https://doi.org/10.1515/9783110289169.

207 Grob, K. and Kaelin, I. (1991). Towards on-line SEC-GC of pesticide residues? The problem of tailing triglyceride peaks. *Journal of High Resolution Chromatography* 14: 451–454.

208 Han, L., Sapozhnikova, Y., and Nuñez, A. (2019). Analysis and occurrence of organophosphate esters in meats and fish consumed in the United States. *Journal of Agricultural and Food Chemistry* 67 (46): 12652–12662. https://doi.org/10.1021/acs.jafc.9b01548.

209 Japan Ministry of Health Labor and Welfare (2006). Annex 3: The Method for Multi-Residue Analysis of Pesticides in Animal Products by GC/MS (Draft). Positive List System for Agricultural Chemical Residues in Foods, no. Annex 3. www.mhlw.go.jp/english/topics/foodsafety/positivelist060228/introduction.html (accessed 5 November 2020).

210 China National Standards (2018). *National food safety standards – determination of 208 pesticide and metabolite residues in plant-derived foods – gas chromatography-mass spectrometry*. GB 23200.113-2018. China National Standards.

211 Che, J., Yu, C., L. Liang et al. (2014). Automated Online GPC/GC-MS for the Determination of Pesticides in Vegetables. *Application Note AN10422*. Shanghai, China: Thermo Fisher Scientific.

212 DongXue, Z., YuLe, Z., ZhiYong, X. et al. (2017). Determination of 41 kinds of pesticide residues in vegetables by QuEChERS automated sample preparation system and online gel permeation chromatography gas chromatography-mass spectrometry. *Journal of Food Safety and Quality* 8 (4): 1376–1382.

213 US Environmental Protection Agency (1994). Method 3640a – Gel Permeation Cleanup, September, pp. 1–24.

214 Showa Denko (2019). Shodex Elution Profile for Pesticides. Application Information. Showa Denko K.K. www.shodex.com/en/dc/09/03/09#! (accessed 16 July 2019).

215 Kerkdijk, H., Mol, H.G.J., and Van Der Nagel, B. (2007). Volume overload cleanup: an approach for on-line SPE-GC, GPC-GC, and GPC-SPE-GC. *Analytical Chemistry* 79 (21): 7975–7983. https://doi.org/10.1021/ac0701536.

216 Grob, K. and Kaelin, I. (1991). Attempt for an on-line size exclusion chromatography-gas chromatography method for analyzing pesticides resdidues in food. *Journal of Agricultural and Food Chemistry* 39: 1950–1953.

217 Hildmann, F., Gottert, C., Frenzel, T. et al. (2015). Pesticide residues in chicken eggs – a sample preparation methodology for analysis by gas and liquid chromatography/tandem mass spectrometry. *Journal of Chromatography A* 1403: 1–20. https://doi.org/10.1016/j.chroma.2015.05.024.

218 Yang, J., Wang, J.-H., Gong, P. et al. (2013). Determination of polycyclic aromatic hydrocarbons (PAHs) in vegetable oil by gas chromatography coupled with triple quadrupole mass spectrometry using isotope dilution. 分析检测食品科学 *(Analysis and Testing Food Science)* 34 (22): 202–207. https://doi.org/10.7506/spkx1002-6630-201322041.

219 Bendig, P. and Vetter, W. (2011). Optimierung der Bestimmung des Flammschutzmittels Decabromdiphenylether (BDE-209). *Lebensmittelchemie* 65: 167–168.

220 Sun, H., Yang, Y., Li, H. et al. (2012). Development of multiresidue analysis for twenty phthalate esters in edible vegetable oils by microwave-assisted extraction–gel permeation chromatography–solid phase extraction–gas chromatography–tandem mass spectrometry. *Journal of Agricultural and Food Chemistry* 60 (22): 5532–5539.

221 Liu, G. (2018). µGPC -GC/MS 联机分析系统快速检测食用油中的19种农药残留 (Rapid detection of 19 pesticide residues in edible oil by µGPC-GC-MS online analytical system). *Application Note.* Beijing: LabTech Instruments Co., Ltd.

222 Liu, G. (2018). µGPC微量凝胶净化-GC/MS联机分析系统快速检测牛油果中的16种有机氯 (Rapid detection of 16 organochlorines in avocado by µGPC microgel purification-GC-MS online analytical system). *Application Note.* Beijing: LabTech Instruments Co., Ltd.

223 Hopfgartner, G., Jahn, S., and Varesio, E. (2014). Integrated Platform Including Bligh and Dyer Extraction and Dual-Column UHPLC-MS/MS Separations for Metabolomics Studies Identification of Endogenous Metabolites from Chlamydomonas Reinhardtii Algae. Baltimore, MD, USA: Life Sciences Mass Spectrometry, School of Pharmaceutical Sciences EPGL, University of Lausanne, Geneva, Switzerland. *ASMS 2014 Poster MO632.*

224 Anderson, N.G. (2012). Solvent selection – II using azeotropes to select solvents. In: *Practical Process Research and Development*, 2e, 169–187. Academic Press. www.sciencedirect.com/topics/chemistry/azeotropic-mixture.

225 Dabrowski, Ł. (2016). Review of use of keepers in solvent evaporation procedure during the environmental sample analysis of some organic pollutants. *TrAC – Trends in Analytical Chemistry* 80: 507–516. https://doi.org/10.1016/j.trac.2015.10.014.

226 GERSTEL (2018). mVAP Option for MPS Robotic. *Spec Sheet.* Mülheim an der Ruhr: GERSTEL GmbH & Co KG.

227 Purschke, K., Heinl, S., Lerch, O. et al. (2016). Development and validation of an automated liquid–liquid extraction GC/MS method for the determination of THC, 11-OH-THC, and free THC-carboxylic acid (THC-COOH) from blood serum. *Analytical and Bioanalytical Chemistry* 408 (16): 4379–4388. https://doi.org/10.1007/s00216-016-9537-5.

228 Sandra, K., Kindt, R., C. Devos, C. Devos et al. (2015). Automated Sample Preparation for Metabolomics Studies Using the GERSTEL MPS Dual Head WorkStation, Part 2: Automated Lipid Fractionation Using Solid Phase Extraction. *Application Note 1/2015*. Mülheim an der Ruhr, Germany: GERSTEL GmbH & Co KG.

229 Lavilla, I., Romero, V., Costas, I. et al. (2014). Greener derivatization in analytical chemistry. *TrAC – Trends in Analytical Chemistry* 61: 1–10. https://doi.org/10.1016/J.TRAC.2014.05.007.

230 Cardinael, P., Casabianca, H., Peulon-Agasse, V. et al. (2015). Sample derivatization in separation science. In: *Analytical Separation Science*, 1725–1755. Weinheim: Wiley-VCH https://doi.org/10.1016/j.trac.2014.05.007.

231 Drozd, J. (1981). *Chemical Derivatization in Gas Chromatography*. Amsterdam, The Netherlands: Elsevier Science Publishers.

232 Pickering, M. (2007). An Overview of Post Column Derivatization Methods from a Pharmaceutical Applications Perspective. *Application Note*. Mountain View, CA, USA: Pickering Laboratories, Inc.

233 Cifuentes Girard, M.F., Ruskic, D., Böhm, G. et al. (2020). Automated parallel derivatization of metabolites with SWATH-MS data acquisition for qualitative and quantitative analysis. *Analytica Chimica Acta* 1127: 198–206. https://doi.org/10.1016/j.aca.2020.06.030.

234 Harding, V.J. and Warneford, F.H.S. (1916). The ninhydrin reaction for amino acids and ammonium salts. *Journal of Biological Chemistry* XXV (2): 319–335.

235 Friedman, M. (2004). Applications of the ninhydrin reaction for analysis of amino acids, peptides, and proteins to agricultural and biomedical sciences. *Journal of Agricultural and Food Chemistry* 52 (3): 385–406. https://doi.org/10.1021/jf030490p.

236 Oulkar, D.P., Hingmire, S., Goon, A. et al. (2017). Optimization and validation of a residue analysis method for glyphosate, glufosinate, and their metabolites in plant matrixes by liquid chromatography with tandem mass spectrometry. *Journal of AOAC International* 100 (3): 631–639. https://doi.org/10.5740/jaoacint.17-0046.

237 Gruening, A., Sander, J., R. Ludwig et al. (2017). Fully automated derivatization and quantitation of glyphosate and AMPA in beer using a standard UHPLC-MS/MS system. *Poster at the RAFA Conference*, Prague, Czech Republic, Shimadzu Europa GmbH, Duiburg, Germany.

238 Pierce, A.E. (1968). *Silylation of Organic Compounds*. Rockford, IL, USA: Pierce Chemical Company.

239 Orata, F. (2012). Derivatization reactions and reagents for gas chromatography analysis. In: *Progress in Agricultural, Biomedical and Industrial Applications* (ed. M.A. Mohd), 83–156. IntechOpen. https://doi.org/10.5772/33098.

240 Pierce Chemical Company (2003). *GC Derivatization – Applications Handbook*. pdfs.semanticscholar.org/4c80/1a0605daa810afa89fed38364b4372278596.pdf (accessed 5 November 2020).

241 Yu, Y., Zhong, S., Su, G.-Y. et al. (2012). Trace analysis of phenolic compounds in water by in situ acetylation coupled with purge and trap-GC/MS. *Analytical Methods* 4: 2156–2161. https://doi.org/10.1039/C2AY05887A.

242 Vila, M., Celeiro, M., Lamas, J.P. et al. (2017). Simultaneous in-vial acetylation solid-phase microextraction followed by gas chromatography tandem mass spectrometry for the analysis of multiclass organic UV filters in water. *Journal of Hazardous Materials* 323: 45–55. https://doi.org/10.1016/J.JHAZMAT.2016.06.056.

243 Buchgraber, M., Androni, S., and Anklam, E. (2007). Determination of cocoa butter equivalents in milk chocolate by triacylglycerol profiling. *Journal of Agricultural and Food Chemistry* 55: 3284–3291. https://doi.org/10.1021/jf063350z.

244 Bligh, E.G. and Dyer, W.J. (1959). A rapid method of total lipid extraction and purification. *The Canadian Journal of Biochemistry and Physiology* 37 (8): 911–917. https://doi.org/10.1139/o59-099.

245 Iverson, S.J., Lang, S.L.C., and Cooper, M.H. (2001). Comparison of the Bligh and Dyer and Folch methods for total lipid determination in a broad range of marine tissue. *Lipids* 36 (11): 1283–1287. https://doi.org/10.1007/s11745-001-0843-0.

246 Suter, B., Grob, K., and Pacciarelli, B. (1997). Determination of fat content and fatty acid composition through 1-min transesterification in the food sample; principles. *Zeitschrift für Lebensmitteluntersuchung und -Forschung A* 204 (4): 252–258. https://doi.org/10.1007/s002170050073.

247 Agilent (2013). Agilent G1676AA *Fiehn GC/MS Metabolomics RTL Library – User Guide*. Manual G1676-90001. Santa Clara, CA 95051 USA: Agilent Technologies, Inc.

248 Miyagawa, H. and Bamba, T. (2018). Comparison of sequential derivatization with concurrent methods for GC/MS-based metabolomics. *Journal of Bioscience and Bioengineering* 127 (2): 160–168. https://doi.org/10.1016/J.JBIOSC.2018.07.015.

249 Soma, Y., Yamashita, T., M. Takahash et al. (2017). Automation of sample preparation for metabolomic analysis using robotic platform. *Poster #362 at the 13th International Conference of the Metabolomics Society*, Brisbane, Australia. Fukuoka, Japan: Kyushu University.

250 Weckwerth, W., Wenzel, K., and Fiehn, O. (2004). Process for the integrated extraction, identification and quantification of metabolites, proteins and RNA to reveal their co-regulation in biochemical networks. *Proteomics* 4: 78–83. https://doi.org/10.1002/pmic.200200500.

251 Rönkkö, T. (2015). Recent developments in solid phase recent developments in solid phase microextraction techniques. MS Thesis. Department of Chemistry, University of Helsinki, Finland.

252 Chiang, J.-S. and Huang, S.-D. (2008). Simultaneous derivatization and extraction of amphetamine and methylenedioxyamphetamine in urine with headspace liquid-phase microextraction followed by gas chromatography–mass spectrometry. *Journal of Chromatography A* 1185 (1): 19–22. https://doi.org/10.1016/j.chroma.2008.01.038.

253 Melchert, H.-U. and Huebschmann, H.-J. (2015). Determination of THC-carbonic acid in urine by NCI. In: *Handbook of GC-MS*, 3rd Ed., 735–741. Weinheim: VCH-Wiley.

254 Faerber, H. and Schoeler, H.F. (1991). Gas chromatographic determination of urea herbicides in water after methylation with trimethylanilinium hydroxide or trimethylsulfonium hydroxide. *Vom Wasser* 77: 249–262.

255 Wang, Q., Ma, L., Yin, C.-R. et al. (2013). Developments in injection port derivatization. *Journal of Chromatography A* 1296: 25–35. https://doi.org/10.1016/j.chroma.2013.04.036.

256 Pan, L. and Pawliszyn, J. (1997). Derivatization/solid-phase microextraction: new approach to polar analytes. *Analytical Chemistry* 692: 196–205. https://doi.org/10.1021/AC9606362.

257 Tsuchiya, F., Hiraishi, T., S. Shimma et al. (2010). Development of a new metabolomics method using GC-TOFMS with automated derivatization system. *Poster at the 58th ASMS Conference*, Salt Lake City, UT, USA, Leco Corporation.

258 Kwon, Y.K. and Sasano, R. (2007). Application of Derivatization in Stomach Shaped Liner. *Technical Report*. Wakayama-city, Japan: Saika Technological Institute Foundation.

259 Takeo, E., Sasano, R. et al. (2017). Solid-phase analytical derivatization for gas-chromatography–mass-spectrometry-based metabolomics. *Journal of Bioscience and Bioengineering* 124 (6): 700–706. https://doi.org/10.1016/j.jbiosc.2017.07.006.

260 Wikipedia (2019). Vortex mixer. http://en.wikipedia.org/wiki/Vortex_mixer (accessed 14 July 2019).

261 Merriam-Webster (2019). Vortex. www.merriam-webster.com/dictionary/vortex (accessed 14 July 2019).

262 Kraft, J.A. and Kraft, H.D. (1962). Apparatus for mixing fluent material. US Patent No. 3, 061, 280.

263 GERSTEL (2019). Source: GERSTEL GmbH Product Information at www.gerstel.com/en/MPS-Agitator-Incubator-Stirrer.htm (accessed 16 July 2019).

264 Tipler, A. (2013). *An Introduction to Headspace Sampling in Gas Chromatography*. PerkinElmer, Inc. www.perkinelmer.com/PDFs/downloads/GDE_Intro_to_Headspace.pdf.

265 Restek (2006). *A Technical Guide for Static Headspace Analysis Using GC*. Bellefonte, PA: Restek Corporation.

266 Huebschmann, H.-J. (2015). Headspace techniques. In: *Handbook of GC-MS: Fundamentals and Applications*, Chapter 2.1.5, 3rd Ed., 26–57. Wiley-VCH. ISBN: 978-3-527-33474-2.

5

Integration into Analysis Techniques

Nichts setzt dem Fortgang der Wissenschaft mehr Hindernis entgegen als wenn man zu wissen glaubt, was an noch nicht weiß.
(translates as: Nothing hinders the progress of science more than the conviction of knowing something one does not know yet.)

Georg Christoph Lichtenberg (1742–1799)
German satirist and physicist [1]

The compatibility of automated sample preparation techniques with the analytical instrumentation is another important factor to consider for a seamless automation from sample preparation to analysis. The offline installation of robots provides samples ready for downstream analysis. Such semi-automated sample preparation workflows can serve several analytical instruments in a laboratory. Preferably, automated sample preparation using robotic systems, in particular with the state-of-the-art micro-methods, get online integrated into various analysis techniques [2]. The hyphenated sample preparation to chromatographic systems with just one integrated software control is the widely preferred solution with gas or liquid chromatography.

5.1 GC Volatiles Analysis

The sampling of gases, vapors, or volatile analytes from samples requires dedicated approaches and tools compared to the application of liquid samples. A major aspect is that an additional clean-up step does not occur, but often the "dynamic" concentration of trace components is required. The analysis of pure, synthetic, liquefied, or atmospheric gases requires specialized equipment, also in the case of onsite sampling, which is beyond the scope of this textbook for automated laboratory sample analysis.

The analysis of a sample headspace is the most automated and well-integrated sample preparation method for GC or GC–MS today. Headspace analyses are easy to handle in the laboratory, require only minimal manual preparation effort, and are run with high sample throughput. However, the volatile extraction techniques with static or dynamic headspace methods differ significantly in instrumentation and application. Automated instrumental solutions are typically dedicated to static or dynamic headspace analysis only. The x,y,z-robotic sampler in open architecture allows options for on-line sampling and the flexible change of the methods, for instance, to move ahead to dynamic extractions, if samples require additional performance.

5.1.1 Static Headspace Analysis

Ron G. Buttery and Roy Teranishi reported in 1961 about the direct GC injection of the "vapor above hot aqueous vegetables and fruit" for aroma analysis [3]. Donald A.M. Mackay et al. published as well in 1961 about the direct GC injection of the flavor of foodstuffs and other materials, in parallel evaluation to a sensory panel, for the "breakdown of odors into their multiple components" [4]. Because the headspace analysis requires thermostatization only as a very simple preparation step, it fulfils all requirements for automation [5]. And in fact, the headspace analysis was the first fully automated instrumental sample preparation and extraction method working online with capillary GC.

Pioneered in the late 1970s by Horst Hachenberg [6], Bruno Kolb, and Leslie S. Ettre [7], the static headspace GC analysis soon became a mature and technically solid sample preparation and analysis method. A first automated system for static headspace analysis was introduced in 1967 by the Bodenseewerk in Überlingen/Germany, the former German subsidiary of the Perkin-Elmer Corporation [8]. The automated static headspace sampling could demonstrate its excellent analytical performance based on the reproducibility of timing, pressure, and temperatures.

Static headspace is an equilibrium technique. In a closed vial, a thermodynamic equilibrium (partition) between the liquid or solid sample and the headspace above is established (Figure 5.1). After reaching equilibrium by the sample incubation over a defined time, an aliquot of the headspace is taken for GC analysis. The knowledge of the partition coefficient of the analytes is helpful to plan the analytical method. The partition coefficient k of a volatile compound describes the tendency of the analyte to leave the matrix into the headspace above. The coefficient k is the ratio of the analyte concentrations in the sample matrix and the gas phase, as of Eq. (5.1), in the equilibrium status. Low k values indicate a high concentration in the gas phase. The equilibrium partition of the analytes depends on temperature and can significantly change with temperature, see Table 5.1. A precise thermostatting and constant temperature must, therefore, be kept for all measurements. Using robotic headspace autosampler, the sample vials are incubated at the set method temperature and shaken, in the case of liquid samples, for a constant time to establish the equilibrium before sampling.

$$k = \frac{c_S}{c_G} = \frac{\text{concentration in the sample}}{\text{concentration in the gas phase}} \qquad (5.1)$$

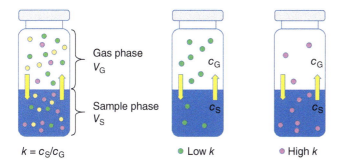

Figure 5.1 Partition of analytes in headspace analysis – compounds with low (green) and high (pink) k factor.

Table 5.1 Partition coefficients of selected compounds in water at different temperatures, sorted by increasing k values at 60 °C.

Substance	40 °C	60 °C	80 °C
n-Hexane	0.14	0.04	< 0.01
Cyclohexane	0.07	0.05	0.02
Tetrachloroethane	1.5	1.3	0.9
o-Xylene	2.4	1.3	1.0
1,1,1-Trichloroethane	1.6	1.5	1.2
Toluene	2.8	1.6	1.3
n-Butyl acetate	31.4	13.6	7.6
Methyl isobutyl ketone	54.3	22.8	11.8
Ethyl acetate	62.4	29.3	17.5
n-Butanol	647	238	99
Isopropanol	825	286	117
Dioxane	1618	642	288

Source: Adapted from Kolb and Ettre [7].

The sample volume applied can have an impact on the method sensitivity. For substances with high partition coefficients (e.g. ethanol, isopropanol, dioxane in water), the sample volume does not affect the peak area. Such substances with high partition coefficients benefit particularly from higher equilibration temperatures. For compounds with low partition coefficients (e.g. cyclohexane, benzene, xylenes in water), the phase volume ratio V_S/V_G determines the method sensitivity, see Table 5.1. In this case, increasing the sample volume leads to higher peak areas. With higher sample volumes, a higher total amount of analytes is present in the vial, which almost quantitatively transfers to the gas phase, for instance well known with the analysis of the BTEX components benzene, toluene, ethylbenzene, and xylene. For all quantitative determinations of compounds with low partition coefficients, an exact filling volume must be maintained.

A robotic x,y,z-autosampler is a recommended solution for the reproducible loading of headspace vials from sample reservoirs for the analysis of highly volatile VOC compounds, e.g. during online analysis from water streams.

5.1.1.1 Overcoming Matrix Effects

For the coupling of headspace GC (HS–GC) with mass spectrometry, the internal standard procedure has proved particularly successful for quantitative analyses. Besides the headspace specific matrix effects, possible variations in the MS detection are also compensated for. It must be noted that the compound partition coefficient is also influenced by the sample matrix. The standard addition method is an effective way to compensate for low recoveries and matrix effects [9, 10]. Several vials of the matrix sample are prepared and measured in series. Known concentrations of the analyte(s) are spiked in a workflow to some of the matrix sample vials and measured together with the not spiked sample. The generated quantitative calibration reflects and compensates the matrix impact. The addition of salt to aqueous samples besides the below described "salting-out effect" serves as well as a standardized matrix.

The full evaporation technique, first published by Michael Markelov and John P. Guzowski, also addresses the known problems with changing and unpredictable sample matrices in static headspace analysis [11]. Besides being independent of the sample matrix, the method offers increased sensitivity in the range of liquid sample injections. Only a small sample aliquot is prepared into standard capped HS vials, using a common workflow with agitator/incubator for equilibration. As only small volumes are dispensed and necessary equilibration times short, the full evaporation technique is well suited for automated high sample throughput applications, e.g. in industrial quality control. The determination of ethanol in fermentation products described by Li et al. [12] uses up to 50 µL of the aqueous sample, dispensed by a micro-syringe into an empty capped HS vial. At an incubation temperature of 105 °C, the small sample is fully evaporated. After a short incubation time of only three minutes, an aliquot of the headspace is taken for GC analysis. Although the sample size is small, the relative GC response is significantly increased, as it is shown in Table 5.2.

Also, the multiple headspace analysis (MHE) is a suitable approach to overcome matrix effects by a matrix-matched calibration in static headspace analysis (refer to Section 5.1.2 below).

Table 5.2 Full evaporation method compared to the conventional static headspace method.

Headspace method	Sample size [µL]	Equilibration temperature [°C]	Equilibration time [min]	Ethanol signal [cts / cts/µL]
Conventional static	2000	70	20	5374/2.7
Full evaporation	10	105	3	780/78

Source: Adapted from Li et al. [12].

5.1.1.2 Measures to Increase Analyte Sensitivity

Agitating the vial with a liquid sample during incubation time supports establishing a reproducible equilibrium (also see Section 4.10.2). Automated headspace samplers are equipped with standard devices for both incubation and mixing the samples and shorten the equilibration time. For liquid samples, vigorous mixing increases the effective surface area by replenishing the phase boundary facilitating the analyte transition into the gas phase. Teflon-coated stirring rods have not proved successful because of losses through adsorption and potential cross contamination. An increased temperature during incubation increases the response of analytes with high partition coefficients, but is practically limited for aqueous samples to 60–80 °C.

For aqueous samples, the ionic strength with the so-called "salting-out effect" increases the analyte concentration in the gas phase right after the start of the incubation and shortens the incubation time significantly [13, 14]. Usually approx. 2 g of sodium chloride, or a saturated NaCl solution, is prepared into the standard headspace vials before the aqueous sample is added [15]. An example is the determination of ethanol in water. The addition of NH_4Cl gives a twofold, and the addition of K_2CO_3 an eightfold increases in the sensitivity of the ethanol detection [8]. Also, a pH adjustment needs to be considered for the analysis of basic or acidic components reducing the dissociation in aqueous phase.

Another efficient but less used technique is the freeze concentration (FC) for water samples. Freezing a water sample from the bottom of a regular headspace vial concentrates the analytes in the still liquid sample above. Already in 1969, Kepner and coworkers reported an up to 40 times concentration increase [16]. The FC method is compatible with the automated use of DI-SPME and the stir-bar sorptive extraction (SBSE), and is described for the composition of green tea, and trace analysis of pesticides, or nitrosamines [17]. Commercial FC devices for current robotic sampling systems are not (yet) available. The realization requires a customized technical solution with a modified sample vial tray.

5.1.1.3 Static Headspace Injection Technique

Robotic samplers mostly take advantage of headspace syringes for GC injection. With the instrument top mount of the robotic sampler the installation of transfer lines and a carrier gas rerouting is avoided. On top, the programmed change of tools for prior sample preparation or to other sampling modes, e.g. to the more sensitive dynamic headspace (see Section 5.1.3) or to liquid injection expands the usability for different and unattended applications significantly. The use of syringes allows flexible injection volumes with a variable in the CDS sequence table of a programmed workflow.

The special headspace syringe is placed in a heated section of the tool (Figure 5.2). Such syringes use a sideport at the upper part of the barrel for carrier gas entry. The dedicated headspace tool provides an inert or carrier gas flow through the side port (Figure 5.3). Flushing of the heated syringe with a pulled up plunger is a recommended cleaning step after each sample injection to avoid any potential carryover in automated workflows.

Samples for headspace analysis are incubated at elevated temperatures. Typically, a gastight syringe is used for the headspace sampling, which may not become a cold

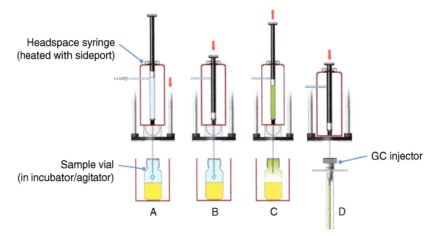

Figure 5.2 Headspace sampling process with a heated syringe. Source: Courtesy CTC Analytics AG. (A) Carrier gas load through syringe sideport after incubation time. (B) Vial pressurization. (C) Headspace aspiration into syringe. (D) GC injection of sample headspace.

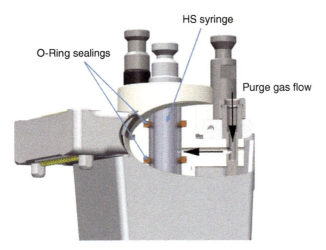

Figure 5.3 Headspace syringe tool with sideport syringe installed, showing purge gas flow for cleaning. Source: Courtesy CTC Analytics AG.

trap with analyte condensation. Hence the syringe needs to be heated well, generally above the incubation temperature. Headspace syringes therefore have to resist the elevated temperatures for repeat analyses. Many conventional headspace syringe needles are glued to the glass body. This limits the useful temperature range to max. 150 °C. Operation above the specified temperature weakens the cement and leads to detached needles when depenetrating the septum of the headspace vial or GC injector. Recent headspace syringe improvements addressed the fixation of the needle. The headspace syringes manufactured by Hamilton Bonaduz AG with glueless fixed needles are specified for operation up to 200 °C, eliminating detached needles also due to contact with organic and chlorinated solvents [18].

Modern GC headspace analysis requires injecting over large temperature ranges. Conventional headspace syringes use a polymer sealed plunger that has a limited sealing performance at high temperatures due to varying thermal expansion between the different materials. Hamilton developed a patented high-temperature syringe, which employs a unique compression spring in the plunger tip that compensates for the materials' different expansion coefficients (see Figure 4.4), creating a better seal over a larger temperature range, and it also extends lifetime [18]. Polymer sealed plungers are usually available for replacement.

The sample headspace aspiration from the sealed vial requires prior pressurization with a corresponding volume of carrier gas, taken via the side port of the syringe, and injected into the septum sealed vial. After several plunger strokes for homogenization and reproducible syringe filling the sample headspace in the programmed volume is transferred to the GC injector.

Headspace syringes use needles with a side hole (Point Style 5, S/Hole), see Figure 4.3. The needle side hole is located a few mm above the tip of the needle. When penetrating vials for headspace collection, it must be made sure the needle penetration depth into the vial is set correctly to position the side hole below the vial septum but well above the sample so that a liquid sample does not reach the needle tip during agitation.

Dedicated headspace samplers with transfer line connection to GC or GC–MS systems use the pressure balanced injection or gas loop technique [19] instead of a headspace syringe. After equilibration and vial pressurization, the sample headspace fills a heated sample loop mounted to a heated 6-way valve. Switching the valve transfers the sample plug via the transfer line into the carrier gas stream to the GC column for separation and detection. The operation sequence of the static headspace analysis using a sample loop is illustrated in Figure 5.4. The pressure-balanced and time-controlled injection method used by Perkin Elmer, Inc., is a third popular automated instrumental solution for a dedicated static headspace analysis [20].

5.1.2 Multiple Headspace Quantification

Quantitative headspace analyses are automated for high sample throughput with the technique of multiple headspace extraction (MHE). The total concentration of an analyte in a sample can be determined from one vial and avoids the incubation of several vials. MHE is the method of choice for the analysis of difficult headspace samples. Matrix effects are practically eliminated with MHE, and the difficulties of quantitative headspace analysis can be overcome [21].

The automated MHE procedure is carried out in a series of consecutive extractions and analyses from one and the same sample vial [19]. This procedure inherently includes a matrix-matched calibration. It is important to note that after the first headspace extraction, the vial is depressurized to the atmosphere and equilibrated again for the next analysis. The depressurization in automated headspace is achieved by using a needle tool for venting the remaining vial pressure to atmosphere. A wide

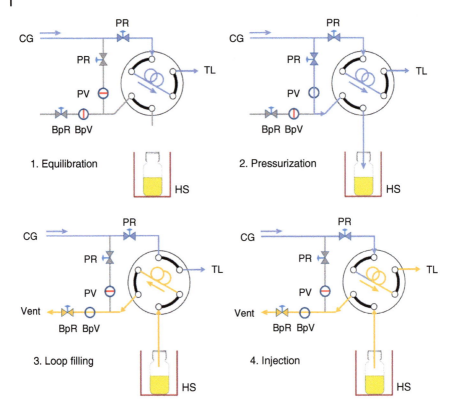

Figure 5.4 Operation sequence of a static headspace system with sample loop for GC injection. (CG carrier gas from GC, PR pressure regulator, PV pressurization valve, BpR backpressure regulator, BpV backpressure valve, TL transfer line to GC injector, HS headspace vial oven. Blue – carrier gas, yellow – sample, gray – not active).

bore needle of gauge 22 is pushed into the vial after aspiration of the headspace sample for a quick pressure equilibration. The steps of multiple headspace analysis with an MHE tool attached are illustrated in Figure 5.5. The shown measurement sequence is repeated for the same sample multiple times.

Repeated analyses from the same vial show a significant drop in the target analyte peak area in the consecutive chromatograms. This decline follows a first-order kinetics if the vial is equilibrated to atmosphere after the first headspace extraction and incubated again as before. Equation (5.2) describes the observed decline of the analyte concentration c. At any time t, the concentration of the analyte depends on the initial concentration c_0 and the constant q in the exponent.

$$c = c_0 \cdot e^{-qt} \tag{5.2}$$

$$A_i = A_1 \cdot e^{-q(i-1)} \tag{5.3}$$

With this procedure, the Eq. (5.2) can be rewritten in (5.3) by replacing the concentration with peak areas (i the number of injections, A_i peak area after i injections, A_1 peak area of the 1st injection) and the time t with the number of analysis. The

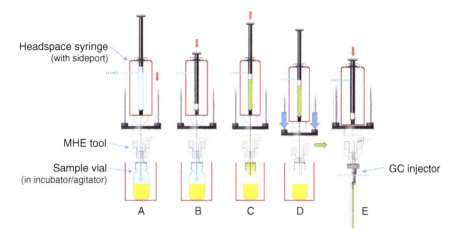

Figure 5.5 MHE workflow with MHE tool attached to a headspace syringe. Source: Courtesy CTC Analytics AG. (A) Carrier gas load through syringe sideport after incubation time. (B) Vial pressurization. (C) Headspace aspiration into syringe. (D) Needle of the MHE tool is pushed into the vial venting remaining headspace to atmosphere. (E) GC injection of sample headspace.

measurements of consecutive extractions will show an exponential decrease of the peak areas. Plotting the peak areas logarithmic against the number of extractions a linear relationship can be expressed in (5.4).

$$\ln A_i = \ln A_1 - q(i-1) \tag{5.4}$$

In practice, a large number of injections from each sample are not needed for achieving quantitative results. Due to the linear relationship in (5.4), the total area can be calculated from the area of the first run if the slope q of a calibration curve is known, which is specific for a given analyte/matrix situation. The slope q, the linear decline in peak areas, is taken from the slope expressed in the regression calculation of the logarithmic graph in Figure 5.6. Finally, in routine, only two data points are required to calculate the slope. If the slope is known and constant for given samples from previous runs only one data point for the following samples is sufficient to calculate the analyte concentration. The total peak area, which is equivalent to the total concentration of the analyte in the sample, is given by the summation of all potential extractions A_i in Eq. (5.5).

$$\sum_{i=1}^{\infty} A_i = \frac{A_1}{1-e^q} \tag{5.5}$$

An example of the MHE quantification is given in the following for the headspace quantification of residual ethylene oxide in PVC. It demonstrates the calculation steps [22]. In this illustration, a series of six MHE measurements are performed from the same sample providing the peak area data of Table 5.3.

The calculation for the total analyte peak area in the sample as of (5.5) with $A_1 = 151\,909$ from Run 1 in Table 5.3 and the slope $q = -0.83$ from the regression

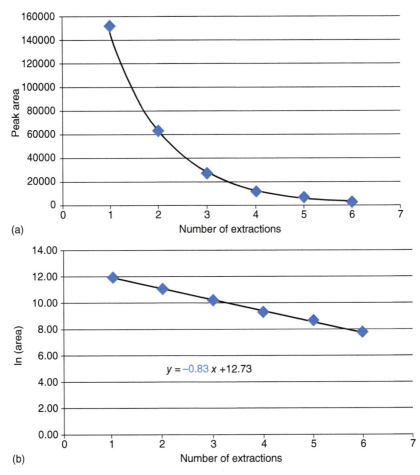

Figure 5.6 Graphical representation of the MHE measurements of Table 5.3. (a) Peak area vs. run number, (b) logarithmic plot, ln (peak area) vs. run number, with regression formula.

Table 5.3 Peak area results of six consecutive MHE headspace analyses.

Run#	Area [cts]	ln (Area)
1	151909	11.93
2	63127	11.05
3	26802	10.20
4	10963	9.30
5	5768	8.66
6	2240	7.71

Source: Adapted from Petersen [22].

Table 5.4 Comparison of the MHE experiment using all six or the first two measurements.

	6 Runs	2 Runs	Delta
Area A_1 [cts]	151909	151909	0
Slope q	−0.8330	−0.8781	−0.0451
exp(q)	0.4347	0.4156	0.0192
Total area A_i [cts]	268743	259928	3.3%

Source: Adapted from Petersen [22].

calculation in Figure 5.6(b), and the calculation of $e^{-0.83} = 0.4347$, finally gives:

$$\sum_{i=1}^{6} A_i = \frac{A_1}{1-e^q} = \frac{151\,909}{1-0.4347} = 268\,743 \text{ cts} \quad (5.6)$$

This result is calculated using all six data points.

With only two measurements, the slope can be calculated with a good and acceptable precision compared to the detailed experiment, as shown in Table 5.4. The difference in the calculated result is low with only 3.3%. For the final determination of the analyte concentration in the sample a regular response calibration with external or internal standards is required.

5.1.3 Dynamic Headspace Analysis

While the static headspace analysis takes an analytical snapshot of the sample after equilibration, the dynamic methods tend to achieve exhaustive extractions. Analytically seen, the advantage of dynamic headspace analysis (DHS) is the more sensitive detection of low-level compounds. The instrumental aspect is very different from static headspace with several different realizations depending on the nature of the sample. Automation of DHS analyses requires significant technical effort with the traditional purge and trap approach. Many equivalent analytical approaches for exhaustive extractions on automated systems became commercially available during recent years and found entrance and acceptance for routine analysis. Automated "dynamic" extraction and sorption methods, foremost SPME and SBSE (see Section 4.4.5), bring along the inherent potential to replace many older DHS methods combined with automated analytical performance and extended usability for gas phase, immersion, and also field sampling. The typical sensitivity ranges achieved by the available static and dynamic headspace methods are shown in Figure 5.7. In comparison of the available dynamic headspace methods, the choice for automated DHS methods is often the preferred option due to their inherently high sensitivity. The DHS methods in-tube extraction (ITEX DHS) and Purge & Trap typically use Tenax™ as sorbent material, which allows a good comparison with literature results [23]. Tenax is a low bleeding polymer resin of poly-2,6-diphenylphenylene oxide (PPPO) with high thermal stability up to 350 °C (product name "Tenax TA", "TA" stands for "trapping agent"). The low

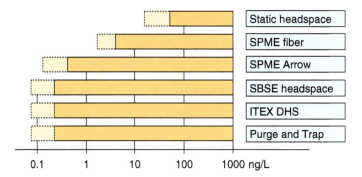

Figure 5.7 Sensitivity range comparison of static and dynamic headspace methods.

affinity for water makes is ideally suitable for analysis of aqueous samples and thermal desorption. "Tenax GR" is the black composite material of Tenax TA and 30% graphite with a higher breakthrough volume for most volatile organics. The strength of Tenax as sorbent material is the general use for a very wide range of analytes and its low affinity for water vapor. In contrast, SPME and SBSE devices use mostly PDMS-based sorbent materials; Tenax as adsorbent is currently not available for these techniques.

5.1.3.1 Purge and Trap

The purge and trap technique (P&T) became a laboratory method already in the 1960s with the analysis of body fluids. In the 1970s, P&T became a popular extraction technique because of the increasing demand of drinking water testing for VOCs [24]. The determination of a large number of VOC compounds in ppq quantities (concentrations of less than 1 µg/L) became possible at that time only with the P&T concentration. In 1974, Tom Bellar from the US EPA in Cincinnati applied P&T successfully for analysing trace levels (ppb) of VOCs in water. P&T gave a 100-fold improvement in sensitivity from existing laboratory equipment. The first commercial purge and trap device was launched in the following by Tekmar with the LSC-1 in 1976 [25]. Many US EPA standard methods like the most applied 502, 524.x, 624, 5030b, or 8260 methods still today demand for an "all-glass purging device", the U-tube sparger as the heart of a purge and trap system, for compliant analysis.

The operation of a P&T system comprises four steps for each sample: (i) the transfer of the water sample into the U-tube, (ii) the purge phase with simultaneous concentration of the extracted analytes on the analytical trap, (iii) the desorption phase of the trap for transfer of the analytes to the GC, and (iv) the bake-out phase to recondition the analytical trap, moisture trap and the drain of the U-tube for the next sample. Each step requires a dedicated parameter set for optimum operation. The analytical trap uses typically Tenax as general sorption material for all of the US EPA methods in the current use. Tenax provides a wide application range, has only a low affinity to evaporated water vapors, and is thermally stable for desorption at high temperatures. The moisture trap is usually a cooled steel capillary or turbulent device to condensate the water vapor. The water layer serves here as a stationary film letting pass non-polar VOCs. Polar compounds can be affected due to retention

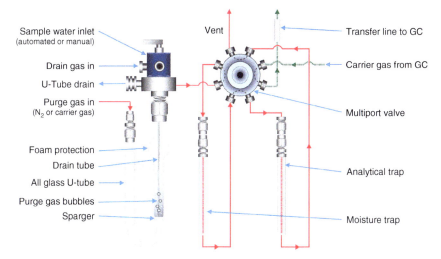

Figure 5.8 Purge & Trap schematics in purging status. Source: Courtesy CDS Analytical LLC.

on the water film causing reduced recoveries, as known for instance for alcohols. A schematics illustrating the purge and trap operation is given with Figure 5.8. The central heated multiport valve is shown in the position for sample purge, using the GC carrier or an auxiliary inert gas like nitrogen. For the desorption phase, the valve is switched so that the GC carrier gas flow then flushes the heated analytical trap against the direction of trapping and transfers the desorbed analytes via the transfer line to the connected GC for analysis. The moisture trap is heated after desorption as well and backflushed to the vent exit. The U-tube is drained by applying gas pressure to the purge and the drain gas lines. The aqueous sample is then pushed out via the drain tube to the drain exit line. Optionally, a wash step of the U-tube can be added at this time.

The automation of the purge and trap process requires additional instrumentation to transfer water samples to the U-tube and optionally to add internal standards. An implementation of a P&T module into a robotic autosampler was developed by CTC Analytics AG for a high degree of automation and analytical flexibility. Due to the automated tool change capability of the employed x,y,z-robot additional methods and workflows like the static headspace, dynamic ITEX DHS, SPME, DLLME, liquid injection, or additional sample preparation procedures can be combined on one unit. With this flexible configuration, VOCs and water dissolved SVOCs can be determined on one GC–MS unit in customized workflows.

The core functions of the automated P&T unit are handled by a multidilutor device with a connected dilutor tool (see also Section 4.1.5). The dilutor tool comprises an exchangeable needle of gauge 22, point style 3 (optional gauge 23, point style AS), and is connected via a flexible transfer tubing of 1.53 m length. This flexibility allows the robotic head to address different vial trays and positions in its working space. Also, attached flow cells for online water analysis can be integrated for a parallel monitoring of different water streams (Figure 6.33). The

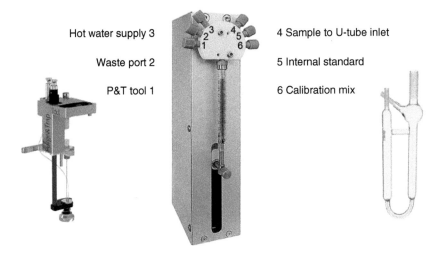

Figure 5.9 Dilutor functions and connection ports for automated purge and trap operation. Source: Courtesy CTC Analytics AG.

dilutor comprises a large volume syringe of up to 10 mL volume with a 6-port ceramic valve. Workflow programming with the various dilutor functions allows customized automated procedures for unattended operation (Figure 5.9). Water samples can be transferred into the sparger U-tube for P&T analysis from different sources, vial sizes, or online streams. Optionally internal standards can be added. The U-tube and sample tubes are optionally rinsed using a hot water supply to eliminate any potential carry over between samples, important for wastewaters. Further, with a standard mix connected to the dilutor, a quantitative calibration can be performed with the required dilution range and data points with each batch of samples. The automated P&T setup is integrated into the CDS of the major instrument manufacturer. Between the samples, blanks and QC standards can be included, controlled by the sequence table of the GC–MS data system.

5.1.3.2 Dynamic Headspace Analysis with In-Tube Extraction

The above described significant effort in P&T analysis with a multifunctional instrumental setup and operation is justified only for US EPA compliant water analysis. Many EPA methods published years ago require the all-glass sparger U-tube. Other dynamic headspace methods for water, sediment, soil, or solid materials do not set out a mandatory use of the classical U-tube.

The dynamic headspace analysis with in-tube extraction (ITEX DHS) facilitates P&T analysis based on Tenax sorbent material with just one tool for an x,y,z-robotic autosampler. The rerouting of carrier gas lines during installation, the need for additional bench space, handling fragile glassware, and most importantly for routine operation, the risk of serious system contamination is entirely avoided. The key to ITEX DHS is a special headspace syringe of approx. 1300 μL volume with a sideport in the upper part of the glass barrel. The needle of the syringe carries about 160 μL Tenax sorbent material in a 50 mm wide section, as shown in Figure 5.10. The syringe

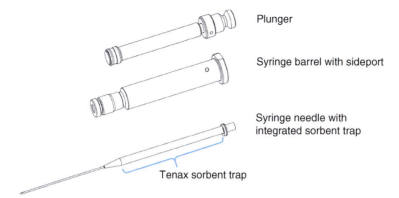

Figure 5.10 Components of the ITEX DHS syringe trap. Source: Courtesy CTC Analytics AG.

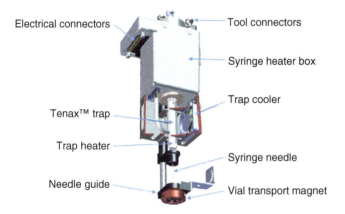

Figure 5.11 ITEX DHS tool cutaway. Source: Courtesy CTC Analytics AG.

with the Tenax trap is enclosed in the ITEX DHS tool displayed with a cutaway in Figure 5.11. Two sections are independently heated. The syringe barrel is heated to a constant temperature. The Tenax syringe trap is located in a fast direct heater, which can be cooled quickly back to room temperature. In addition to the popular Tenax TA and GR sorbent materials, the syringe needle traps are available filled with different sorbents like Carboxen, Carbopack, Carbosieve, or combinations of those (see Table 5.7 for the sorbent material performance characteristics). The ITEX DHS syringe is a closed system and does not require auxiliary purge gas flows, or a transfer line to a GC injector. The GC injection is a regular syringe injection and does not require any modifications on the GC side. An invaluable benefit is the low risk of serious contamination. In a worst case, only a syringe needs to be cleaned.

For ITEX DHS operation, the sample vial is incubated a defined time as for regular headspace analysis in a temperature-controlled agitator. For analyte purging and trapping the needle of the ITEX DHS tool penetrates the vial septum while the vial is kept in the agitator, or moved to the stronger mixing heatex stirrer device before (see Section 4.10.5). The typical operation steps with ITEX DHS are illustrated in

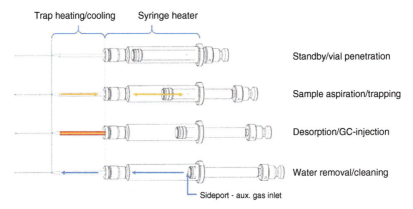

Figure 5.12 ITEX DHS syringe operation principle.

Figure 5.12. In the purge phase, the vial is pressurized and the sample headspace is pulled by several plunger up/down movements through the fan-cooled sorbent trap. Analytes get extracted by trapping on the sorbent material. The number of plunger movements determines the desired method sensitivity. This parameter allows a unique method flexibility and customization for low or high analyte concentrations in a sample. A next step is the optional removal of residual moisture from the Tenax trap. A gentle gas flow can be used to remove moisture from the sorbent material by pulling up the plunger beyond the side hole while keeping the trap still cold at room temperature. After sampling the ITEX DHS tool moves to the GC injector for desorption. After the needle penetrated the GC injector septum the trap heater rapidly heats the sorbent material. A plunger stroke down finally completes the injection. Syringe and trap cleaning follow as necessary.

The analytical performance of ITEX DHS compares well with the classical P&T technology. Since the first publications in 2006 [26] the application range extended from water, beverages, beer and wine, to plant materials, forensic analyses from blood and urine, flavour compounds in cosmetics, and more. Results obtained already in 2008 demonstrate that the ITEX DHS technique provides excellent precision, accuracy, and detection levels comparable to standard P&T methods [27]. For VOCs the linear dynamic ranges were from 0.1 to 50 µg/L, with an average correlation coefficient of 0.999 and a precision of 11% RSD, respectively. The achieved MDLs were between 0.01 and 0.05 µg/L in water. In particular, it was noted the ease-of-use and maximum up-time compared to the complex P&T systems.

The ITEX DHS performance concerning the analytical efficiency for high boiling components is significantly improved by applying a vacuum to the sample vial during the extraction phase. Pascal Fuchsmann and co-workers demonstrated an up to 450 times signal improvement for selected compounds compared to regular pressure ITEX DHS or SPME conditions [28].

The vacuum method, named "Dynamic Headspace Vacuum Transfer in Trap" (DHS-VTT) extraction, works with the same equipment as HS-ITEX DHS. The only modification is a vacuum pump with a pressure control interface connected to an additional workflow controlled three-way solenoid for the instrument gas line, also becoming the vacuum supply line. "The DHS-VTT technique enriches the gas phase rapidly by limiting the matrix effects because of the reduced pressure in the vial." The response increase is particularly valuable and intended for full scan MS analysis for flavor compound identification. Best results were achieved, in particular for low volatile sulfur compounds, using a Tenax TA/Carbosieve SIII blend as trapping material in the ITEX DHS sorbent trap (for sorbents performance characteristics see Table 5.7).

5.1.3.3 Dynamic Headspace Analysis Using Sorbent Tubes

Another solution for dynamic headspace routine analysis was developed by GERSTEL and is available as the dynamic headspace module (DHS) based on sorbent tubes. The DHS system is shown in Figure 5.13 [29]. Manual and automated extensions are optionally available with large sample containers up to 1 L volume (DHSLarge).

In the automated DHS mode, the samples were weighed into 20 mL screw cap vials and placed into the racks of the x,y,z-robotic sampler. For analysis, the vials are transferred using the transport magnets of a tool into the DHS unit. The sample is thermostated and agitated while the headspace is purged with a controlled flow of inert gas using a double-needle penetrating the sample vial septum, as shown in Figure 5.14. Extraction temperatures can be selected up to 200 °C, or cooled down to 10 °C. The sample headspace continuously passes a sorption tube during the extraction process. Typical sorption tubes packed with sorbent material are shown in Figure 5.19. Purged VOCs can be trapped on different kinds of sorbent material, as mentioned in Table 5.7, the most often used is the universal Tenax TA sorbent material. The temperature of the sorbent tubes can be programmed from 20 to 70 °C for optimum trapping conditions and to reduce water retention.

After the extraction process, the loaded sorbent tubes are transferred from the DHS module to the "Thermal Desorption Unit" (TDU) [30] for analysis, which is mounted on top of the "Cooled Injection System" (CIS), a PTV-type inlet, as shown in Figure 4.62. As necessary, a dry purge step for water removal from the sorbent

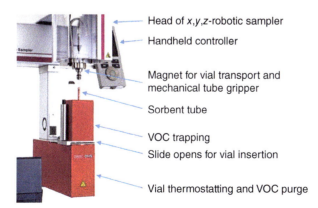

Figure 5.13 Dynamic headspace unit using sorbent tubes. Source: Courtesy GERSTEL GmbH & Co KG.

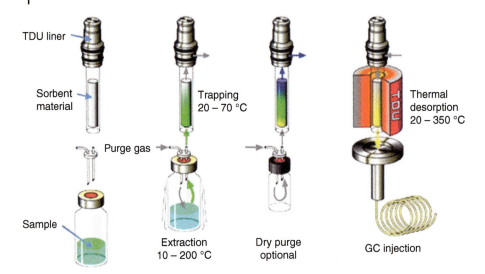

Figure 5.14 DHS process using sorbent tubes. Source: Courtesy GERSTEL GmbH & Co.

material can be added to the workflow. The volatile analytes are cryofocused in the temperature programmable inlet (PTV). For analysis, the inlet is then heated to transfer the analytes in a sharp band to the analytical column. Using a rack of sorbent tubes on the x,y,z-robot ensures that individual sorption tube can be used for each sample to avoid any potential cross-contamination, for example, from highly concentrated samples. The time-consuming extraction and trapping process can be run in the "prep-ahead" mode as outlined in Section 3.3.2 for enhanced productivity so that a continuous sample analysis without wait times for the GC–MS is achieved.

5.1.3.4 Needle Trap Microextraction

In response to the demand for more robust SPME systems, the needle trap device (NTD) technique has been introduced in 2001 [31–33]. Blunt tip needles (point style 3) with a side hole were packed in a length of 2–3 cm from the tip with sorbent material of up to 4 μL sorbent volume (compare at Table 4.10). Single layers or combinations of PDMS, DVB, or Carboxen are in use. A schematic of the NTD is shown in Figure 5.15. During sampling, the side hole is closed by a short 10 mm polymer tube fitting the needle diameter. The analytes are trapped on the sorbent by passing a low volume gaseous sample of 5–100 mL through the sorbent material plug in the needle. An active sampling can be achieved manually utilizing 1–20 mL single-use sterile syringes, which are fitted to the top Luer connector of the NTD, using several plunger strokes. Alternatively, passive and active onsite sampling can be achieved [34]. The sample gas is drawn through the sorbent material and the target analytes are concentrated on the sorbent material. After sampling, the needle is placed into a dedicated rack of an autosampler and inserted into the hot injector of a GC for thermal desorption [33]. During the insertion process, the short sealing polymer tube is pushed upward by the injector cap revealing access to the side hole inside of the injector. When the NTD needle is inserted and touches the narrow-neck inlet liner

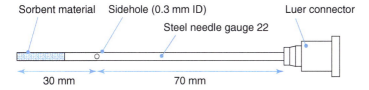

Figure 5.15 Needle trap schematics with approx. dimensions (Source: Adapted from Risticevic et al. [33]).

Figure 5.16 Desorption of the needle trap device using a tapered liner in the GC injector.

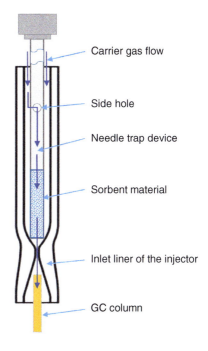

the carrier gas flow is then directed through the side hole into the needle passing the sorbent material and transferring the analytes by thermally desorption into the column, as illustrated in Figure 5.16.

The needle-trap microextraction (NTME) is a green, solvent-free, easy-to-handle, and low-cost method with significant advantages in terms of sensitivity, reusability, and portability of the NTDs [35]. Before usage, the needle trap sorbents are conditioned in a heating device, e.g. with Tenax for 30 minutes at 195 °C and under permanent nitrogen flow to eliminate any contaminations. The conditions need to be set as of the employed sorbent material. NTDs containing DVB, Carbopack X, and Carboxen 1000 or DVB and Carboxen 1000 were conditioned at 250 °C. NTDs containing PDMS, Carbopack X, and Carboxen 1000 or PDMS and Carboxen 1000 can be conditioned up to 290 °C. Afterwards, both ends of the needles were sealed with Teflon caps. After onsite sampling the Teflon caps can be reattached for transport [36]. Applications are mostly reported for at-site breath gas analysis. The analysis of breath VOCs represents a noninvasive diagnostic tool. The sampling and pre-concentration can be done directly at the patient. No additional steps for sample preparation are required. The automated VOC analysis follows with dedicated x,y,z-robotic autosamplers via GC–MS [37].

5.1.4 Tube Adsorption

The adsorption of volatile compounds on sorbent material filled tubes followed by thermal desorption (TD) is the most versatile and widely used analytical method for gaseous sample concentration and introduction to GC. It provides access to a large variety of analytical solutions from air monitoring to solid material thermal extraction covering organic gases, volatile, and semi-volatile analytes. These methods are well automated with different commercial instrumental solutions. Thermal desorption is inherently less labor intensive than solvent extraction, avoiding solvent use completely, and requiring no manual sample preparation. A selection of applications using automated TD is listed in Table 5.5. A large number of international standard methods are based on tube adsorption/desorption collected in Table 5.6.

Monitoring VOCs from air or gaseous samples requires prior trapping of the analytes on sorbent-packed tubes. Permanent gases, such as nitrogen, oxygen, argon, also sulfur dioxide, hydrogen chloride, carbon mono and dioxide, and methane pass through the sorbent materials and cannot be analyzed. The concentration levels can vary from low ppt from a high dilution in ambient air to even the percentage level at emission sites. Compounds of different chemical nature, polarity, boiling point, and high humidity can occur with air samples making sample collection a particular challenge. The compounds of interest typically get actively or passively

Table 5.5 Applications using automated thermal desorption for analysis.

Analytes	Analytical method	Sample
VOCs	Active or passive air sampling	Sorbent material
VOCs	Active or passive air sampling	Stir bar
VOCs	Sample headspace	Stir bar
VOCs	Sample headspace	Monotrap
VOCs	Sample outgassing	Solid material
VOCs	Pyrolysis	Solid material
VOCs	Dynamic headspace sampling	Stir bar
VOCs	Dynamic headspace sampling	Monotrap
Organic gases, freons	Active or passive air sampling	Sorbent material
SVOCs	Sample outgassing	Solid material
SVOC additives	Thermal extraction	Solid material
Aerosols	Active or passive air sampling	Sorbent material
Dissolved polars	Immersion	Stir bar
Reaction products	Pyrolysis	Solid material

Table 5.6 Official standard methods and applications applying thermal desorption.

Method	Title	References
AgBB/DIBt-Mitteilungen	Health-related Evaluation Procedure for Volatile Organic Compounds Emissions (VOC and SVOC) from Building Products	[38]
ANSI/ASHRAE/USGBC/IES Standard 189.1	2018 International Green Construction Code – Powered by Standard 189.1-2017	[39]
ANSI/BIFMA M7.1 – 2011(R2016) FES Test	Standard Test Method for Determining VOC Emissions from Office Furniture Systems, Components and Seating	[40]
ASTM D5197 – 16	Standard Test Method for Determination of Formaldehyde and Other Carbonyl Compounds in Air (Active Sampler Methodology)	[41]
ASTM D5466 – 15	Standard Test Method for Determination of Volatile Organic Compounds in Atmospheres (Canister Sampling Methodology)	[42]
ASTM D6196 – 15e1	Standard Practice for Choosing Sorbents, Sampling Parameters and Thermal Desorption Analytical Conditions for Monitoring Volatile Organic Chemicals in Air	[43]
ASTM D7648 / D7648M – 18	Standard Practice for Active Soil Gas Sampling for Direct Push or Manual-Driven Hand-Sampling Equipment	[44]
ASTM D7706 – 17	Standard Practice for Rapid Screening of VOC Emissions from Products Using Micro-Scale Chambers	[45]
ASTM D7758 – 17	Standard Practice for Passive Soil Gas Sampling in the Vadose Zone for Source Identification, Spatial Variability Assessment, Monitoring, and Vapor Intrusion Evaluations	[46]
California Specification 01350	Standard Method for the Testing and Evaluation of VOC Emissions from Indoor Sources Using Environmental Chambers	[47]
CEN/TS 13649:2014	Stationary source emissions. Determination of the mass concentration of individual gaseous organic compounds. Sorptive sampling method followed by solvent extraction or thermal desorption	[48]
ISO 12219-1:2012	Interior air of road vehicles – Part 1: Whole vehicle test chamber – Specification and method for the determination of volatile organic compounds in cabin interiors	[49]
ISO 12219-2:2012	Interior air of road vehicles – Part 2: Screening method for the determination of the emissions of volatile organic compounds from vehicle interior parts and materials – Bag method	[50]

(Continued)

Table 5.6 (Continued)

Method	Title	References
ISO 12219-3:2012	Interior air of road vehicles – Part 3: Screening method for the determination of the emissions of volatile organic compounds from vehicle interior parts and materials – Micro-scale chamber method	[51]
ISO 12219-4:2013	Interior air of road vehicles – Part 4: Method for the determination of the emissions of volatile organic compounds from vehicle interior parts and materials – Small chamber method	[52]
ISO 12219-5:2014	Interior air of road vehicles – Part 5: Screening method for the determination of the emissions of volatile organic compounds from vehicle interior parts and materials – Static chamber method	[53]
ISO 16000-3:2011	Indoor air – Part 3: Determination of formaldehyde and other carbonyl compounds in indoor air and test chamber air – Active sampling method	[54]
ISO 16000-6:2011	Indoor air – Part 6: Determination of volatile organic compounds in indoor and test chamber air by active sampling on Tenax TA sorbent, thermal desorption and gas chromatography using MS or MS-FID	[55]
ISO 16200-1:2001	Workplace air quality – Sampling and analysis of volatile organic compounds by solvent desorption/gas chromatography – Part 1: Pumped sampling method	[56]
ISO 16200-2:2000	Workplace air quality – Sampling and analysis of volatile organic compounds by solvent desorption/gas chromatography – Part 2: Diffusive sampling method.	[57]
JSA – JIS A 1901 (Japan Industry Standard)	Determination of the emission of volatile organic compounds and aldehydes by building products – Small chamber method	[58]
SW-846 Test Method 8260B	Volatile Organic Compounds by Gas Chromatography/Mass Spectrometry (GC/MS)	[59]
US EPA Compendium Method TO-14A	Determination Of Volatile Organic Compounds (VOCs) In Ambient Air Using Specially Prepared Canisters With Subsequent Analysis By Gas Chromatography	[60]
US EPA Compendium Method TO-15	Determination Of Volatile Organic Compounds (VOCs) In Air Collected In Specially-Prepared Canisters And Analyzed By Gas Chromatography/Mass Spectrometry (GC/MS)	[61]
US EPA Compendium Method TO-17	Determination of Volatile Organic Compounds in Ambient Air Using Active Sampling Onto Sorbent Tubes	[62]
US EPA Method 0030	Volatile Organic Sampling Train	[63]
US EPA Method 0031	Sampling Method for Volatile organic compounds (SMVOC)	[64]
US EPA Method 5041A	Analysis for Desorption of Sorbent Cartridges from Volatile Organic Sampling Train (VOST)	[65]

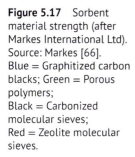
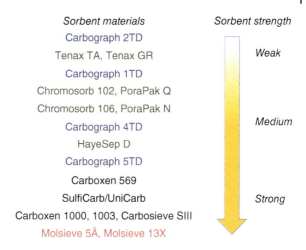

Figure 5.17 Sorbent material strength (after Markes International Ltd). Source: Markes [66]. Blue = Graphitized carbon blacks; Green = Porous polymers; Black = Carbonized molecular sieves; Red = Zeolite molecular sieves.

collected using standardized tubes filled with sorbent materials [67]. The thermal desorption of the loaded sample tubes is performed reversed to the sampling direction. This allows the combination of sorbent material layers of increasing strength in one tube. An overview of the different sorbent materials, strengths, performance characteristics, and typical applications are given with Figure 5.17 and Table 5.7. As a rule of thumb, the boiling points of the compounds to be sampled give first directions for the choice of sorbent materials, or for using multilayer sorbent tubes. Analytes with boiling points >100 °C go well with a weak sorbent strength (e.g. Tenax TA/GR, Carb2TD). Boiling points from 30 to 100 °C require the medium sorbent strength (e.g. Carbograph1TD), and compounds with low boiling points in the range of −30 to 50 °C must be collected using strong sorbent materials (e.g. CarbX, Unicarb, Carboxen1000, CS III).

The sample tubes get loaded externally as of the required sampling project. A selection of commonly performed TD applications is given in Table 5.5. While the sample collection is usually at site and offline, the analysis in the laboratory is performed with automated thermal desorption systems connected to GC–MS as the method of choice for analysis of VOC compounds. The dimensions of the used sorbent-packed tubes with a length of 3½ in. and 0.25 in. OD became an industry standard defined in the late 1970s by the Working Group 5 of the UK Health & Safety's Committee on Analytical Requirements. Basic functionality requirements for thermal desorption were specified including a two-stage operation, splitting with a recollection of samples, trap-cooling, standard system checks, and automation [68]. The thermal desorption of the loaded sorbent tubes in a carrier gas stream is the necessary instrumental link from the collected sample to GC or GC–MS analysis.

Thermal desorption is a dynamic process. With the rising temperature, the absorbed analytes get increasingly desorbed from the sorbent material. The GC carrier gas directed through the sorbent tube is purging and diluting the trapped compounds from the sorbent or sample matrix in a gas volume of around 100–200 mL. With the two-stage operation, the carrier gas stream from the heated sorbent tube is first directed through a narrow electrically cooled focusing trap collecting the analytes in a sharp band. Flash heating the focusing trap releases

Table 5.7 Performance characteristics of common sorbent materials (after Markes [67]).

Graphitised carbon blacks

Non-specific carbon sorbents for trace-level applications, less suitable for some labile or highly reactive species.

Carbograph 2TD	Very weak sorbent, suitable for alkylbenzenes, C8 hydrocarbons to C20 SVOCs
Carbograph 1TD	General-purpose sorbent for C6 to C14, often used in two- or three-bed sorbent tubes.
Carbograph 4TD	Medium-strength sorbent suitable for light hydrocarbons.
Carbograph 5TD	Medium to strong sorbent suitable for light hydrocarbons, C3 to C7 volatility range.

Porous polymer sorbents

Inert and hydrophobic sorbent materials suitable for labile and reactive compounds, and humid conditions.

Tenax TA	Most used TD sorbent, most suitable from hexane (C6) to C30.
Tenax GR	Offers slightly higher breakthrough volumes than Tenax TA for some compounds.
PoraPak Q	Slightly polar general-purpose PoraPak sorbent.
PoraPak N	More polar than PoraPak Q, and typically used for monitoring volatile nitriles.
HayeSep D	Primarily used for monitoring volatile chemical warfare agents.

Carbonised molecular sieves

Strongest sorbents ideally suited for trapping the most volatile compounds.

Carboxen 569	More hydrophobic than most carbonised molecular sieves.
SulfiCarb	For reactive sulfur species and light VOCs, C3 to C8 volatility range. Replaces UniCarb
Carbosieve SIII	For ultra-volatile hydrocarbons, C2 to C5 range.
Carboxen 1003	For air monitoring methods, C2 to C5 range. Replacing Carboxen 1000 or Carbosieve SIII.
Carboxen 1000	For ultra-volatile hydrocarbons.

Zeolite molecular sieves

Selective hydrophilic sorbents for specific TD applications, such as monitoring nitrous oxide.

Molecular sieve 5Å	Strong sorbent typically used for nitrous oxide monitoring.
Molecular sieve 13X	Strong sorbent typically used for buta-1,3-diene monitoring.

Table 5.7 (Continued)

Other non-carbon sorbents

Silanised glass wool	Replaces quartz wool in sorbent tubes and cold traps, for very labile compounds, degrading >275°C.
Unsilanised glass wool	Used in sorbent tubes where inertness is not an issue.
Quartz wool	First 'sorbent' in a multi-bed sorbent tube for high-boiling compounds C30 to C40.
Quartz beads	Used in cold traps when monitoring high-boiling, chemically active compounds.

Activated charcoal

NOT suitable for TD sampling tubes - Requires solvent desorption.

Source: Courtesy Markes International Ltd.

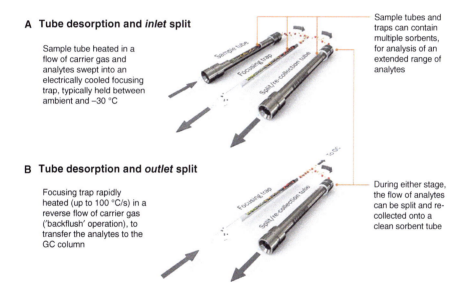

A Tube desorption and *inlet* split

Sample tube heated in a flow of carrier gas and analytes swept into an electrically cooled focusing trap, typically held between ambient and −30 °C

Sample tubes and traps can contain multiple sorbents, for analysis of an extended range of analytes

B Tube desorption and *outlet* split

Focusing trap rapidly heated (up to 100 °C/s) in a reverse flow of carrier gas ('backflush' operation), to transfer the analytes to the GC column

During either stage, the flow of analytes can be split and re-collected onto a clean sorbent tube

Figure 5.18 Principle of the two-stage thermal desorption with backflush of the internal focusing trap. Source: Courtesy Markes International Ltd. (A) Sample thermal desorption into a cold focusing trap. (B) GC injection of the trapped sample from a fast heated focusing trap.

the analytes all at once for injection onto the GC column. A small gas volume of typically 100–200 µL only transfers the analytes to the GC with a sharp peak shape for good sensitivity. The two-stage process is illustrated in Figure 5.18.

Alternatively, a solvent extraction of the sorbent material can be performed [69, 70]. The solvent extraction ends up usually in analyte dilution in the applied solvent volume and lower sensitivity compared to online TD. A pre-concentration of the sample solvent volume is required to yield similar detection limits [71]. The solvent extraction procedure is a static equilibrium procedure. The analytes are partitioning between the sorbent, the solvent, and a headspace phase. This limits desorption efficiency to the partition coefficients of the compounds. For this reason, the standard methods for solvent extraction specify only 75% recovery [72, 73]. The evaporation and concentration of the desorption solvent are critical causing losses of trapped VOCs.

Charcoal-based sorbent materials cannot be thermally desorbed (usually coconut-shell-activated carbon). Solvent desorption from charcoal sampling often involves hazardous solvents like CS_2 or DCM in manual operation [74]. Alternatively, as nonhazardous solvents, for instance, pentane [75], ethylenglycol-monoethylether [69], or anionic surfactant solutions [76], can be used for solvent desorption from charcoal sorbent materials. The eluted sample can be analyzed several times, also with different detectors, but is diluted in solvent, a limitation for trace analyses. Analytical limitations to be considered are a potential contamination from the solvent itself, also a quenching of early eluters by the chromatographic elution of the broad solvent peak in a similar retention area.

The solvent desorption from charcoal trapping material can be automated in robotic sampling systems as well. The Canada Ministry of the Environment and Climate Change, Laboratory Services Branch, issued a standard method in the drinking water surveillance program using micro-SPE cartridges filled with coconut charcoal using an in-line automated sample preparation with an x,y,z-robotic sampler and GC–HRMS detection [77]. About 10 mL of sample water is passed slowly at 50 µL/s through the micro-SPE cartridge by the robotic sampler. After a drying step, the sorbent material is eluted with 500 µL DCM, of which 10 µL are injected to GC–HRMS. Compared to Tenax material, charcoal absorbs substantial amounts of water, which needs to be considered requiring the drying step before GC analysis.

An alternative sorbent material to charcoal, and applicable to thermal desorption, is provided with a hybrid material based on a monolithic silica structure with activated carbon modification (or graphite carbon) providing a large surface area with bonded octadecylsilane (C18, ODS) [78]. Referencing the monolithic silica network with through pores and mesopores the analyte collection process is named monolithic material sorptive extraction (MMSE) [79].

Thermal desorption into a GC injector is a single snapshot, a one sample-one run analysis. Often a sample is required to be repeated, analyzed with different conditions, but cannot after injector desorption. As the thermal desorption is mostly a split injection anyway (see Figure 5.18), the logical step is the collection of the split flow onto a second internal sorbent tube. During sample tube desorption, a heated internal valve of the autosampler directs the flow to the internal cold trap and at the same time a split- flow to the second sorbent tube. The collection of the split flow on the second sorbent tube allows a repeat analysis, and, if necessary, the tube with the split sample can be archived for future reference [80].

Instrumental solutions for automated thermal desorption are commercially available as dedicated instruments for handling the standard sorbent tubes. Such automated thermal desorption devices are coupled as external units via transfer lines, or are integrated with GC or GC–MS systems. The GC-integrated thermal desorption

(TD) uses multifunctional GC injectors, which are served by robotic autosamplers. A common concept here is the transfer of sample tubes from dedicated racks of the autosampler into the GC injector designed to serve as the TD unit. This concept allows flexible use for a variety of applications without modifications of the installed hardware using appropriate workflows for the robotic autosampler.

The thermal desorption unit (TDU) developed by the GERSTEL GmbH & Co KG is coupled to a cooled injection system (CIS) mounted in the GC, shown in Figure 4.62 for the thermal desorption of sorbent filled tubes or SBSE stir bars. The CIS serves both as a cryo-focusing internal trap and as a temperature programmable GC inlet. A selection of the thermal desorption tubes, which are uniquely equipped with a transport adapter on the top side with optional silicon inlet septum for liquid injection into the tube, is shown in Figure 5.19. The septum at the top is located inside the transport adaptors, not visible in the graphics. The loaded tubes are transferred using a gripper tool from the sample rack into the TDU. A ball activated closure mechanism firmly holds the injector O-ring in place and seals the injector for operation. The capacity of desorption tubes is only dependent on the installed number of tray holders and racks on the x,y,z-robot.

GL Sciences introduced a thermal desorption device capable of handling capped tubes [81]. The tubes are manually capped after the onsite sampling and placed with caps attached into the racks of the x,y,z-robotic sampler, hence a decapping unit is required before the loaded sorbent tubes can be inserted into the GC injector for desorption. A pneumatically activated gripper is used for the transfer of the capped desorption tubes to an automated decapping station, as shown in Figure 5.20. After taking off the top and bottom caps of the tube, the gripper transfers the tubes into the GC injector for thermal desorption. A specially designed GC injector is

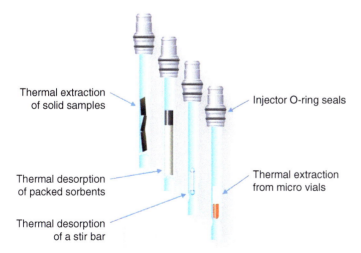

Figure 5.19 Thermal desorption tubes for the thermal desorption unit (TDU). The top sealing is optionally equipped with a silicon septum for injection of liquids. Besides this a TD 3.5+ module for standard 3½ in. sorbent tubes is available. Source: Courtesy of GERSTEL GmbH & Co.; KG [30].

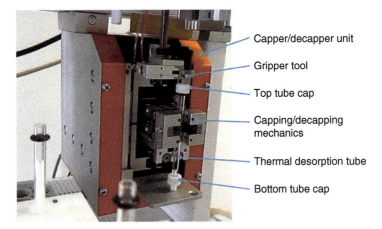

Figure 5.20 Device for automated capping/decapping of desorption tubes. Source: Courtesy GL Sciences B.V.

Figure 5.21 Desorption tube insertion sequence with gripper tool into the open OPTIC injector, then closed for operation. Source: Courtesy GL Sciences B.V.

used. The head of the OPTIC GC injector opens pneumatically. After insertion of the desorption tube, the injector head closes and the tube is purged with carrier gas. This sequence of operation is shown in Figure 5.21. For thermal extraction or desorption, the injector is then rapidly heated and the collected analytes transferred to the GC column for analysis [82]. After the GC run is completed the tube is moved back to the capping station, recapped and moved back into the tube tray. A next sample tube is taken for a new analysis cycle. The OPTIC injector works with the standardized thermal desorption tubes of $1/4$ in. × $3\frac{1}{2}$ in. (6.35 mm × 82.55 mm) dimension, but also with regular liquid injection inlet liners (5 mm × 81 mm). In the same way, using the same gripper, a programmed automated GC inlet liner exchange can be accomplished for regular liquid injections in a workflow for large sample series.

A proven example for a fully automated thermal desorption system installed as a separate unit for the analyses of up to 100 sorbent tubes and a sample overlap for enhanced sample throughput is the TD 100 system by Markes International Ltd. The unit is targeted to high-throughput analytical laboratories. The sorbent tubes remain capped at all times, making automation mechanically simple while avoiding sample contamination and analyte loss. A proprietary diffusion-locking technology

Figure 5.22 Diffusion lock tube cap. Source: Courtesy Markes International Ltd.

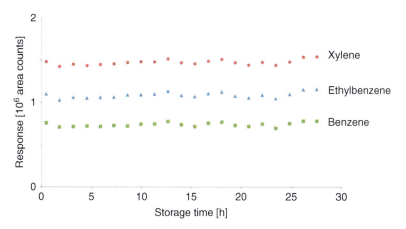

Figure 5.23 Diffusion lock performance for > 24h storage. Source: Courtesy Markes International Ltd.

eliminates analyte diffusion into and out of the tube, see Figure 5.22. Pumped sampling and automated thermal desorption is made possible without taking caps off [83]. The integrity of the sample is given even for long storage times of more than 24 hours, see Figure 5.23. In addition, the sorbent tubes can be made traceable using barcodes or RFID tube tags. The desorption units features an onboard tube tag read/write functionality and updates tube and sample information during analysis, see also Figure 3.17.

An up-to-date unique solution for combining sorptive extraction, thermal desorption, static and dynamic headspace, and SPME in an automated concept is provided with the multimode robotic platform Centri™ [84]. It is designed to connect to GC–MS instruments for the automated sampling, pre-concentration, and injection of VOCs and SVOCs from liquids and solids, and the thermal desorption of the standard packed sorbent tubes. Analyte flows from all sampling modes are concentrated on a sorbent-packed, cryogen-free focusing trap for enhanced GC performance, as illustrated in Figure 5.18.

Another analytical solution for manual and automated application of sorbent-based adsorption was developed by Entech Instruments with the Sorbent Pens™ [85, 86]. This technique connects the headspace extraction under reduced pressure with a trapping of the volatiles and a subsequent thermal desorption

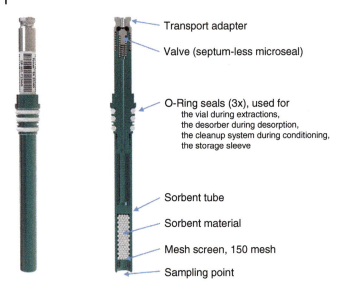

Figure 5.24 Sorbent pen cross section. Source: Courtesy Entech Instruments.

(Figure 5.24). The vacuum-assisted headspace extraction further expands HS–GC to lower LODs for analytes of lower volatility, also provides shorter extraction times. The technique is emerging also as vacuum-assisted solid-phase micro-extraction (Vac-HSSPME) [87].

The solution developed by Entech Instruments is called "vacuum-assisted sorbent extraction" (VASE™). The low pressure reduces the boiling point of analytes providing an increased response for higher boiling analytes [88]. The sorbent pens get inserted into headspace vials, as shown in Figure 5.25, or can be used for active and passive sampling. A vacuum provided by a diaphragm pump is applied through the pen to the headspace sample vial. Sorbent materials are available in the typical range of materials, as outlined in Table 5.7. The pens can be filled with different kind of sorbent materials, also use multilayers from weak trapping material like PDMS-coated glass beads as a first layer, followed by a medium-strength sorbent such as Tenax TA, then finally use Carboxen materials for strongest interaction.

VASE is used to perform compositional analysis of the sample, often providing exhaustive extraction of small test portions. The majority of compounds that are recovered, providing better sensitivity and reproducibility compared to other techniques that perform a partial extraction, and where these other techniques are not operating under equilibrium conditions. VASE typically uses vials of either 20 or 40 mL size. Only 1–5 g of sample is typically required to reach low part per trillion levels using full scan GC–MS, or even sub-ppt sensitivity using the SIM mode.

The sorbent pens are of a standard ¼ in. OD with 3.5 in. length manufactured from AISI 316 type stainless steel and are Silonite™ coated, a silica-based surface treatment for additional inertness. The pen comprises three functional sections, as

Figure 5.25 Sorbent pen inserted to a 20 mL headspace vial with vacuum line connected. Source: Courtesy Entech Instruments.

shown in Figure 5.24. The top is designed as the transport adapter for the *x,y,z*-robot, a 1/16″ vacuum line connection, and a septumless micro-valve [89]. The valve keeps the inside vacuum condition once the vacuum line is removed during extraction or the water removal step. In the center of the pen body are the O-rings for sealing with several devices like the vial during extraction, the desorber during desorption, the clean-up system during conditioning, and the storage sleeve during transport. The bottom part is filled up with sorbent material or several sorbent layers. The large phase loading of the sorbent pens makes them less susceptive to competitive matrix interferences. The opening at the sampling point is covered by a silonite-treated metal mash screen. Sorbent pens can be reused for several hundred analyses when operated within specifications.

For analysis, the sample test portion is placed into a headspace vial, capped with the sorbent pen inserted, and connected to a vacuum line. The extraction takes place under vacuum conditions. The sample vial is controlled heated during the headspace extraction process. The pen works as a closed system, so a "breakthrough" during trapping does not occur. Collected water is removed in a next step by diffusion from the tube using a cold tray. Diffusion happens fast as the vials are still under vacuum, which gets stronger by volume reduction due to the condensation of the water vapor in the cooled tray. For thermal desorption an *x,y,z*-autosampler transfers the pens from sealed tubes in a customized rack and places them into a desorption unit which is built as an additional injector into a GC. The automated sampling system uses the top transport adapter with a ball-lock mechanism as it is illustrated in Figure 3.20. The desorption unit is installed as an additional GC injector, with a mounting kit available for many GC brands. The workflow is semi-automated. The x,y,z-autosampler allows the automated processing of up to 120 pre-loaded sample pens with desorption for GC or GC-MS analysis.

5.2 GC Liquid Injection

The automated injection to gas chromatography systems (GC) is most probably the widest use of the so-called "liquid autosamplers" and the hyphenation to mass spectrometry (GC–MS). While simple autosamplers, also known as autoinjectors, do sequence controlled injections only, the usage of robotic sample preparation systems integrated to GC and even more to GC–MS is strongly on the rise.

5.2.1 Sandwich Injection

A very suitable way for automated systems of injection from different solutions, solvent, or vials is the so-called "sandwich injection" technique (Figure 5.26). Sandwiching a sample between volumes of solvents helps to handle small sample volumes also for LC injections. Sample carryover is minimized since the solvent behind the sample works as a wash solvent. For GC injections air gaps are used between different liquid plugs. Using the sandwich technique solvent plugs with internal standard, analyte protectants, or derivatization agents can be injected at once [90]. With the hot needle GC injection, the maximum tolerated solvent volume for the used inlet liner needs to be taken into account (Section 5.2.2). Due to the increased solvent volume, the fast liquid band GC injection (Section 5.2.3) is most suitable when using the sandwich technique with automated sampling systems [91].

5.2.2 Hot Needle Injection

The "hot needle injection" was coined by Koni Grob together with many of his great basic publications on GC injection techniques [92, 93]. This vaporizing injection

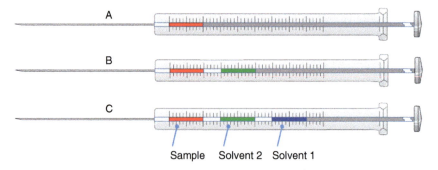

Figure 5.26 Sandwich injection principle: Solvents 1 and 2 can be optionally pulled up before the sample as additional functional agents like internal standard, derivatization reagent, analyte protectant, or a flushing solvent, separated by air gaps. (A) Regular sample injection. (B) Most used for an additional internal standard (ISTD). (C) Optionally added another solvent, reagent, or additional ISTD plug.

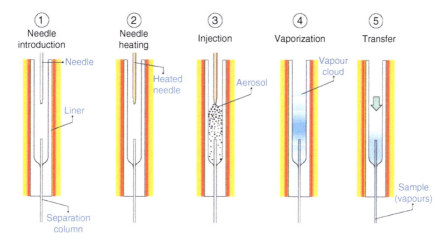

Figure 5.27 "Hot Needle Injection" with thermospray formation. Source: Image used with permission of Thermo Fisher Scientific Inc. (1) Pull up the sample into the syringe barrel. Introduce the syringe needle to approx. the center of the evenly heated inlet liner. (2) Wait for needle heat-up, approx. 3 to 5 s. (3) Inject with "normal" speed, corresponds to manual injection speed. Aerosol formation occurs. (4) Vaporization takes place inside the liner. The vapor cloud stays well within the liner. (5) Transfer of the vapor to the GC column with carrier gas speed. Recondensation at the beginning of the GC column takes place in splitless mode.

technique is also known as the "Normal", "Grob type", "Thermospray", or simply applied as the "Manual" injection technique. As the name already implies, the hot needle technique requires heating the syringe needle before injection. This needs an evenly heated injector, in particular at the septum top, called the "Grob type" injector. The injection steps are illustrated in Figure 5.27.

The liquid sample must be pulled up into the syringe barrel before penetrating the injector septum. It works well with an empty straight or single baffled GC inlet liner. Caution must be taken for the expansion volume of the liquid volume to be injected. The resulting solvent vapor volume must be significantly lower than the inlet liner volume [94]. A pressure pulse (or surge pressure) can be applied during vaporization to control the volume of the vapor cloud. Otherwise poor recovery, severe injector contamination and carry-over will occur. The hot needle technique works for split or splitless injection. For splitless injection, the GC oven must be set to a temperature below the actual solvent boiling point at the chosen headpressure to achieve the recondensation and a narrow band formation by the solvent effect in the GC column during the injection process. The initial isothermal phase should be kept for about two minutes at a general flow rate of around 1 mL/min before ramping the oven temperature.

For the hot needle injection, the empty syringe needle is pushed down into the injector. The needle and bottom of the syringe barrel must become hot. This can take up to five seconds after insertion of the needle through the septum. Modern GCs keep the septum cool for longer septa lifetime. So less hot injector tops require a little longer wait time. Upon pushing down the syringe plunger with a normal manual injection speed, typically in the range of 20 μL/s, solvent vaporizes already in the hot needle. Aerosol formation happens in the hot needle and a quick desolvation for the contained analytes occurs in the inlet liner. This technique is known to be the most gentle injection technique for sensitive compounds, protecting analytes well from an inlet liner glass wall contact. After injection, a short waiting time of up to two seconds is used and the syringe needle can be withdrawn. Limitations of the hot needle injection technique are seen with analytes of a wide boiling point range due to potential discrimination effects of high boilers. Also note: The effective liquid injection volume is the sum of the pulled up sample volume plus the needle volume!

For quantitative work, the hot needle injection provides the best precision results with lowest RSDs.

5.2.3 Liquid Band Injection

In contrast to the above hot needle injection the injection mode called "liquid band injection" became the typical GC autoinjector mode, also known as the "fast injection" mode. In 1986, a patent for a fast injection technique to GC was granted to Francis M. DiNuzzo and James S. Fullemann from the Hewlett-Packard Company, Palo Alto (CA), US [95]: " ... the steps of inserting the syringe needle into the vaporization chamber, operating the plunger of the syringe, and withdrawing the needle from the vaporization chamber are all done in such a short time that the needle does not get hot enough to vaporize a significant amount of the solvent or any solute. Thus, a sample volume leaves the needle as a liquid, ... , no components are vaporized from the solvent solution that pushes the sample volume out of the needle, and there is little or no discrimination ... for solvents such as hexane that boil at 69 °C by making the total time that the top of the needle dwells in the vaporization chamber 500 milliseconds or less. For solvents that boil at lower temperatures, the dwell time should be reduced. Satisfactory operation is attained when pentane is used as a solvent if the dwell time is 76 milliseconds."

The fast liquid band injection mode illustrated in Figure 5.28 works very much different from a manual or automated injection with a hot needle (see Figure 5.27). The fast mode has several advantages concerning the recovery of high boiling compounds as the unaltered liquid sample is transferred from the syringe into the inlet liner without prior vaporization in the syringe needle. It works equally well for split and splitless injection modes. A few special requirements make the liquid band injection work. The usage of inlet liners with a glass wool plug is mandatory, available commercially for all GC brands and inlet types. The syringe needle must have a conical tip to allow the solvent beam to be directed straight downward into the glass wool plug. The liquid beam may not hit the column entry or bottom of the liner. The size of the glass wool plug should be dimensioned this way that the planned liquid volume can be accepted by the plug without rinsing down. This can be tested best outside of the GC, just manually with the solvent of choice. Injection volumes from 1 μL to several 10th of microliters are possible [96]. The glass wool plug is positioned close to the bottom of the liner for splitless injections but may not cover the column. For working in split mode the glass wool plug has to be closer to

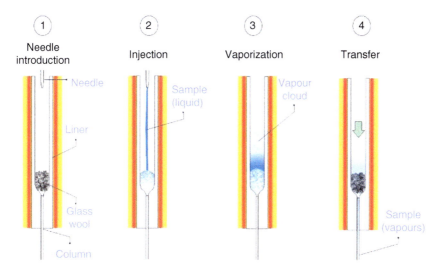

Figure 5.28 Fast Injection with Liquid Band Formation into a glass wool fitted inlet liner. Source: Image used with permission of Thermo Fisher Scientific Inc. (1) Introduce the syringe needle only to approx. the top of the inlet liner (depth is injector type dependent). (2) Inject fast. The liquid band formation occurs. The injected liquid sample volume is shot into the glass wool plug. (4) Vaporization takes place from the glass wool plug, cooling the glass wool. The vapor cloud stays well within the liner. (5) Fast transfer of the vapor to the GC column. Recondensation at the beginning of the GC column takes place in splitless mode and accelerates the transfer by volume reduction.

the top of the liner to provide sufficient mixture space of the evaporated analytes with carrier gas before reaching the split point. Before injection, the sample solvent plug can but is not required to be pulled back from the needle into the syringe barrel.

This fast injection mode is only available with fast autosamplers or robotic sampling using typical injection speeds of about 100 µL/s. It is not possible to inject that fast and precise manually. Upon injection, the syringe needle penetrates the septum down just to the entrance of the liner. For different brand injector designs, the penetration depth has to be adjusted accordingly as it is very much different from the "normal" mode. The injection takes place immediately without a wait time and fast into the hot injector. A liquid band is shot down the liner into the glass wool. Also, there is no wait time necessary after injection. The syringe needle leaves the injector immediately. The duration of the whole injection process is completed with modern sampling systems in the range of 100 ms only!

The evaporation starts slowly from the glass wool. First the solvent is evaporated, then followed by the more and more concentrated analytes. This process is slow as the glass wool has very limited heat capacity. Evaporation of the solvent cools the glass wool and prevents an explosive solvent evaporation at once. Radiation from the inner inlet liner walls supports the ongoing gentle evaporation until completed. A liner overflow with solvent vapor does not occur, hence larger liquid volumes can be injected, and no surge pressure (or pressure pulse) is required.

Also with the fast injection in the splitless mode, the GC oven is set during evaporation below the actual solvent boiling point to assure recondensation and the focusing solvent effect. For larger volume injections to make use of the full capacity of a 10 µL syringe, the recommended isothermal time should be set to 3–5 minutes before start ramping the GC column oven. For example, the obtained chromatograms from 1, 5, and 10 µL injections of a phthalate standard solution in hexane are demonstrated in Figure 5.29 with the GC oven setting shown in Figure 5.30. The retention times and peak shapes are identical and independent from the injection volume. The peak height corresponds well with the injection volume.

The liquid band injection mode is also recommended for large volume injections (LVI). LVI often makes sample extract concentration needless. This is of great advantage for automated analyses as time-consuming evaporation steps can be avoided for improved sample integrity and increased laboratory productivity [97]. For injections above 10 µL sample volume, the use of a 1–2 m long pre-column of 0.32 or 0.53 mm ID is recommended. Please note: The fused silica columns of 0.53 ID are prone to breaking when bent. In such applications for routine analysis, a corresponding flexible metal column, e.g. Siltec™ deactivated is recommended [98]. The concurrent solvent recondensation effect (CSV) [99] pulls the solvent vapor quickly down into the pre-column due to the strong volume reduction caused by the solvent recondensation at the beginning of the pre-column, as shown in Figure 5.31 [100]. The initial isothermal temperature phase should be set appropriate to the injected liquid volume to allow the solvent peak to pass the column before ramping the GC oven.

The strong analytical advantage of the liquid band injection is the discrimination-free injection for high boiling compounds, shown for an alkane mixture in Figure 5.32. Note the excellent peak shape and separation of the early eluting C8 alkane, even from the hexane solvent peak, and the tailing free uniform peak height with low discrimination up to C40. These features are of importance for all matrix extracts of real-life samples. Although the glass wool used in the inlet liner is usually deactivated, this inertness does not last long. During the analyses of real life samples, significant amounts of involatile sample matrix are injected and build up as a dark involatile residue in the glass wool. This protects the analytical column for an extended usage and requires a frequent liner change as a preventive maintenance measure. For a preventive inlet liner exchange automated solutions are available for a customized configuration of robotic x,y,z-systems (see Section 5.2.4).

5.2.4 Automated Liner Exchange

In GC analysis sample extracts get injected into a glass inlet liner, which, depending on the injection mode, is typically equipped with a short plug of deactivated glass wool. The GC inlet liner is the point of sample evaporation, but also serves to protect column performance. The glass wool plug becomes increasingly contaminated with a deposit of nonvolatile matrix and needs regular change in a preventive maintenance plan. Matrix deposit in the inlet liner leads to analyte degradation and adsorption, especially for sensitive, polar, and high boiling analytes, leading to a

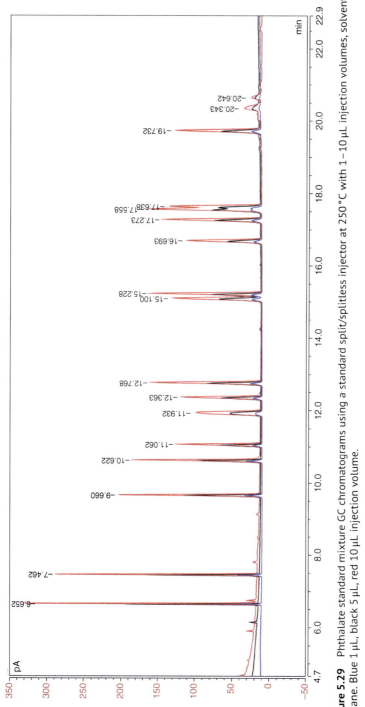

Figure 5.29 Phthalate standard mixture GC chromatograms using a standard split/splitless injector at 250 °C with 1–10 μL injection volumes, solvent hexane. Blue 1 μL, black 5 μL, red 10 μL injection volume.

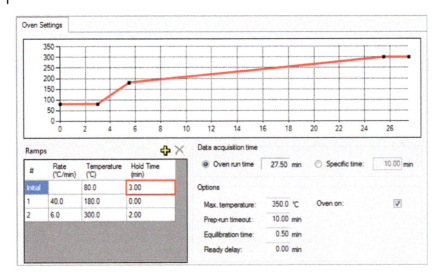

Figure 5.30 GC oven temperature program for up to 10 μL injection volume – Note the extended initial isothermal hold time. Source: Hans-Joachim Hübschmann.

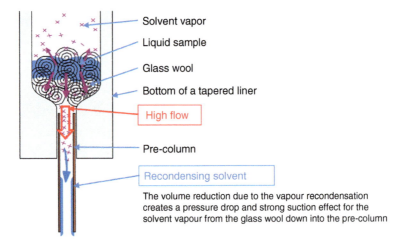

Figure 5.31 Concurrent solvent recondensation (CSR) principle. Source: Image used with permission of Thermo Fisher Scientific Inc.

notable drop in response and ultimately inaccurate quantitative results. The GC inlet liner is the cheapest part in the sample flow path and should not become the root cause for costly consequences. Replacing the inlet liner saves the laboratory time and money, otherwise spent on time-consuming column clipping, retention time adjustments, or an even more costly early column exchange.

The frequency of a required inlet liner change depends on the application, the type of analytes, and the sample clean-up. For instance, in pesticide analysis, using the QuEChERS extraction the useful liner lifetime can vary from only a few injections for honey or pollen analysis (acetonitrile dissolves sugars!) to as many as several

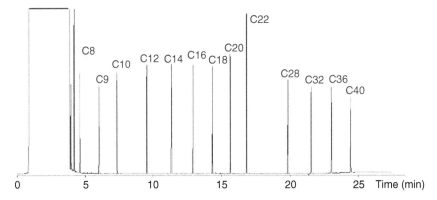

Figure 5.32 Chromatogram of 35 µL alkane standard in hexane, CSV injection mode. Conditions: 35 µL CSV injection, level 500 pg/µL in hexane, pre-column 5 m × 0.32 mm ID deactivated, analytical column Rtx-1, 15 m × 0.32 mm × 0.5 µm. Source: Image used with permission of Thermo Fisher Scientific Inc.

hundred injections after µSPE clean-up. The inlet liner should be changed to maintain performance after a certain number of real-life sample injections to prevent analyte degradation and a drop in performance.

The inlet liner performance needs to be checked for degradation regularly. In routine analysis operation and even more important in automated workflows, a known standard containing the sensitive compounds of interest is analyzed within the sample queue. A preventative maintenance schedule needs to be established changing the septum, liner and O-ring seal before performance degrades. The frequency can be "every day, every week, every month, every sample queue, depending on sample type and cleanliness" [101]. In automated routine analysis, the liner exchange can be executed by x,y,z-robotic systems. The robotic GC inlet liner exchange solutions, combined with a dedicated GC inlet, are prepared for a workflow driven unattended liner change.

The automated liner exchange (ALEX, developed by GERSTEL GmbH & Co KG) uses the proprietary Cooled Injection System (CIS) injector. A specially designed tray with clean individual liner compartments holds up to 40 GC inlet liners fitted to a liner adapter with a septum and O-ring seals, depicted in Figures 5.33 and 5.34. The adapter is kept in place and seals pneumatically in the CIS injector. With this design, the liner and septum are exchanged together automatically for safe routine operation [102]. The liner exchange is controlled by the analysis sequence table and can be customized as needed for the particular application. Also, the change to different liner types is supported. The same tool and operation principle is used for the automated insertion of thermal desorption tubes, as it is demonstrated in Figure 5.19.

Another device for automated GC inlet liner exchange named LINEX is offered by GL Sciences B.V. for the proprietary OPTIC injector. For the automated exchange, the liners are transported between a dedicated liner tray and the inlet by an x,y,z-robot equipped with a pneumatic gripper tool. For liner insertion to the injector, the head of the OPTIC injector is pneumatically opened, and the new clean

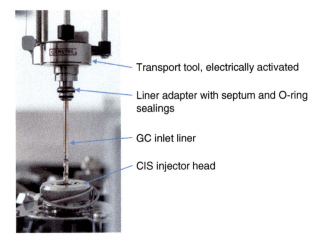

Figure 5.33 Robotic liner exchange with the MPS ALEX System. Source: Courtesy GERSTEL GmbH & Co KG.

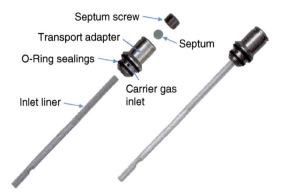

Figure 5.34 Exploded view of the transport adapter for the automated GC inlet liner exchange ALEX. Source: Courtesy GERSTEL GmbH & Co KG.

liner is placed into the inlet. Any of the standard OPTIC liners with or without a glass wool plug can be replaced. The head is closed again and the inlet is purged with carrier gas. The LINEX device is used for inserting thermal desorption tubes into the injector as well. The operation sequence illustrating the LINEX gripper tool with the insertion of a liner into the open OPTIC injector is demonstrated in Figure 5.21.

5.3 LC–GC Online Injection

The on-line coupling of LC with GC is not new. Already in 1980, Ronald E. Majors reported in his review about multidimensional LC about the potential of online LC–GC with the determination of a herbicide in sorghum [103]. In 1982, Dori J. Luzbetak and Joseph J. Hoffmann used an automated LC–GC method for the analyses of biomass-derived hydrocarbons as renewable sources of energy [104].

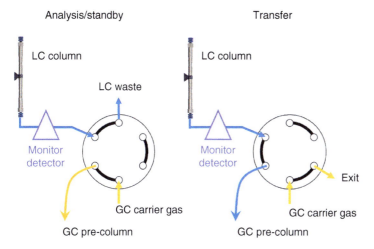

Figure 5.35 LC fraction transfer to GC (Transfer valve).

A loop-type LC–GC interface was presented in 1987 by Fausto Munari and Koni Grob at the International Symposium for Capillary Chromatography (ISCC) in Riva del Garda, Italy, for the automated analysis of drugs in urine [105]. A basic publication on the loop-type LC–GC interface and its usage finally paved the way for multiple applications in the later years [106]. First commercial systems were developed starting in 1989 by the Brechbühler AG, Schlieren, Switzerland, in collaboration with Koni Grob and Maurus Biedermann of the Food Control Authority of the Canton of Zürich, Switzerland, for the analysis of the saturated and aromatic mineral oil hydrocarbons (MOSH/MOAH) in food packaging products [107]. Also with the "Dualchrom", a commercial LC–GC system was launched in 1990 by the former Carlo Erba Strumentazione in Italy, today part of Thermo Fisher Scientific. Typical automated LC–GC transfers are in use with online LC, online GPC/µGPC, or online SPE for the trace analysis of complex mixtures in food and environmental samples, for instance, the online clean-up of pesticides [108], phthalates [109], PAHs [110], or mineral oil hydrocarbons [111, 112] or petrochemical analysis [113].

Today, the formerly described sample loop, which defines a preset transfer volume, is often replaced by a direct connection of the LC flow via a 6-way valve as shown in Figure 5.35, installed on top of the GC oven roof. The valve switching allows the flexible retention time-based control and calibration of the transferred fraction volume with µL/min flows of the LC side. The transfer volume of the analyte fraction is determined by the valve switching points on the retention time scale. This way the transfer volume can be adjusted by the automated method for different applications without a mechanical adjustment of a sample loop. The volume of the collected fraction is finally a controlled parameter in the workflow programming. A monitor detector signal, required for calibration of the switching times, can also be used for triggering the valve switching.

In a typical configuration, the flow from the LC column passes the monitor detector, then the 6-way valve with an outlet to waste (Figure 5.35 Analysis/Standby).

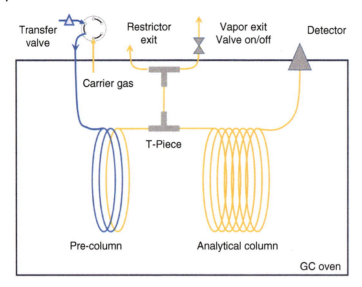

Figure 5.36 On-line LC-GC configuration. The analyte LC fraction is transferred into the pre-column.

The valve is connected to the carrier gas regulation of the GC, and a pre-column installed below in the GC oven. This valve inlet replaces the regular injector of the GC. The pre-column is connected directly to the valve routed from the GC oven through a spare inlet slot in the roof of the GC oven. The LC–GC inlet valve is not heated and operates at room temperature.

Inside of the GC oven, the pre-column outlet is connected to a capillary T-piece, as shown in Figure 5.36. The T-piece not only connects the pre-column to the analytical column but also with a short capillary to a solvent vapor exit valve (SVE). This valve must be heated to prevent solvent condensation during the injection process.

In regular LC operation, the 6-way valve is switched to the standby position. The LC effluent is directed to waste. The GC columns are provided with carrier gas from the regular GC instrument carrier gas regulation using the built-in electronic pressure control (EPC). For the transfer of the LC analyte fraction, the temperature of the GC oven is set close above the boiling point of the solvent at the applied carrier gas pressure [114] and stays isothermal during the injection process. The GC *Ready* signal triggers the sample injection to the LC separation. The SVE valve is open at this point. For transfer of an LC fraction, the 6-way valve is switched so that the LC flow is directed during a previously calibrated retention time window into the pre-column of the GC as shown in Figure 5.35 Transfer. At the end of the transferred fraction, the valve is switched back into the standby/analysis position. A *Start* signal is sent to accomplish the GC analysis with FID or MS detection as directed by the particular application.

During the transfer phase, the solvent plug of the LC sample fraction is pushed into the pre-column, first by the LC flow, after switching the transfer valve by the carrier gas. At the set oven temperature close to the corrected boiling point, the solvent plug is concurrently evaporating from its front end. At the T-piece, the solvent vapor leaves the pre-column via the open solvent vapor exit. The analytical column is at this point a strong restrictor for the vapor stream and remains in this phase with a negligible flow only, if a mass spectrometer detector is directly connected. The concurrent solvent evaporation allows the transfer of solvent volumes significantly larger than the nominal inner volume of the pre-column [115]. After the transfer and evaporation of the major part of the solvent, the SVE valve will be closed. Only a capillary restrictor with low flow is left open for venting the SVE line avoiding potential solvent vapor back stream, see Figure 5.36. After closing the SVE valve, all analyte compounds and remaining solvent move into the analytical column and get recondensed by the solvent effect at the beginning of the stationary phase of the separation column. This process is comparable to a regular GC liquid injection in splitless mode. The GC oven temperature program accomplishes the compound separation on the analytical column for the final analyte detection. In the automated setup, the LC column is usually backflushed during the GC analysis, accomplished with an additional backflush valve on the LC side, awaiting the *Ready* signal of the GC to start the next sample run.

5.4 LC Injection

Six-port valves with sample loops are the standard for sample extract delivery to LC or LC–MS analysis, as shown in Figure 5.37. In automated systems, the valve is switched by an electrical drive mounted on the rail of the x,y,z-robot and is controlled by the programmed analysis workflow.

The loop volume should correspond to the column capacity and required detection sensitivity. The LC pump and column connection gets installed this way that the transfer from the sample loop is achieved by loop backflush, see Figure 5.38. The loop can be operated in filled-loop or partial-loop mode. It is a good practice in the filled-loop mode to overfill the loop at least twofold to ensure uniform filling. If smaller volumes need to be injected, either the change of sample loop is advised or the partial-loop mode can be used, with some caveats. The injection volume the partial-loop mode should be less than half of the nominal loop volume, and the loop must be injected in the reverse direction (backflush). Then the partial-loop filling allows workflow-driven flexibility for injection of different sample volumes. For good precision, the injection volume must be reproducibly delivered, which is the typical strength of automated robotic systems. The precise dosage of volumes below $10\,\mu L$ is achieved with automated partial-loop injections. Syringes for LC injection must have the blunt needle tip (Point Style 3, for LC), see also Figure 4.3. Also, pipetting tools can be used by robotic samplers for liquid injection using the filled-loop technique (see also Section 5.4.2). It is advised to use the autosampler bottom sensing function during LC injection to secure a reliable leak-tight fit to the injection

234 | *5 Integration into Analysis Techniques*

Figure 5.37 LC Injection ports with sample loops for accessible automated liquid sample injection to LC and column switching. Source: Courtesy CTC Analytics AG.

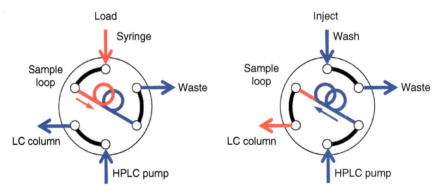

Figure 5.38 LC Injection valve – load with partial loop filling – inject in backflush.

port seal. After loading the sample to the sample loop, the valve is switched to the inject position so that the flow of the mobile phase takes the sample plug to the pre- or separation column.

5.4.1 Dynamic Load and Wash

For fast and mostly direct MS injections, a low carryover and rapid cleaning tool is required. This is the typical automated application requirement in high-throughput applications. With a dynamic load and wash injection tool, the sample is no longer

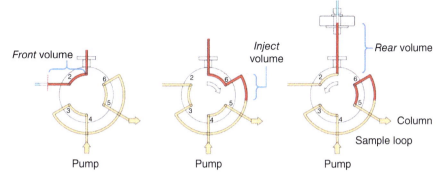

Figure 5.39 Dynamic load and wash injection sequence. Source: CTC Analytics AG.

in contact with the syringe but instead is aspirated partially into a holding loop as part of the injection tool. The sample solution does not get in contact with a syringe at all. The holding loop and sample lines are actively flushed after injection with solvents. Hence, no time-consuming syringe cleaning is necessary. The syringe acts like a dilutor syringe for sample aspiration and dispensing in order to pull-up and inject the correct volume.

In a typical workflow, the syringe and holding loop are preloaded with wash solvent at the start of the injection process. Before the sample is aspirated, a back air gap is drawn up separating the sample plug from the wash solvent, maintaining the sample integrity. After the sample, a front air gap is used. The total sample volume aspirated comprises a rear volume, the inject volume, and front volume, as shown in Figure 5.39. Right after injection, while keeping the needle still in the injection port, the complete sample path including the needle and the valve are flushed actively from the backside with up to two different solvents in sequence. The selected wash solvents are pumped from the back to the front to intensely flush all parts that were in contact with the sample plug, as shown in Figure 5.40 with the wash step after sampling. Two self-priming micro-pumps deliver the required solvents in the programmed volume. The active cleaning procedure is up to four times faster compared to a standard syringe washing process. At the end of the injection cycle, all parts which were in contact with the sample before are completely cleaned for the next sample in sequence. A built-in sensor at the inlet of the holding loop monitors the aspiration of air gaps. This ensures the reliable operation and logging, respectively, a feedback in cases of sample foaming, missing sample, or needle clogging (Figure 5.41).

5.4.2 Using LC Injection Ports with a Pipette Tool

Pipette tools are used for injection into LC or LC–MS systems as well. For the injection of samples with the pipette tool into an LC system, an injection port with a backflush system for the injection valve and the port is recommended. Injection ports for pipette tools can comprise of an active backflush system for the injection valve

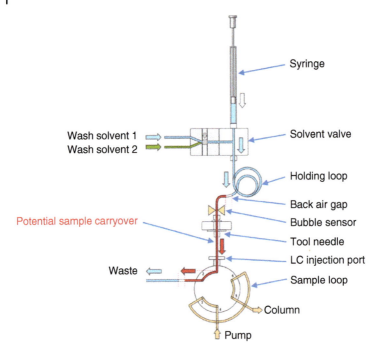

Figure 5.40 Dynamic load and wash tool for high-throughput and low carryover in LC-MS applications. The wash step is shown with active pumping of wash 1 solvent (blue) through the sample holding loop and valve pushing potential sample contaminations (red) to waste. Source: Courtesy CTC Analytics AG.

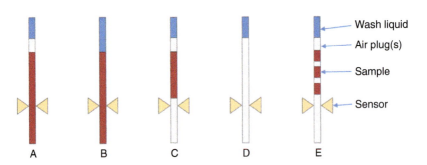

Figure 5.41 Liquid monitoring with the sensor of an LC-MS injection tool. (A) Correct operation. (B) Missing rear air gap. (C) Not enough sample. (D) Needle clogging. (E) Sample foaming. Source: Courtesy CTC Analytics AG.

and port using a separate diaphragm pump with a solvent reservoir. It is advised to perform sample loop overfill injections with pipette tips only. Using the bottom sensing function with controlled force ensures the leak tightness of the pipette tip with the inlet seal (Figure 5.42).

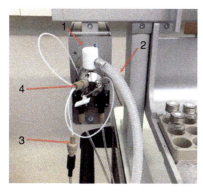

1 Injection port
2 Waste tube
3 Check valve
4 Tubing with check valve

Figure 5.42 LC injection with backflush for pipette injections. Source: Courtesy CTC Analytics AG.

References

1 Lichtenberg, G.C. (1889–1793). Sudelbücher II (Scrapbooks II) Heft J (Booklet J). In: *Georg Christoph Lichtenberg Schriften und Briefe*, Band Zwei (ed. W. Promies). Carl Hanser Verlag München Wien [1438].
2 Ramos, L. (2012). Critical overview of selected contemporary sample preparation techniques. *Journal of Chromatography. A* 1221: 84–98. https://doi.org/10.1016/j.chroma.2011.11.011.
3 Buttery, R.G. and Teranishi, R. (1961). Gas–liquid chromatography of aroma of vegetables and fruit. Direct injection of aqueous vapors. *Analytical Chemistry* 33: 1439.
4 Mackay, D.A.M., Lang, D.A., and Berdick, M. (1961). The objective measurement of odor. *Analytical Chemistry* 33 (10): 1369–1374.
5 Kolb, B. (1982). Anwendung der headspace analysis. *LaborPraxis*: 156–167.
6 Hachenberg, H. and Schmidt, A.P. (1977). *Gas Chromatographic Headspace Analysis*. London, UK: Heyden and Son.
7 Kolb, B. and Ettre, L. (2006). *Static Headspace-Gas Chromatography: Theory and Practice*, 2nd ed. Hoboken, New Jersey: John Wiley & Sons, Inc.
8 Perkin, Elmer (2005). The evaluation of gas chromatographic instrumentation at Perkin Elmer. *50th anniversary of gas chromatography at Perkin Elmer*. Perkin Elmer Life and Analytical Sciences, Shelton, CT, USA. https://books.google.com.sg/books?id=nGPmpb4VvEgC&printsec=frontcover.
9 Thompson, M. (2009). *Standard additions: myth and reality*. AMC Technical Briefs, AMCTB No. 37 March 2009, Royal Society of Chemistry, Analytical Methods Committee.
10 CVUA Stuttgart. (2017). *Workflow to perform quantification by standard addition procedure*. EU Reference Laboratory for Pesticide, Stuttgart, Germany. www.eurl-pesticides.eu/userfiles/file/EurlSRM/StdAdd_Workflow_EurlSRM.pdf.
11 Markelov, M. and Guzowski, J.P. (1993). Matrix independent headspace gas chromatographic analysis. This full evaporation technique. *Analytica Chimica Acta* 276 (2): 235–245. https://doi.org/10.1016/0003-2670(93)80390-7.

12 Li, H., Chai, X.S. et al. (2009). Rapid determination of ethanol in fermentation liquor by full evaporation headspace gas chromatography. *Journal of Chromatography. A* 1216 (1): 169–172. https://doi.org/10.1016/j.chroma.2008.11.024.

13 Herrington, J.S. (2019). SPME fundamentals – don't forget the salt for HS VOCs. In: *Restek Chromatography Blog*, 1. Restek Corp.: Bellefonte, PA blog.restek.com/?p=62001 (accessed 24 November 2019).

14 Majors, R.E. (2009). Salting out liquid–liquid extraction. *LCGC North America* 27 (7): 526–533. www.chromatographyonline.com/saltingoutliquidliquidextractionsalle.

15 Zuba, D., Parczewski, A., and Różańska, M. (2001). Effect of salt addition on sensitivity of HS-SPME-GC method of volatiles determination. In: *Paper #46 presented at the 39th TIAFT Triennial Meeting 2001*. Prague, Czech Republic.

16 Kepner, R.E., Van Straten, S., and Weurman, C. (1969). Freeze concentration of volatile components in dilute aqueous solutions. *Journal of Agricultural and Food Chemistry* 17 (5): 1123–1127. https://doi.org/10.1021/jf60165a023.

17 Logue, B.A., Skaggs, C. et al. (2020). Sample preparation goes subzero: ice concentration linked with extractive stirrer (ICECLES). *LCGC Europe* 33 (1): 37–41.

18 Hamilton (2016). Innovative HDHT-Type Headspace Syringe. Giarmata, Romania: Hamilton Bonaduz AG, *Specification Sheet*. www.hamiltoncompany.com (accessed 31 October 2019).

19 Huebschmann, H.-J. (2015). Chapter 2.1.5 Headspace techniques. In: *Handbook of GC/MS: Fundamentals and Applications*, 3rd Ed. Wiley-VCH Verlag GmbH & Co. KGaA.

20 Hinshaw, J.V. (2012). Headspace Sampling. Part II Instrumentation. *LCGC Europe*, 40–47. New York, NY, USA: Advanstar Communications Inc. http://www.advanstar.com/, http://www.speciation.net/Database/Journals/LCGC-Europe-;i2108.

21 Kolb, B., Pospisil, P., and Auer, M. (1984). Quantitative headspace analysis of solid samples; a classification of various sample types. *Chromatographia* 19 (1): 113–122. https://doi.org/10.1007/BF02687726.

22 Petersen, M.A. 2008. *Quantification of volatiles in cheese using multiple headspace extraction (MHE)*. Presentation, University of Copenhagen, Department of Food Science, Frederiksberg, Denmark.

23 LeRoy, S., Villière, A. et al. (2018). Headspace solid phase microextration vs. dynamic headspace extraction to explore breast milk volatile fraction. In: *Flavour Science – Proceedings of the XV Weurman Flavour Research Symposium* (eds. B. Siegmund and E. Leitner). Graz, Austria: Verlag der Technischen Universität Graz https://doi.org/10.3217/978-3-85125-593-5-99.

24 Seibel, B. 2015. "*Fundamentals of Purge and Trap*". Teledyne Tekmar. blog.teledynetekmar.com/fundamentals-of-purge-and-trap.

25 Teledyne Tekmar. (2020). *The History of Teledyne Tekmar*. Web presentation at https://www.teledynetekmar.com/about-us/about-us (accessed 12 October 2019).

26 Jochmann, M.A. 2006. Solventless extraction and enrichment methods for compound-specific isotope analysis. PhD Thesis. Eberhard-Karls University, Tübingen, Germany.

27 Wang, H. and George, E. (2008). Determination of volatile organic compounds in water by in-tube extraction gas chromatography ion trap mass spectrometry. In: *Application Note 01320*. Sunnyvale, CA: Varian Inc.

28 Fuchsmann, P., Stern, M.T. et al. (2019). Development and performance evaluation of a novel dynamic headspace vacuum transfer 'In Trap' extraction method for volatile compounds and comparison with headspace solid-phase microextraction and headspace in-tube extraction. *Journal of Chromatography. A* 1601: 60–70. https://doi.org/10.1016/j.chroma.2019.05.016.

29 Gil, C., O. Lerch, et al. 2007. Automated dynamic headspace sampling using replaceable sorbent trap. *Application Note 1/2007*, GERSTEL Gmbh & Co KG, Mülheim an der Ruhr, Germany.

30 GERSTEL (2016). Thermal desorption unit TDU 2. In: *Product Information*, 1–4. Mülheim, Germany: GERSTEL GmbH & Co.KG www.gerstel.com/pdf/s00135-042-02_TDU2_Flyer_en.pdf.

31 Koziel, J.A., Odziemkowski, M., and Pawliszyn, J. (2001). Sampling and analysis of airborne particulate matter and aerosols using in-needle trap and SPME fibre devices. *Analytical Chemistry* 73: 47–54. https://doi.org/10.1021/AC000835S.

32 Lord, H.L. and Zhan, W. (2010). Fundamentals and applications of needle trap devices: a critical review. *Analytica Chimica Acta* 677 (1): 3–18. https://doi.org/10.1016/J.ACA.2010.06.020.

33 Risticevic, S., Niri, V.H. et al. (2009). Recent developments in solid-phase microextraction. *Analytical and Bioanalytical Chemistry* 393: 781–795. https://doi.org/10.1007/s00216-008-2375-3.

34 Asl-Hariri, S., Gómez-Ríos, G.A. et al. (2014). Development of needle trap technology for on-site determinations: active and passive sampling. *Analytical Chemistry* 86 (12): 5889–5897. https://doi.org/10.1021/ac500801v.

35 Mieth, M., Schubert, J.K. et al. (2010). Automated needle trap heart-cut GC/MS and needle trap comprehensive two-dimensional GC/TOF-MS for breath gas analysis in the clinical environment. *Analytical Chemistry* 82 (6): 2541–2551. https://doi.org/10.1021/ac100061k.

36 Hein, D. (2017). *Controlled sampling on packed micro extraction needles (needletrap)*. Poster Presentation, PAS Technology Deutschland GmbH, Singapore. www.pastec.com/fileadmin/files/dokumente/concept_nt/Praesentationen/NeedleTrap-EN.pdf (accessed 12 October 2019).

37 Miekisch, W., Trefz, P. et al. (2014). Microextraction techniques in breath biomarker analysis. *Bioanalysis* 6 (9): 1275–1291. https://doi.org/10.4155/bio.14.86.

38 Umweltbundesamt. (2018) Anforderungen an die Innenraumluftqualität in Gebäuden: Gesundheitliche Bewertung der Emissionen von flüchtigen organischen Verbindungen (VVOC, VOC und SVOC) aus Bauprodukten, translated: Health-related Evaluation Procedure for Volatile Organic Compounds

Emissions (VVOC, VOC and SVOC) from Building Products. Ausschuss zur gesundheitlichen Bewertung von Bauprodukten AgBB – August 2018, Berlin, Germany, pp. 1–26.

39 International Code Council, Inc. (ICC) and ASHRAE (2018). *2018 International Green Construction Code*. Country Club Hills, USA: ICC Publications.

40 ANSI/BIFMA (2016). *M7.1 – 2011(R2016) FES Test Method, Standard Test Method for Determining VOC Emissions from Office Furniture Systems, Components and Seating*. Grand Rapids, USA: BIFMA.

41 ASTM International (2016). *ASTM D5197-16, Standard Test Method for Determination of Formaldehyde and Other Carbonyl Compounds in Air (Active Sampler Methodology)*. West Conshohocken, PA, USA: ASTM International https://doi.org/10.1520/D5197-16.

42 ASTM International (2015). *ASTM D5466-15, Standard Test Method for Determination of Volatile Organic Compounds in Atmospheres (Canister Sampling Methodology)*. West Conshohocken, PA, USA: ASTM International https://doi.org/10.1520/D5466-15.

43 ASTM International (2015). *ASTM D6196-15e1, Standard Practice for Choosing Sorbents, Sampling Parameters and Thermal Desorption Analytical Conditions for Monitoring Volatile Organic Chemicals in Air*. West Conshohocken, PA, USA: ASTM International https://doi.org/10.1520/D6196-15E01.

44 ASTM International (2018). *ASTM D7648 / D7648M-18, Standard Practice for Active Soil Gas Sampling for Direct Push or Manual-Driven Hand-Sampling Equipment*. West Conshohocken, PA, USA: ASTM International https://doi.org/10.1520/D7648_D7648M-18.

45 ASTM International (2017). *ASTM D7706-17, Standard Practice for Rapid Screening of VOC Emissions from Products Using Micro-Scale Chambers*. West Conshohocken, PA, USA: ASTM International https://doi.org/10.1520/D7706-17.

46 ASTM International (2017). *ASTM D7758-17, Standard Practice for Passive Soil Gas Sampling in the Vadose Zone for Source Identification, Spatial Variability Assessment, Monitoring, and Vapor Intrusion Evaluations*. West Conshohocken, PA, USA: ASTM International https://doi.org/10.1520/D7758-17.

47 California Department of Public Health (2017). *Standard Method for the Testing and Evaluation of Volatile Organic Chemical Emissions from Indoor Sources Using Environmental Chambers, Version 1.2*. Sacramento, CA, USA: California Department of Public Health.

48 European Committee for Standardization (2014). *Stationary source emissions – Determination of the mass concentration of individual gaseous organic compounds – Sorptive sampling method followed by solvent extraction or thermal desorption*. CEN/TS 13649:2014. Brussels, Belgium: CEN-CENELEC Management Centre https://standards.cen.eu/dyn/www/f?p=204:110:0::::FSP_PROJECT:31874&cs=1FFD2CB9B2C56E33D9FE0C96686F0E8EA (accessed 29 September 2019).

49 International Organization for Standardization. (2012). Interior air of road vehicles – Part 1: Whole vehicle test chamber – Specification and method

for the determination of volatile organic compounds in cabin interiors. ISO 12219-1:2012, International Organization for Standardization, ISO Central Secretariat, Vernier, Geneva, Switzerland. www.iso.org/standard/50019.html (accessed 29 September 2019).

50 International Organization for Standardization. (2017). Interior air of road vehicles – Part 2: Screening method for the determination of the emissions of volatile organic compounds from vehicle interior parts and materials – Bag method. ISO 12219-2:2012, International Organization for Standardization, ISO Central Secretariat, Vernier, Geneva, Switzerland. www.iso.org/standard/54865.html (accessed 29 September 2019).

51 International Organization for Standardization. (2017). Interior air of road vehicles – Part 3: Screening method for the determination of the emissions of volatile organic compounds from vehicle interior parts and materials – Micro-scale chamber method. ISO 12219-3:2012, International Organization for Standardization, ISO Central Secretariat, Vernier, Geneva, Switzerland. www.iso.org/standard/54866.html (accessed 29 September 2019).

52 International Organization for Standardization. (2018). Interior air of road vehicles – Part 4: Method for the determination of the emissions of volatile organic compounds from vehicle interior parts and materials – Small chamber method. SO 12219-4:2013, International Organization for Standardization, ISO Central Secretariat, Vernier, Geneva, Switzerland. www.iso.org/standard/54867.html (accessed 29 September 2019).

53 International Organization for Standardization. (2019). Interior air of road vehicles – Part 5: Screening method for the determination of the emissions of volatile organic compounds from vehicle interior parts and materials – Static chamber method. ISO 12219-5:2014, International Organization for Standardization, ISO Central Secretariat, Vernier, Geneva, Switzerland. www.iso.org/standard/56876.html (accessed 29 September 2019).

54 International Organization for Standardization. (2011). Indoor air – Part 3: Determination of formaldehyde and other carbonyl compounds in indoor air and test chamber air – Active sampling method. ISO 16000-3:2011, International Organization for Standardization, ISO Central Secretariat, Vernier, Geneva, Switzerland. www.iso.org/standard/51812.html (accessed 29 September 2019).

55 International Organization for Standardization. (2011). Indoor air – Part 6: Determination of volatile organic compounds in indoor and test chamber air by active sampling on Tenax TA sorbent, thermal desorption and gas chromatography using MS or MS-FID. ISO 16000-6:2011, International Organization for Standardization, ISO Central Secretariat, Vernier, Geneva, Switzerland. www.iso.org/standard/52213.html (accessed 29 September 2019).

56 International Organization for Standardization. (2020). Workplace air quality – Sampling and analysis of volatile organic compounds by solvent desorption/gas chromatography – Part 1: Pumped sampling method. ISO 16200-1:2001, International Organization for Standardization, ISO Central Secretariat, Vernier,

Geneva, Switzerland. www.iso.org/standard/30187.html (accessed 29 September 2019).

57 International Organization for Standardization. (2020). Workplace air quality – Sampling and analysis of volatile organic compounds by solvent desorption/gas chromatography – Part 2: Diffusive sampling method. ISO 16200-2:2000, International Organization for Standardization, ISO Central Secretariat, Vernier, Geneva, Switzerland. www.iso.org/standard/30188.html (accessed 29 September 2019).

58 Japanese Standards Association. (2015). Determination of the emission of volatile organic compounds and aldehydes by building products – Small chamber method. JSA – JIS A 1901, Japanese Standards Association, Tokyo, Japan. https://standards.globalspec.com/std/9964318/jis-a-1901 (accessed 29 September 2019).

59 US Environmental Protection Agency. (1996). Volatile Organic Compounds by Gas Chromatography/Mass Spectrometry (GC/MS). SW-846 Test Method 8260B Rev.2, US Environmental Protection Agency, Washington, D.C., USA. https://19january2017snapshot.epa.gov/hw-sw846/sw-846-test-method-8260b-volatile-organic-compounds-gas-chromatographymass-spectrometry_.html (accessed 29 September 2019).

60 Center for Environmental Research Information, Office of Research and Development, U.S. Environmental Protection Agency. (1999). Determination Of Volatile Organic Compounds (VOCs) In Ambient Air Using Specially Prepared Canisters With Subsequent Analysis By Gas Chromatography. Compendium Method TO-14A, U.S. Environmental Protection Agency, Cincinnati, OH, USA. www3.epa.gov/ttnamti1/files/ambient/airtox/to-14ar.pdf (accessed 29 September 2019).

61 Center for Environmental Research Information, Office of Research and Development, U.S. Environmental Protection Agency. (1999). Determination Of Volatile Organic Compounds (VOCs) In Air Collected In Specially-Prepared Canisters And Analyzed By Gas Chromatography/Mass Spectrometry (GC/MS). Compendium Method TO-15, 2nd ed., U.S. Environmental Protection Agency, Cincinnati, OH, USA. https://www3.epa.gov/ttnamti1/files/ambient/airtox/to-15r.pdf (accessed 29 September 2019).

62 Center for Environmental Research Information, Office of Research and Development, U.S. Environmental Protection Agency (1999). *Determination of Volatile Organic Compounds in Ambient Air Using Active Sampling Onto Sorbent Tubes*. Compendium Method TO-17, 2nd Ed. Cincinnati, OH, USA: U.S. Environmental Protection Agency www3.epa.gov/ttnamti1/files/ambient/airtox/to-17r.pdf (accessed 29 September 2019).

63 US Environmental Protection Agency. (1986). Volatile Organic Sampling Train. SW-846 Test Method 0030, US Environmental Protection Agency, Washington, D.C., USA. www.epa.gov/sites/production/files/2015-12/documents/0030.pdf (accessed 29 September 2019).

64 US Environmental Protection Agency. (1996). Sampling Method for Volatile Organic Compounds (SMVOC). SW-846 Test Method 0031, US Environmental

Protection Agency, Washington, D.C., USA. www.epa.gov/sites/production/files/2015-12/documents/0031.pdf (accessed 29 September 2019).

65 US Environmental Protection Agency. (1996). Analysis for Desorption of Sorbent Cartridges from Volatile Organic Sampling Train (VOST). SW-846 Test Method 5041A, US Environmental Protection Agency, Washington, D.C., USA. www.epa.gov/sites/production/files/2015-12/documents/5041a.pdf (accessed 29 September 2019).

66 Markes (2012). Advice on Sorbent Selection, Tube Conditioning, Tube Storage and Air Sampling. In: *Application Note TDTS 5*. Llantrisant, UK: Markes International Ltd.

67 Woolfenden, E. (2010). Sorbent based sampling methods for volatile and semi-volatile organic compounds in air – Part 1: Sorbent-based air monitoring options. *Journal of Chromatography. A* 1217: 2674–2684.

68 Markes. (2012). Analytical thermal desorption: history, technical aspects and application range. *Application Report TDTS 12*, Markes International Ltd., Llantrisant, UK.

69 Krebs, G., Schneider, E., and Schumann, A. (1991). Headspace GC analysis of volatile, aromatic, and halogenated hydrocarbons out of soil gas. *GIT Fachzeitschrift Labor* 35: 19–22.

70 International Organization for Standardization. (2014). Workplace air quality – Sampling and analysis of volatile organic compounds by solvent desorption/gas chromatography. ISO 16200-1, International Organization for Standardization, ISO Central Secretariat, Vernier, Geneva, Switzerland. www.iso.org/standard/30187.html (accessed 29 September 2019).

71 Ramírez, N., Cuadras, A. et al. (2010). Comparative study of solvent extraction and thermal desorption methods for determining a wide range of volatile organic compounds in ambient air. *Talanta* 82 (2): 719–727. https://doi.org/10.1016/J.TALANTA.2010.05.038.

72 Lepera, J.S. and Colacioppo, S. (2002). Rapid determination of desorption efficiency, and analysis of solvent mixtures for occupational exposure studies. *Chromatographia* 55 (7–8): 463–466. https://doi.org/10.1007/BF02492278.

73 International Organization for Standardization (2001). Workplace air quality – Sampling and analysis of volatile organic compounds by solvent desorption/gas chromatography. Part 1: Pumped sampling method. ISO 16200-1:2001 https://www.iso.org/standard/30187.html.

74 Cucciniello, R., Proto, A. et al. (2015). An improved method for BTEX extraction from charcoal. *Analytical Methods* 7 (11): 4811–4815. https://doi.org/10.1039/c5ay00828j.

75 Guo, D., Shi, Q. et al. (2011). Different solvents for the regeneration of the exhausted activated carbon used in the treatment of coking wastewater. *Journal of Hazardous Materials* 186 (2–3): 1788–1793. https://doi.org/10.1016/J.JHAZMAT.2010.12.068.

76 Hinoue, M., Ishimatsu, S. et al. (2017). A new desorption method for removing organic solvents from activated carbon using surfactant. *Journal of Occupational Health* 59 (2): 194–200. https://doi.org/10.1539/joh.16-0214-OA.

77 Korenková, E., and M. Cepeda-Leucea. 2017. *"The determination of 1,4-dioxane in water by automated spe gas chromatography-high resolution mass spectrometry (GC-HRMS)"*. Ministry of the Environment and Climate Change-E3534. 30 January 2017, Revision 1.

78 Martens, M., Hogekamp, H. et al. (2011). *Evaluation of Monolithic Material Sorptive Extraction (MMSE) as an Alternative Aroma Extraction Technique*. Deventer, The Netherlands: Friesland Campina Innovation https://doi.org/10.1016/b978-0-12-398549-1.00077-5.

79 Sato, A., K. Sotomaru, and M. Takeda. 2009. A novel approach for aroma components analysis using a monolithic hybrid adsorbent as a new generation medium. "MonoTrap". *Poster presented at the International Symposium on Essential Oils (ISEO)*, GL Sciences, Saitama, Japan.

80 Woolfenden, E. (2001). Optimising analytical performance and extending the application range of thermal desorption for indoor air monitoring. *Indoor and Built Environment* 10 (3–4): 222–231. https://doi.org/10.1159/000049240.

81 GL Sciences (2020). Automated LINer EXchanger for gas chromatography. In: *Product Information*, 1–4. GL Sciences: B.V., Eindhoven, The Netherlands.

82 Berger-Karin, C., Hendriks, U., and Geyer-Lippmann, J. (2008). Comparison of natural and artificial aging of ballpoint inks. *Journal of Forensic Sciences* 53 (4): 989–992. https://doi.org/10.1111/j.1556-4029.2008.00770.x.

83 Markes (2012). Diffusion-locking technology. In: *Technical Note TDTS 61*, 1–3. Llantrisant, UK: Markes International Ltd.

84 Markes (2018). Centri – a breakthrough in sample automation and concentration for GC–MS. In: *Product Information*. Llantrisant, UK: Markes International Ltd.

85 Cardin, D.B. (2017). Vacuum assisted sample extraction device and method. US patent 2017/0261408 A1, filed 6 March 2017 and issued 2017. patents.google.com/patent/US20170261408A1/en.

86 Entech Instruments (2019). Sorbent pens headspace analysis. In: *Product Information*, 1–12. Simi Valley, CA: Entech Instruments www.entechinst.com/download/entech-sorbent-pens/.

87 Zhakupbekova, A., Baimatova, N., and Kenessov, B. (2019). A critical review of vacuum-assisted headspace solid-phase microextraction for environmental analysis. *Trends in Environmental Analytical Chemistry* 22: e00065. https://doi.org/10.1016/j.teac.2019.e00065.

88 Psillakis, E. (2020). The effect of vacuum: an emerging experimental parameter to consider during headspace microextraction sampling. *Analytical and Bioanalytical Chemistry*, Document No. Silonite MicroValves –180118-52: 1–8, https://doi.org/10.1007/s00216-020-02738-x.

89 Entech Instruments (2018). Silonite microvalves. In: *Product Information*, 1–8. Simi Valley, CA: Entech Instruments.

90 Wylie, P.L., and K. Chen. 2016. Using Sandwich Injections to Add Matrix, Internal Standards and/or Analyte Protectants for the GC/Q- TOF Analysis of Pesticide Residues. *EPRW 2016 Poster PV 027*, Wilmington, DE USA, and Santa Clara, CA USA: Agilent Technologies Inc.

91 Klee, M.S. (2019). Sandwich injections. *Separation Science* 2019 blog.sepscience.com/gaschromatography/sandwich-injections (accessed 29 September 2019).

92 Grob, K. (1994). Injection techniques in capillary GC. *Analytical Chemistry* 66 (20): 1009A–1019A.

93 Grob, K. (2001). *Split and Splitless Injection for Quantitative Gas Chromatography. Concepts, Processes, Practical Guidelines, Sources of Error*, 4th Ed. D-69469 Weinheim (Federal Republic of Germany): WILEY-VCH Verlag GmbH.

94 Huebschmann, H.-J. (2015). *Handbook of GC-MS: Fundamentals and Applications*, Chapter 2.2 Gas Chromatography. 3rd Ed. Weinheim, Germany: Wiley-VCH. ISBN: 978-3-527-33474-2.

95 DiNuzzo, F.M., and Fullemann, J.S. (1986). Apparatus and method for introducing solutes into a stream of carrier gas of a chromatograph. US Patent 4,615,226, filed 4 February 1985.

96 Cochran, J. (2010). Large volume injection for gas chromatography using a commercially-available, unmodified splitless injector. Poster at the 34th ISCC Conference Riva del Garda, Italy.

97 Krumwiede, D., Munari, F., and Münster, H. (2008). The application of large volume injection techniques for increased productivity and sensitivity in routine POPs analysis with GC-HRMS and TripleQuad GC-MS. *Organohalogen Compounds* 70: 466–468.

98 Restek. (2019). MXT Capillary Columns. Product Information. Restek Bellefonte, PA, USA. www.restek.com (accessed 08 December 2019).

99 Magni, P. (2005). Method and device for vaporization injection of high volumes in gas chromatographic analysis. US Patent No. US06955709, filed 19 December 2002.

100 Biedermann, M., Fiscalini, A., and Grob, K. (2004). Large volume splitless injection with concurrent solvent recondensation: keeping the sample in place in the hot vaporizing chamber. *Journal of Separation Science* 27: 1157–1165. https://doi.org/10.1002/jssc.200401847.

101 Cochran, J. (2014). *Change the GC Inlet Liner – When?* Bellefonte, PA, USA: Restek blog.restek.com/?p=11288.

102 Lerch, O. (2010). GERSTEL automated linear exchange (ALEX) and its benefits in GC pesticide analysis. *Application Note 7/2010*, Mülheim an der Ruhr, Germany GERSTEL GmbH & Co KG.

103 Majors, R.E. (1980). Multidimensional high performance liquid chromatography. *Journal of Chromatographic Science* 18 (10): 571–579. https://doi.org/10.1093/chromsci/18.10.571.

104 Luzbetak, D.J. and Hoffmann, J.J. (1982). On-line HPLC/GC techniques for the analyses of biomass derived hydrocarbons. *Journal of Chromatographic Science* 20 (3): 132–135. https://doi.org/10.1093/chromsci/20.3.132.

105 Munari, F. and Grob, K. (1988). Automated on line HPLC-HRGC. Instrumental aspects and application for the analysis of heroin and metabolites in urine. *HRC & CC* 11: 172.

106 Munari, F. and Grob, K. (1990). Coupling LC to GC: Why? How? With what instrumentation? *Journal of Chromatographic Science* 28: 61–66.

107 Mottay, P., Hofstetter, U., and Pichler, P. (1989). *Mineral Oil Determination in Food and Food Packaging by LC-GC Coupling*. Brechbühler AG, Schlieren, Switzerland: Product Information.

108 Kitagawa, M., Hori, S., Miyagawa, H. et al. (2002). Analysis of residual pesticides in agricultural products by on-line GPC-GC/MS. *Shimadzu Review* 59 (1/2): 15–23.

109 Munari, F., and A. Fankhauser-Noti. 2006. Reducing blank problems for the analysis of phthalates in fatty foods by online LC–GCMS. Thermo Electron Corp. *Poster at the ISCC Riva del Garda*.

110 Schulz, C.M., F. Janusch, et al. 2017. Simple, Fast, Innovative, and Automated Determination of 27 Polycyclic Aromatic Hydrocarbons (PAHs) in Oils and Fats by LC-LC-GC-MS. *Poster at the RAFA Conference 2017*, Prague. Eurofins WEJ Contaminants GmbH, Hamburg.

111 Weber, S., Schrag, K. et al. (2018). Analytical methods for the determination of mineral oil saturated hydrocarbons (MOSH) and mineral oil aromatic hydrocarbons (MOAH) – a short review. *Analytical Chemistry Insights* 13: 1–16. https://doi.org/10.1177/1177390118777757.

112 Nestola, M. and Schmidt, T.C. (2017). Determination of mineral oil aromatic hydrocarbons in edible oils and fats by online liquid chromatography–gas chromatography–flame ionization detection – evaluation of automated removal strategies for biogenic olefins. *Journal of Chromatography. A* 1505: 69–76. https://doi.org/10.1016/j.chroma.2017.05.035.

113 Kelly, G.W. and Bartle, K.D. (1994). The use of combined LC–GC for the analysis of fuel products – a review. *Journal of High Resolution Chromatography* 17 (6): 390–397. isi:A1994NY08200003.

114 Grob, K. and Läubli, T. (1987). Minimum column temperature required for concurrent solvent evaporation in coupled HPLC–GC. *Journal of High Resolution Chromatography* 10 (8): 435–440. https://doi.org/10.1002/jhrc.1240100803.

115 Grob, K., Walder, C., and Schilling, B. (1986). Concurrent solvent evaporation for on-line coupled HPLC–HRGC. *Journal of High Resolution Chromatography* 9 (2): 95–101. https://doi.org/10.1002/jhrc.1240090208.

6

Solutions for Automated Analyses

> *If you define the problem correctly, you almost have the solution.*
> Steve Jobs (1955–2011)
> Co-founder of Apple Inc.

This chapter goes into real-life applications. The selected examples provide details of proven practical solutions for automated workflows. From the many different analytical methods used in daily routine in laboratories worldwide, a selection of often used sample preparation methods is presented. The workflows cover the general liquid handling with different solutions for dilution and derivatization and approaches for the analysis of volatile compounds (VOCs), as well as the corresponding non-volatile analytes (SVOCs). The workflows are presented with the background and operation principle of the application, required chemicals, solvents, and consumables, followed by a description of the suggested configuration of a generic x,y,z-robotic rail system with a note on the function of the required modules. For all applications, polar and non-polar wash solvents for syringes are required, which fit best the particular application, the involved matrix or chemicals, following a general best practice. The wash solvents are referenced as solvents 1 and 2 in workflows, just to indicate the difference of polarity, without specifying a particular solvent. Finally, the workflows covered in this section are commented with initial pre-treatment steps and described stepwise for implementation on different x,y,z-platforms. The sequence of activities of the robotic platform is illustrated for many applications in a graphical display for ease of reproduction using one of the programming tools mentioned in Section A.6 about the robot system control.

The comprehensive description of the application and workflow allows the installation of the automated solution in a laboratory for the same or equivalent analytical tasks on many robotic platforms. The described sample preparation and analysis solutions are of general applicability. The detailed robotic system configurations refer to the rail-based x,y,z-robotic samplers available from many instrument companies and vendors with different product names, system integration, and capability of using a range of standard and customized tools and modules, usually with equivalent or comparable functionality. Consumables and accessories are offered commercially from many local and international distributors.

First About Safety

For the safe handling and responsible disposal of solvents and chemicals mentioned in this document refer to the safety data sheets (SDS) available from the supplier, or as download from the internet, for instance via http://www.chemspider.com.

6.1 Dilution

Dilutions are daily tasks in every laboratory, a regularly repeating preparation workload for all quantitative assays. As such, it is an always constant workflow demanding for automation, not only to release hands but also for improved analytical quality and traceability. Generally, commercial analytical standards are available for trace analysis at high ppm level (µg/mL) and require appropriate dilution to laboratory stock solutions, working standards and further dilution to fit the method quantitative calibrations according to regulations. Often internal standards (ISTD), analyte protectants (APs) or "keeper" solvents need to be considered as well.

Three approaches to automated dilution workflows can be distinguished using different strategies and instrument configurations. Often, the first step is the dilution of ppm concentrated commercial standards down to ppb levels, then the preparation of suitable calibration dilutions according to the required decision limits for quantification. The most challenging task is the composition of multi-compound standards from individual standard solutions or even solid reference materials with the required quality control management and documentation.

6.1.1 Geometric Dilution of Reference Standards

Application

The geometric series or serial dilution uses a constant dilution factor between successive dilutions, for instance by a constant factor of 10. This way with a few steps a dilution of several orders of magnitude can be achieved, without using much solvent. For further calibration tasks, it may be necessary to create such intermediate solutions, which then are diluted further to achieve the required calibration concentrations. The mechanical precision of the automated robotic sampler is critical to achieve a decent precision of the required dilution levels.

Figure 6.1 Automated dilution scheme for a geometrical dilution of a concentrated stock solution (constant dilution factor 1:10 as an example).

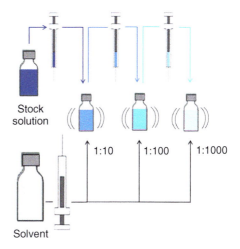

Scope and Principle of Operation

The geometric dilution is required for most of the commercial standard solutions to dilute from the usually delivered high µg/mL level down to a required level of pg/µL or below. Constant volumes of the standard, starting with the stock solution, are diluted with solvent using a constant factor for dilutions over several orders of magnitude. A dilution factor of for instance 1:10 is using a dilution ratio of 1:9. A series of dilutions with the set dilution factor are executed to achieve the desired concentration level, illustrated in Figure 6.1.

Solvents and Chemicals
- Stock standard solutions
- Suitable solvents for dilution and syringe washing

Consumables
- 2 mL vials with magnetic caps (for trace analysis applications).
- Liquid syringe of suitable volume, e.g. 1000 µL, needle length 57 mm, polytetrafluoroethylene (PTFE) plunger, gauge 22, point style flat.

System Configuration

Typically, a dilution task is integrated with sample preparation configurations and optional automated tool change. The system can be installed as a standalone configuration as well. The minimum hardware requirement for a robotic sampling system for benchtop prepared dilutions is shown in Figure 6.2. For large volume dilutions into 10 or 20 mL vials, the use of a dilutor or pipetting tool instead of syringes is

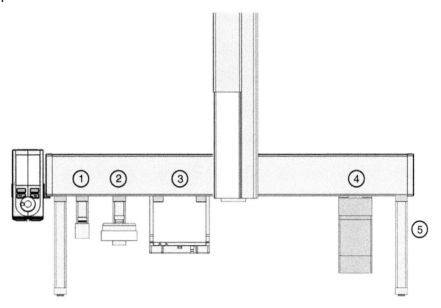

Figure 6.2 System configuration for geometric dilutions (benchtop installation). 1 Standard wash station, with position for stock solution; 2 Solvent reservoirs; 3 Trayholder for 2 mL vial racks; 4 Vortex mixer; and 5 Benchtop mounting legs.

recommended. In case a septum penetration needs to be avoided, the use of two decapper units is recommended (not shown here, refer to Section 6.1.3):

System – Tool – Module	Task
PAL RSI System	x,y,z-Robotic system with manual tool change for sample preparation. The robot x-rail length can be short, if used as a standalone solution
1 pc Park Station	Park station optional for an automated tool change system only
1 pc D8/57 Liquid Tool	Tool for a 1000 µL sample preparation syringe
1 pc Tray Holder	Tray holder for 3× vial racks
3 pc Rack 54× 2 mL	54 pos. for 2 mL vials
1 pc Solvent Module	For 3× 100 mL dilution solvent reservoirs, or use for separate syringe wash solvents
1 pc Standard Wash Module	Wash module for up to 5× syringe wash solvents
2 pc Adapter	Standard Wash Station Adapter to place 2 mL vials instead of the 10 mL wash vial
1 pc Vortex Mixer	For vial mixing
1 pc Mounting kit	Installs the x,y,z-robot instrument top, or benchtop

Workflow

Initial Manual Steps

- The standard stock solution is placed into a dedicated slot of the standard wash station. 10 mL vials, or 2 mL vial sizes with adapter, can be used. If necessary, the standard ampule is transferred to a 10 mL respectively 2 mL vial and capped.
- Screw-capped vials with magnetic caps according to the planned dilution levels are placed into the vial rack.
- Sufficient solvent(s) for dilution and syringe cleaning are provided with the solvent reservoirs.

Automated Workflow

- The automated dilution sequence starts with the cleaning of the syringe using the dedicated wash solvent, which is usually identical to the dilution solvent, but taken from a separate reservoir of the solvent station.
- After each standard and solvent dispensing, the vial is moved by magnetic transport to the Vortex mixer for short mixing and then returned for the next dilution step in the sequence. After each dilution step, the syringe is washed several times with solvent.
- The resulting dilution of a six-magnitude sequence with a constant dilution factor 1:10 of an aqueous blue dye solution is shown in Figure 6.3. In each dilution step, 100 µL standard of the previous dilution (initially of the stock solution) is diluted with 900 µL solvent each into 2 mL vials.

6.1.2 Dilution for Calibration Curves

Application

In contrast to the geometric dilution described in Section 6.1.1, the dilution for quantitative calibrations requires several data points in the range of the expected sample concentrations, the working range of the analytical method, often down to the limit of quantification (LOQ) in trace analysis [1, 2]. One or more suitable dilution levels from a geometric dilution are used as intermediate solutions to generate the calibration curve (Figure 6.4). The described workflow for calibration dilution is mainly used for the commercially available multi-compound standards. Weighing the data points in unequal distance with accumulation around the required decision level (e.g. the legislative maximum residue level (MRL) limits) and the incorporation of a blank value are typical measures in gas chromatography-mass spectrometry

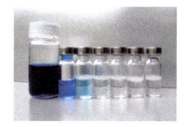

Figure 6.3 Automated geometrical dilution result (using a blue dye).

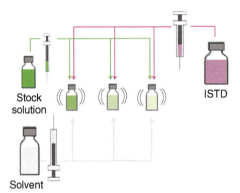

Figure 6.4 Automated calibration dilution scheme with individual syringes for stock solution, ISTD, and solvent.

(GC-MS) and liquid chromatography-mass spectrometry (LC-MS) quantification for improved precision in residue analysis. Wide quantitative calibrations use equidistant calibration points to minimize the impact on the regression curve from a potential offsets of data points at the high concentration levels. In general, the difference between calibration levels should not be greater than a factor of 10 [3].

Scope and Principle of Operation

The setup of a quantitative calibration is typically different for different methods but constant as long as the analytical method is in use. The automation of such dilution workflows is straightforward, but requires for method flexibility a procedure allowing different concentration values and dilution factors. Alternatively, fixed concentration levels can be "hard coded" in the programmed workflow for ease of use and reduced error, with less operational variability. Another aspect is the introduction of one or more ISTD for a relative response calibration.

For the setup of the calibration dilution scheme, it is advisable to prepare a calculation spreadsheet as illustrated for instance in Table 6.1 for a range from 0.1 to 100 pg/µL, typically applied for persistent organic pollutants (POPs) or pesticides trace analysis. For optimum precision of the dilution, it is essential that the employed syringes are used only in the range of 10–100% of the nominal volume (refer to Section 4.1.2.1). Multiple dispensing from syringes should be avoided, but can technically be incorporated into workflows for the reduction of time-consuming syringe change procedures. The calculation also informs about the minimum required solvent volumes for dilution and syringe wash. An optional ISTD, AP, or keeper solvent need to be considered in the dilution ratio with a reduced solvent volume according to the ISTD volume added. Individual syringes are used for stock/standard solutions, ISTD and solvents to prevent any potential carryover. Sufficient syringe wash steps need to be considered. Wash cycles and volumes need to be individually set as required by the particular application. For dispensing larger solvent volumes, the use of a dilutor tool is recommended.

The standard addition method, often applied for but not limited to quantitative headspace (HS) analysis, follows the same concept with the sample material, or a representative matrix. Multiple additions of the standard concentrations in proper spacing are performed with analysis for the same sample. It needs to be considered that the method requires linearity throughout the range for a valid extrapolation and, the calibration must go through zero. The standard addition method is

Table 6.1 Calculation scheme for a calibration dilution: Dn dilution levels from stock solutions. Cn planned calibration dilutions.

Dilution/calibration level	Conc.	Units	Aspirate volume ... [µL]	from dilution level	ISTD volume [µL]	Added solvent volume [µL]	Total volume in vial [µL]	Dilution ratio	Wash to waste [µL]	Total solvent volume [µL]
D0	100	ng/µL	./.	./.	./.	./.	1000	./.	1000	1000
D1	10	ng/µL	100	D0	./.	900	1000	1/10	1000	1900
D2	1	ng/µL	100	D1	./.	900	1000	1/10	1000	1900
C100	100	pg/µL	100	D2	100	800	1000	1/10	1000	1800
C50	50	pg/µL	200	C100	100	100	400	1/2	1000	1100
C25	25	pg/µL	100	C100	100	200	400	1/4	1000	1200
C10	10	pg/µL	100	C100	100	800	1000	1/10	1000	1800
C5	5.0	pg/µL	200	C10	100	100	400	1/2	1000	1100
C2	2.5	pg/µL	100	C10	100	200	400	1/4	1000	1200
C1	1.0	pg/µL	100	C10	100	800	1000	1/10	1000	1800
C05	0.50	pg/µL	200	C1	100	100	400	1/2	1000	1100
C02	0.25	pg/µL	100	C1	100	200	400	1/4	1000	1200
C01	0.10	pg/µL	100	C1	100	800	1000	1/10	1000	1800
Total solvent volume, single calibration run					1000	5900			13000	18900

recommended for confirmatory quantitative analyses in cases of MRL exceedances. An alternative practical approach to compensate for matrix effects in GC analyses is the use of APs as a standard matrix for calibrations as well [4, 5].

Solvents
- Solvents to be used must be compatible with the intended GC or LC application.
- Suitable wash solvents need to be provided in sufficient volume, see as example Table 6.1.

Consumables
- 10 mL amber capped vials for stock solutions.
- 2 mL amber vials with magnetic screw cap closure.
- Liquid syringe, 1000 µL, for liquid handling, needle gauge 22 or 23, point style 3/LC.
- Liquid syringe, 100 µL, for liquid handling, needle gauge 22 or 23, point style 3/LC.
- Liquid syringe, 25 µL, for liquid handling, needle gauge 22 or 23, point style 3/LC, or
- Liquid syringe, 10 µL, for liquid handling, needle gauge 22 or 23, point style AS/conical.

System Configuration
The minimum hardware requirements for a robotic dilution system, shown in Figure 6.5, are:

System – Tool – Module	Task
PAL RTC System	x,y,z-Robotic system with automated tool change for sample preparation and liquid injection, 850-mm rail length
1 pc Park Station	Park station for tools not in use during workflow
2 pc D7/57 Liquid Tool	Tools for the 100 µL and a 10 µL GC injection syringe (optional for liquid injections)
1 pc D8/57 Liquid Tool	Tool for a 1000 µL sample preparation syringe
1 pc Dilutor	Optional instead of a large volume syringe, for solvent dispensing from reservoirs
1 pc Tray Holder	Tray holder for 3× vial racks
3 pc Rack 15× 10/20 mL	15 pos. for 10/20 mL vials (optional in case of HS standard addition applications)
3 pc Rack 54× 2 mL	54 pos. for 2 mL vials
1 pc Rack 60× 10/20 mL vial	60 pos. for 10/20 mL vials
1 pc Solvent Module	For 3× 100 mL solvent reservoirs
1 pc Standard Wash Module	Wash module for up to 5× syringe wash solvents
2 pc Adapter	Standard Wash Station Adapter to place 2 mL vials instead of the 10 mL wash vial
1 pc Fast Wash Module	Wash module for 2× active wash solvents, recommended for large syringe wash steps
1 pc Vortex Mixer	For intense vial mixing
1 pc Mounting kit	Installs the x,y,z-robot instrument top, or benchtop

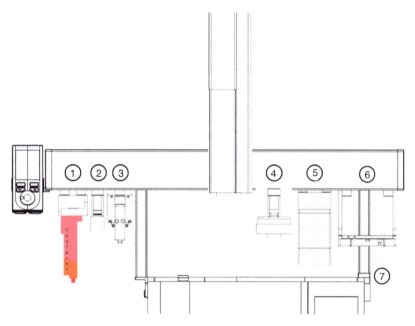

Figure 6.5 Minimum configuration for the calibration dilution workflow. 1 Tool park station; 2 Standard wash station, positions for stock solution, ISTD vials; 3 Fast wash station for two solvents and waste; 4 Large solvent station (optional dilutor not shown); 5 Vortex mixer; 6 Sample vial tray; and 7 Instrument top mounting legs.

Workflow

The described workflow is kept flexible for the required number and concentration of the dilution levels. It is used for the automated dilution of standards up to nine levels from the stock solution. The amount of stock solution and diluent is calculated automatically from the user-defined dilution ratio. The program automatically selects the suitable tool based on the amount to be transferred. For volume calculations, the formulas (1) and (2) are applied:

(1) $Stock\ Solution\ Volume = \dfrac{Total\ volume}{Dilution\ ratio}$

(2) $Solvent\ Volume = Total\ Volume - Internal\ Standard\ Volume - Stock\ Solution\ Volume$

If the required volume is more than $100\,\mu L$, a $1000\,\mu L$ syringe is selected for the liquid transfer. If necessary, the same tool can perform multiple solvent transfers if the required volume exceeds the tool capacity. Multiple transfers are then performed until the required volume is obtained, though not recommended due to error propagation. The addition of ISTD is considered in the volume calculation within the workflow program.

The workflow steps are graphically illustrated in Figure 6.6. The input of the desired dilution ratios and additional workflow parameters are requested at the start of the program. Variables are the final volume, number of calibration levels, ISTD use, and ISTD volume. The total volume is defined by the use of 10 mL vials. Default, minimum, and maximum parameter settings are taken care of with appropriate entry checks. The outlined workflow is intended for benchtop stand-alone operation, but can be extended to online GC and LC injection.

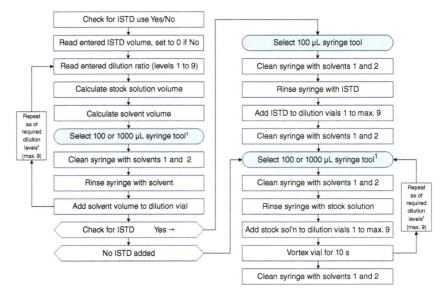

Figure 6.6 Workflow steps for the dilution of standards for calibration curves. [1]For volumes >100 µL, a 1000 µL syringe is recommended; [2]Number of calibration levels.

Figure 6.7 Automated quantitative calibration dilution result (using a green dye).

Measurements

An equidistant dilution example using a green dye is shown in Figure 6.7 illustrating the stepwise reduction of the colorant intensity. Figure 6.8 shows a weighted calibration curve for GC-MS trace analysis with the dilution of 1,2,3-trimethylbenzene (1,2,3-TMB, CAS No. 526-73-8) in the range of 0.08–80 ng/µL. The curve fit with the determination coefficient R^2 0.9998 demonstrates the excellent precision of the automated preparation for the calibration dilution.

6.1.3 Preparation of Working Standards

Application

Significant additional workload for a laboratory arises from the typical situation that standard compounds are available individually and need to be combined in one multi-compound working standard for quantification. The manual preparation and standard administration require additional manpower and are not free from human error and variability. Also, the volume only control is prone to error. Here,

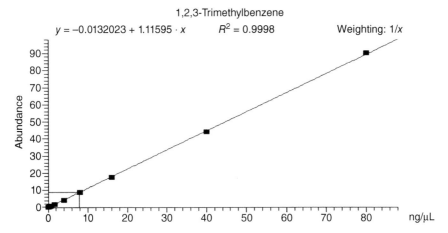

Figure 6.8 Quantitative calibration after automated dilution in the range of 0.08–80 ng/μL 1,2,3-trimethylbenzene.

the individual concentrations, necessary dilution steps, and resulting factors need to be considered, making such a dilution scheme a complex but regularly reoccurring task. Often a dedicated person in the laboratory is entrusted to prepare working standards for different methods and occupied full time.

For the preparation, use, and storage of pesticides stock solutions and working standards, the European Commission demands a record that must be kept with the identity and amount of all reference standards and solvents employed. Logging the reference standards with concentration, amount, and expiration date together with the procedural steps, dosing and weighing results in a database allow traceability, for instance, and in particular with the transfer of such data to a laboratory management and information system. "The date of preparation, the identity and mass (or volume, for highly volatile analytes) of the reference standard and the identity and volume of the solvent (or other diluents) must be recorded" [3].

Scope and Principle of Operation
Methods have been developed taking the task of working standard preparation to the automated level [6]. Solutions include the storage of reference standards in cooled trays, decapping/recapping vials to prevent septum puncture, barcode reading for standard identification, use of pipetting instead of syringes to avoid cross contaminations, and the weighing of reference standard vials as well as the target vials for gravimetric control. This comprehensive approach complies with the requirements of the European Commission SANTE for a laboratory quality control and management system [3].

Solvents
Solvents to be used must be compatible with the intended GC or LC application.

Consumables

- 10 mL screw-capped amber vials for stock solutions.
- 2 mL amber vials with magnetic screw cap closure.
- Pipette tips, 200 µL and 1000 µL for liquid handling.

System Configuration

The minimum hardware requirements for an x,y,z-robotic system for combining standards and dilution as shown in Figure 6.9 are:

System – Tool – Module	Task
PAL DHR RSI/RTC System	x,y,z-Robotic system with dual head and automated tool change for sample preparation, 1600 mm rail length
1 pc Park Station	Park station for tools not in use during workflow for the right head
1 pc Dilutor and Tool	Dispensing of solvent(s) as programmed from reservoirs to target vials
2 pc Pipetting Tool	For liquid handling with 1000 µL and 200 µL pipette tips
2 pc Tray Holder	Tray holder for 3× vial racks
2 pc Cooled Stack	Cooled tray for 3 × 30 10 mL vials (2 mL vials optional)
12 pc Rack 15× 10/20 mL	15 pos. racks for 10 mL vials in cooled storage
3 pc Rack 54× 2 mL	54 pos. racks for 2 mL target vials
3 pc Rack 15× 10/20 mL	15 pos. racks for optional 10/20 mL target vials
1 pc Vortex Mixer	For intense vial mixing
1 pc Barcode Reader	Scanning barcode reader for vial identification
1 pc Balance	Analytical balance, top feeding, serial interface, benchtop mounted
1 pc Mounting kit	Installs the x,y,z-robot benchtop

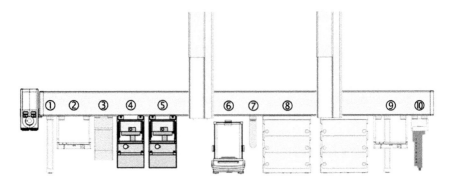

Figure 6.9 Dual-head x,y,z-robotic system for analytical multi-compound standards preparation. 1 Benchtop mounting legs; 2 Trayholder for target vial racks (2/10/20 mL); 3 Vortex mixer; 4 Decapper 1; 5 Decapper 2; 6 Analytical balance (benchtop-mounted); 7 Barcode reader; 8 Cooled storage for standards (2×); 9 Trayholder for pipette tips; 10 Tool park station; and 11 Dilutor with tool (backside-mounted, not shown).

Workflow

Initial Manual Steps

- Analytical standards are placed in barcode-labeled vials of up to 10 mL size into the cooled stack. The vials must be equipped with magnetic screw caps for transport.
- Barcode-labeled empty vials with magnetic caps are placed into the trayholder racks as required.
- Initially, the weight of the reference/ISTD vials is determined and logged on to the database with additional information from the barcode, as compound ID, expiry date, etc.
- The required reference standard vial positions, volumes of standards and solvent, and target vial positions are entered via the user interface of the program.

Automated Workflow

- The standard vial is moved by the right robot head from the cold stack through the barcode reader to the analytical balance. The barcode and weight are recorded. Then moved on to the decapper 2 and decapped.
- An empty target vial is moved by the left head from the trayholder through the barcode reader to decapper 1, decapped, and weight empty. Barcode and weight are recorded. The target vial stays in the balance.
- The right head uses a pipetting tool and transfers the programmed volume from the standard in the decapper to the target vial in the balance. The weight is recorded, and the pipette tip disposed.
- The standard vial gets capped and returned to the cooled stack.
- The target vial is returned from the balance and capped.
- A next standard vial is moved by the right head from the cold stack through the barcode reader to the analytical balance. The barcode and weight are recorded. Then moved on to the decapper 2 and decapped.
- The process is repeated for the required number of standards.
- The target vial after final capping is moved to the vortexer for mixing, then back to the rack position in the trayholder.
- A next target vial is processed using the programmed parameters.

6.2 Derivatization

Derivatization reactions are common sample preparation steps in many chromatographic methods. GC and GC-MS are still the methods of choice for the quantification even of polar compounds, which require derivatization before GC injection. Typical examples are the quantification of anabolic steroids or drugs of abuse, in particular for forensic applications. The alternative LC-MS methods suffer from the often reported analyte "ion suppression" in matrix samples. Also in LC analysis, derivatization reactions are common for the introduction of chromophores, fluorophores, or increase of the molecular weight for improved separation and detectivity in LC-MS. Usually, there are several manual steps involved in handling hazardous reactive agents. Robotic autosamplers installed online with the analytical instrumentation offer workflows for in-time sample derivatization before injection to GC or LC.

6.2.1 Silylation

Application

The silylation of polar analytes is probably the most used derivatization technique for GC analyses. As "silylation" we understand the substitution of an acidic hydrogen atom (–OH, =NH, –SH) by a trimethylsilyl (TMS) group $(CH_3)_3Si$–. The general reaction for the formation of the TMS derivatives is:

$$(CH_3)_3Si–X + R–H \rightarrow (CH_3)_3Si–R + HX$$

Scope and Principle of Operation

N,O-Bis(trimethylsilyl)trifluoro-acetamide (BSTFA) and N-methyl-N-trimethylsilyl-trifluoroacetamide (MSTFA) became the most commonly used silylating agents for GC applications. Within the numerous silylation reagents, MSTFA is the most important as the reagent itself and its by-product, N-methyltrifluoroacetamide, are even more volatile than BSTFA and its by-products. It was Manfred Donike of the Institute for Biochemistry of the Cologne University (today "German Sport University Cologne") who paved the way for the application of MSTFA as a strong silylating agent [7]. It can be used after evaporation of extracts to dryness without solvent as it can dissolve even highly polar substances. The silylation power of MSTFA can be further increased in the presence of catalysts like TMCS or, as introduced by Manfred Donike as well, ammonium iodide (NH_4I). MSTFA reacts with NH_4I to form trimethyliodosilane (TMSI), which has been reported to be the most powerful TMS donor available [8]. Ethanethiol is added here to reduce the generated iodine to hydrogen iodide under the formation of diethyl disulfide. The MSTFA derivatization became the standard for quantitative anabolic steroids analysis [9, 10], and many other GC analyses for instance the metabolomics workflow (see also Section 6.2.3), for biodiesel analysis [11], or drugs [12], just to mention a few of the most important routine GC and GC-MS applications.

General procedure for silylation with BSTFA or MSTFA [13]:

- Use up to 10 mg sample in 2 mL vials.
- Add 100–500 µL BSTFA or MSTFA with 1% TMCS, resp. MSTFA with NH_4I, ethanethiol (1000:2:3).
- Add 1.0 mL solvent (pyridine; acetonitrile is recommended for amino acids; not required for MSTFA).
- Shake for 30 seconds to dissolve the analytes.
- Heat for 15 minutes at 60–70 °C to facilitate dissolution and silylation (or one hour reaction time at room temperature.
- Inject to GC.

Solvents and Chemicals
- BSTFA, ≥99%, CAS No. 25561-30-2, usually with 1% TMCS.
- MSTFA, ≥98.5%, CAS No. 24589-78-4, for GC derivatization, also available activated with NH_4I.
- NH_4I, ≥99.5%, CAS No. 12027-06-4.
- Ethanethiol, CAS No. 75-08-1.
- Pyridine, ≥99.9%, CAS No. 110-86-1.
- Acetonitrile, for HPLC, CAS No. 75-05-8.

Consumables
- 2 mL screw-top clear vials with magnetic screw caps.
- Liquid syringe, 1000 µL, for reagent and solvent handling, needle gauge 22 or 23, point style 3/LC.
- Liquid syringe 10 µL, for GC injection, needle gauge 23s, point style AS/conical.

System Configuration
A recommended system configuration for a robotic system equipped for derivatization reactions is shown in Figure 6.10. The minimum hardware requirements are:

System – Tool – Module	Task
PAL RTC System	x,y,z-Robotic system with automated tool change for derivatization and GC injection, 850 mm rail length
1 pc Park Station	Park station for tools not in use during the workflow
1 pc D7/57 Liquid Tool	Tool for a 10 µL GC injection syringe (optional for liquid injections)
1 pc D8/57 Liquid Tool	Tool for 1000 µL sample preparation syringe
1 pc Headspace Tool	For solvent evaporation with pre-heated nitrogen stream
1 pc MHE Module	Double-needle for solvent evaporation
1 pc Tray Holder	Tray holder for 3× vial racks
3 pc Rack 54× 2 mL	54 pos. racks for 2 mL vials
1 pc Solvent Module	For 3× 100 mL solvent/reagent reservoirs
1 pc Standard Wash Module	Wash module for up to 5× syringe wash solvents and reagents
2 pc Adapter	Standard Wash Station Adapter to place 2 mL vials instead of the 10 mL wash vial
1 pc Vortex Mixer	For intense vial mixing
1 pc Agitator/Incubator	Sample Incubation of up to 6× 20 mL vials
1 set Vial inserts 2 mL	Inserts for the use of 2 mL vials in the Agitator
1 pc Mounting kit	Installs the x,y,z-robot GC instrument top
1 pc GC system	With split/splitless, or PTV type injector, with detector as of application
1 pc MS system	Single or triple quadrupole analyzer, electron ionization (EI) ion source, as of the intended application

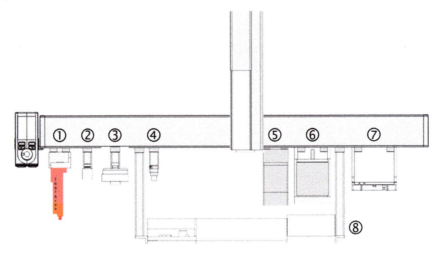

Figure 6.10 Recommended system configuration for a derivatization workflow with GC injection. 1 Tool park station; 2 Standard wash station for solvents and reagents; 3 Solvent reservoirs; 4 Double-needle device; 5 Vortex module; 6 Agitator/incubator; 7 Trayholder for sample racks; and 8 GC top mounting legs.

Workflow
Initial Manual Steps

- Place the capped sample vials after evaporation into the vial racks.
- Provide the silylation agent in 2 or 10 mL vials into an unused position of the "standard wash station".
- If solvents are required, they a provided with the solvent reservoirs.

Automated Workflow

- The workflow steps are graphically illustrated in Figure 6.11.
- Vials with the sample extract are placed into the racks of the robotic sampler for evaporation, derivatization, and online analysis.

Conclusion
The automated derivatization step can be employed in a variety of sample preparation procedures before GC injection. A large number of silylating agents are commercially available addressing specific derivatization requirements. The outlined automated workflow can be adjusted with volumes, solvents, reaction temperatures, and times to fit the optimum conditions for the particular silylation agents and analytes.

6.2.2 SPME On-Fiber Derivatization

Application
The solid phase micro-extraction (SPME) of polar compounds may require derivatization prior to GC analysis. The sample extraction, derivatization, and desorption in the GC injector can be integrated into one automated workflow.

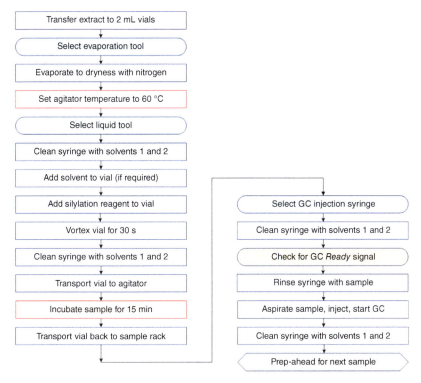

Figure 6.11 Workflow steps for the MSTFA derivatization with GC injection.

Scope and Principle of Operation

The described workflow that can be used for a variety of derivatization agents within those for silylation is the most common, see Section 4.4.5.4. For the on-fiber derivatization, the SPME sorption phase is exposed into the headspace of the derivatization agent. As a derivatization strategy, the exposure to the reagent can take place before extraction, after extraction, or both, before and after extraction. Choosing the derivatization strategy for a particular assay, it should be considered that silyl derivatives are sensitive to hydrolysis limiting the efficiency of a pre-extraction derivatization from aqueous samples. And, polydimethylsiloxane (PDMS) as a non-polar absorbent material and highly viscous liquid dissolves TMS-derivatized analytes well.

The derivatization agent is placed into a dedicated 10 or 20 mL vial, which is placed into a reserved position of the agitator/incubator for the derivatization reaction. If different temperatures for extraction and derivatization are required, the use of two incubator modules with the required temperature setting is recommended.

Solvents and Chemicals
- Derivatization agent of choice.
- Polar and non-polar wash solvents.

Consumables

- 20 mL headspace screw-top clear vials with magnetic screw caps.
- 10 mL headspace screw-top clear vials with magnetic screw caps.
- SPME Arrow or SPME Fiber with sorbent phase of choice.
- GC inlet liner, 1.3 or 1.7 mm ID (fitting to the chosen SPME Arrow OD).
- GC column: DB-5MS UI, 30 m × 0.25 mm × 0.25 µm, or equivalent.
- Syringe 10 µL, for GC injection, needle gauge 23s, point style AS/conical (optional for liquid injections of standards).

System Configuration

The minimum hardware requirements for an x,y,z-robotic system for SPME sampling and derivatization in Figure 6.12 are:

System – Tool – Module	Task
PAL RSI System	x,y,z-Robotic system with manual tool change for SPME operation, 850 mm rail length
1 pc Park Station	Park station, optional for an automated tool change system only
1 pc D7/57 Liquid Tool	Tool for a 10 µL GC injection syringe, optional for liquid injections of standards
1 pc Tray Holder	Tray holder for 3× vial racks
2 pc Rack 15× 10/20 mL	15 pos. for 10/20 mL vials
1 pc Rack 54× 2 mL	54 pos. for 2 mL vials
1 pc Rack 60× 10/20 mL vial	60 pos. for 10/20 mL vials
1 pc Standard Wash Module	Wash module for up to 5× syringe wash solvents
1 pc SPME Arrow tool	Tool for SPME Arrow operation (alternative)
1 pc SPME Fiber tool	Tool for SPME Fiber operation (alternative)
1 pc Heatex Stirrer	Heated module for stirring during SPME Arrow extraction
1 pc SPME conditioning module	Conditioning of SPME Fibers/Arrows before and after analysis
1 pc Agitator/Incubator	Sample incubation of up to 6× 20 mL vials
1 pc Mounting kit	Installs the x,y,z-robot GC instrument top
1 pc GC system	With split/splitless, or PTV type injector
1 pc SPME Arrow kit	Arrow inlet adaptor (required for Arrow operation)
1 pc MS system	Single or triple quadrupole analyzer

Workflow

The automated workflow follows the regular SPME extraction procedure with the additional handling of a derivatization reagent. The steps for on-fiber derivatization are graphically illustrated in Figure 6.13. A 10 mL vial with a small amount of the derivatization reagent is placed into a high index position of the sample rack. For the derivatization step, the reagent vial is transferred into a reserved position of the

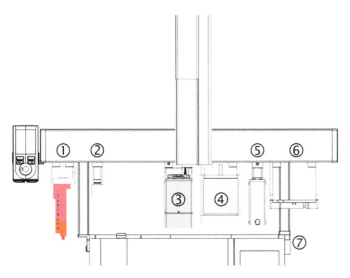

Figure 6.12 Robotic sampler configuration for SPME derivatization workflows. 1 Tool park station (optional for automated tool change); 2 Standard wash station; 3 Heatex stirrer (for optional SPME Arrow operation); 4 Agitator/Incubator; 5 SPME conditioning station; 6 Trayholder; and 7 GC top mounting legs.

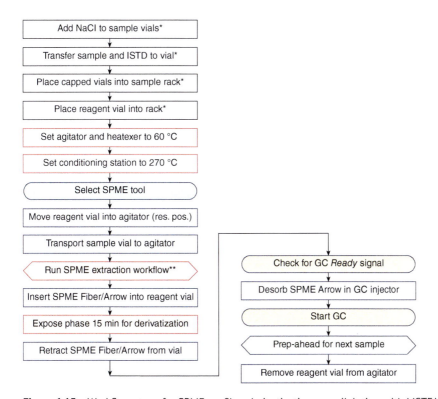

Figure 6.13 Workflow steps for SPME on-fiber derivatization, e.g. silylation with MSTFA (*initial manual steps, **see e.g. Figure 6.39).

agitator/incubator module. This position is reserved for the reagent vial only and not used for overlapping the sample vial incubation. It is recommended to use the same temperature setting for extraction and derivatization, e.g. 60 °C for a silylation with MSTFA. In such applications with derivatization reactions, which require a different temperature setting, a configuration using two incubator modules with the required temperature setting is recommended.

The on-fiber derivatization step can be integrated into the workflow before, after, or before and after the SPME extraction, depending on the intended application. For derivatization, the SPME device penetrates the heated reagent vial and exposes the sorbent phase into the reagent headspace of the vial, and stays exposed in the vial for the programmed derivatization time. After all samples are processed, the reagent vial is moved back to the initial rack position.

Sample Measurements
A fully automated procedure for the analysis of short-chain amines by SPME on-fiber derivatization with pentafluorobenzaldehyde (PFBAY) and GC-MS detection was developed by Emanuela Gionfriddo et al. using a PDMS/DVB SPME phase [14]. The results describe two derivatization approaches, prior- and post-extraction. The response of the prior load of the derivatization agent onto the SPME fiber, leading to a simultaneous extraction and derivatization, provided the higher response compared to the post-extraction derivatization. The PFBAY load time on the coating was 30 minutes in both batches. The SPME extractions were performed at 60 °C for 10 minutes and 500 rpm agitation using an x,y,z-robotic autosampler. The desorption in the GC injector was carried out for five minutes at 250 °C.

Conclusion
Using SPME extraction for the analysis of polar analytes by GC-MS is facilitated by the automated compound derivatization before GC desorption in a seamless workflow. A wide variety of derivatization reagents can be employed using on-fiber pre- or post-extraction derivatization schemes.

6.2.3 Metabolite Profiling by Methoximation and Silylation

Application
GC-MS analysis is extremely powerful for metabolite profiling due to the high chromatographic separation power of hundreds of compounds in one run, and the unique capability of their identification by library search. The analysis of the mostly polar compounds by GC-MS requires a two-step derivatization procedure (*aka* MeOx-TMS derivatization) to reflect the in-vivo state without any artefact formation [15].

Scope and Principle of Operation
The first methoximation reaction (MeOx) protects carbonyl functions. Pyridine as a solvent serves as a catalyst in the reaction. The following silylation step delivers the TMS derivatives of the polar compounds [16].

The MeOx reaction incubates the sample with the methoxyamine solution in pyridine for 60–90 minutes at 28 °C. MSTFA or BSTFA with 1% TMCS as a catalyst for silylation is then added to the mixture and incubated at 37 °C for an additional 30 minutes. After completion of the reaction, the vial is cooled down to RT in the sample rack before injecting into GC-MS for analysis. The complete sequence of liquid handling steps, derivatization, and incubation at different temperatures, finally the GC-MS injection, is carried out in one comprehensive workflow using the robotic sampling system.

Solvents and Chemicals
- Methoxyamine solution ($CH_3ONH_2 \cdot HCl$), CAS No. 593-56-6, 20 mg/mL in pyridine.
- MSTFA, CAS No. 24589-78-4, for GC derivatization.

Consumables
- 2 mL screw-top clear vials with magnetic screw caps.
- Liquid syringe 10 µL, for GC injection, needle gauge 23s, point style AS/conical, used in the workflow for derivatization and GC injection.
- Liquid syringe 100 µL, needle gauge 23s, point style AS/conical, can optionally be used for the sample derivatizations steps.
- GC column: DB-5MS, 30 m × 0.25 mm × 0.25 µm, or equivalent.

System Configuration
The standard wash station in this configuration is used to place the methoximation and silylation reagents. Adapters are used to place 2 mL vials here to replace the standard 10 mL vials.

The minimum hardware requirements of the robotic system for the two-step MeOx-TMS derivatization as shown in Figure 6.14 are:

System – Tool – Module	Task
PAL RTC System	x,y,z-Robotic system with automated tool change for sample preparation and liquid injection, 850 mm rail length (automated tool exchange optional)
1 pc Park Station	Park station for tools not in use during workflow
1 pc D7/57 Liquid Tool	Tool for a 10 µL GC injection syringe
1 pc Tray Holder	Tray holder for 3× vial racks
3 pc Rack 54× 2 mL	54 pos. for 2 mL vials
1 pc Standard Wash Module	Wash module for up to 5× syringe wash solvents
2 pc Adapter	Standard Wash Station adapter to place 2 mL vials instead of the 10 mL wash vial
1 pc Fast Wash Module	Wash module for 2× active wash solvents
2 pc Agitator/Incubator	Sample incubation of up to 6× 20 mL vials (second incubator necessary for a second incubation temperature only)
1 set Vial inserts 2 mL	Inserts for the use of 2 mL vials in the Agitator
1 pc Mounting kit	Installs the x,y,z-robot GC instrument top
1 pc GC system	With split/splitless, or PTV type injector.
1 pc MS system	Single or triple quadrupole analyzer, EI ion source

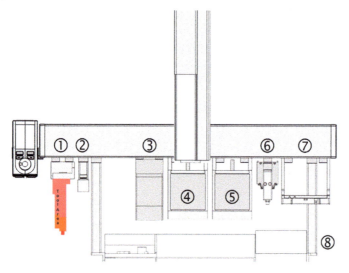

Figure 6.14 x,y,z-Robot configuration for the metabolomics two-step MeOx-TMS derivatization. (1) Tool park station (required for use of a second large volume syringe), (2) Standard wash and reagent station, (3) Vortex mixer, (4) Agitator/Incubator 1, (5) Agitator/Incubator 2, (6) Fast wash station, (7) Trayholder, (8) GC top mounting legs.

Analysis Parameter

The method parameters for GC-MS analysis are provided in Table 6.2. For compound identification, the GC-MS system is run in full scan mode generating mass spectra for library search. For individual compound quantification, a triple quadrupole system is used in targeted multiple reaction monitoring (MRM) mode. High resolution and accurate mass (HR/AM) mass spectrometer facilitate the compound identification and quantification.

Table 6.2 Selected analysis parameters for metabolite analysis.

GC	
Inlet temperature	250 °C
Inlet mode	Splitless, 2 min (or with split 10 : 1)
Injection volume	1.0 µL
Carrier gas	He, constant flow at 1.5 mL/min
Oven program	60 °C (1 min), 10 °C/min to 325 °C (10 min).
Transfer line temperature	340 °C
MS	
System, Ionization	Single/triple quadrupole, EI mode
Ion source temperature	250 °C
Scan mode	Full scan
Scan m/z	m/z 40–650
Scan time	0.25 s

Workflow

The standard workflow uses one syringe of 10 µL volume only for reagent additions and GC injection. Optionally, a second larger volume syringe can be employed for the sample preparation. In this setup a tool park station is required in the configuration, as shown in Figure 6.14. Accordingly, a syringe exchange and additional cleaning steps are programmed in the workflow before and after the GC injection.

As a general procedure, 10 µL of methoxyamine solution is added to the dried sample extract in 2 mL vials. The vials are transferred to the agitator/incubator for 90 minutes reaction time at 28 °C. Then, 20 µL silylating agent is added to the vial while in the agitator, allowing for an additional 30 minutes reaction time at 37 °C. In case the derivatization steps are required to be executed at the temperatures of 28 and 37 °C, the use of two agitator units set to the desired temperatures is required. Setting different temperatures at one agitator is not recommended due to time-consuming cool-down times between the samples. After placing the vial back to the trayholder, the samples are injected to GC-MS. The programmed workflow operates in a prep-ahead and the full overlapping mode of the two derivatzation steps to ensure equal derivatization conditions for all samples, and a seamless duty cycle of the GC-MS system (Figure 6.15).

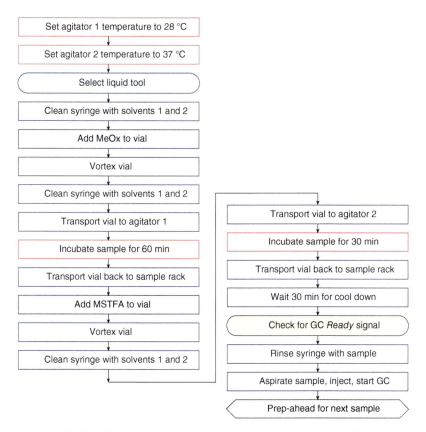

Figure 6.15 Workflow steps for the automated two-step MeOx and TMS derivatization procedure. The incubation times are handled in overlapping mode for both derivatization reactions.

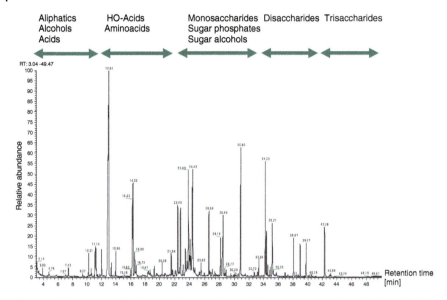

Figure 6.16 A metabolomics profile after MeOx-TMS derivatization. The eluting chemical compound classes are assigned to the corresponding retention regions (after, Source: Redrawn from Fragner et al. [17]).

Sample Measurements

The chromatogram in Figure 6.16 shows a full scan mode analysis of a complex sample with assignments of the elution regions of typical metabolite groups. The elution of compounds starts with small molecules and ranges up to high molecular saccharide structures. Important to note is the high dynamic range required to cover both, the major components as well as low abundance metabolites.

For the overlapping of a large series of analyses, the scheduler of the robot automatically synchronizes cycle times of sample preparation and GC runtime. The robot

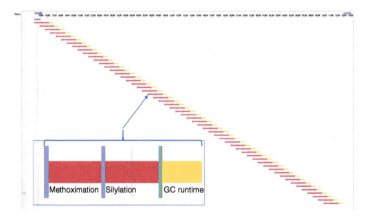

Figure 6.17 Overlapped MeOx-TMS workflow for high sample throughput. Source: Courtesy CTC Analytics AG.

starts preparation of the next samples in time to inject right after a *Ready* signal is received from the GC-MS system. In 24 hours approx. 26 sample preparations and analyses can be completed as it is shown in the course of the overlapped MeOx-TMS workflow in Figure 6.17.

6.3 Taste and Odor Compounds Trace Analysis

Application

Several organic compounds contribute to the taste and odor (T&O) of water and beverages even at low ng/L concentrations. Well-known examples are geosmin (GSM, Figure 6.37) and 2-methylisoborneol (MIB, Figure 6.38) as off-odor contaminants in drinking water, see also Section 6.7. Several additional compounds add to the so-called "off-odor" of drinking water. The group of methoxypyrazines creates an undesirable note at very low detection levels of only 1 ng/L, also known for instance from wine with 3-isobutyl-2-methoxypyrazine (IBMP, Figure 6.18) and 3-isopropyl-2-methoxypyrazine (IPMP, Figure 6.19). The haloanisole compounds 2,4,6-trichloroanisole (TCA, Figure 6.20) and 2,4,6-tribromoanisole (TBA, Figure 6.21), causing a musty or moldy taint, can be present in many products, known from wines, or found even in coffee. TCA and TBA contribute to the known "cork taint" in wine. They are also responsible for the sometimes unpleasant musty and moldy smell of packaging materials [18].

Figure 6.18 3-Isobutyl-2-methoxypyrazine (IBMP).

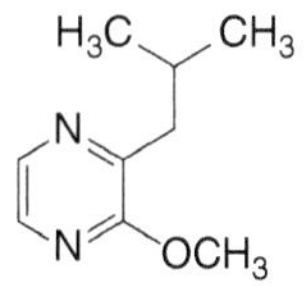

Figure 6.19 3-Isopropyl-2-methoxypyrazine (IPMP).

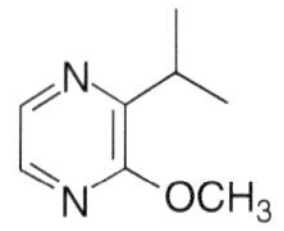

Figure 6.20 2,4,6-Trichloroanisole (TCA).

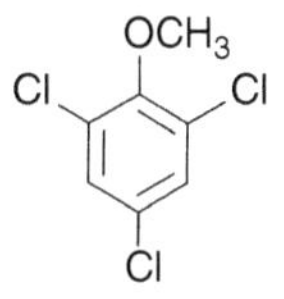

Figure 6.21 2,4,6-Tribromoanisole (TBA).

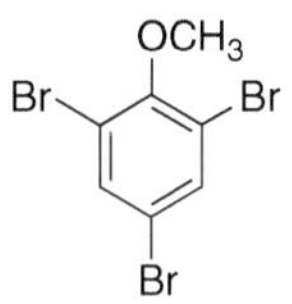

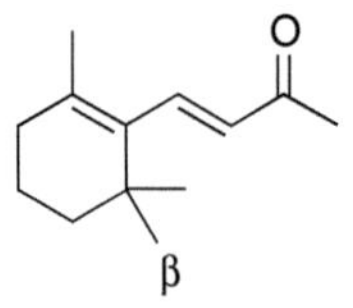

Figure 6.22 β-Ionone (BIN).

Odor thresholds in wine go down 2 ng/L. Even TCA levels below can still impact the wine appearance with "muted" aromas and flavors. TBA has an even lower odor threshold down to 0.08–0.3 ng/L in water and 2–6 ng/L in wine. While TBA and TCA are responsible for off-odors, the here included β-ionone (BIN, Figure 6.22) contributes with a raspberry scent and odor threshold of 90 ng/L positively to the wine aroma [19] and is used as a parameter in authenticity control. Elevated concentrations of α-ionone and BIN can serve as indicators for a potential adulteration [20].

Scope and Principle of Operation

SPME is established as the standard sample preparation method for analysis of the T&O compounds due to the ease of automation, the solvent-free green character, and an online direct desorption into GC or GC-MS. It provides the potential for high extraction sensitivity in the headspace and the immersion mode likewise required for low-level T&O analysis.

Also, the trace analysis of the various halophenols as by-products from disinfection processes can be included in this SPME method. In this case, the prior in-situ derivatization with acetic acid is required [21].

The described method uses the SPME Arrow technique with optimized conditions for the choice of the sorbent phase, salting-out effect, incubation temperature, and extraction time. The SPME headspace extraction of T&O compounds is demonstrated for water samples, to be applied for different beverages analogously. Carbon wide range-polydimethylsiloxane (CWR-PDMS), divinylbenzene-carbon wide range (DVB-CWR), and divinylbenzene-polydimethylsiloxane (DVB-PDMS) were used in this workflow to evaluate the sensitivity for the individual T&O compounds of interest [22].

Solvents and Chemicals

- Standard solutions of GSM, MIB, IPMP, IBMP, TCA, TBA, and BIN.
- Methanol, for HPLC, ≥99.9%, CAS No. 67-56-1, for preparing standard solutions.
- Water, HPLC quality, CAS No. 7732-18-5.
- NaCl, puriss. p.a., ≥99.5%, CAS No. 7647-14-5.

Consumables

- GC inlet liner: 2 mm ID × 5 mm × 95 mm, splitless liner.
- GC column: Rxi-624 sil 60 m × 0.25 mm × 1.4 μm, or equivalent.
- SPME Arrows, 1.1 mm OD:
  - CWR/PDMS, phase thickness 120 μm × 20 mm.
  - DVB/CWR/PDMS, phase thickness 120 μm × 20 mm.

– DVB/PDMS, phase thickness 120 µm × 20 mm.
- 20 mL headspace amber vials with magnetic screw caps.

System Configuration

The minimum robotic system hardware requirements for the trace analysis of T&O compounds as shown in Figure 6.23 are:

System – Tool – Module	Task
PAL RSI System	x,y,z-Robotic system for manual tool change, length of the rail 1200 mm
1 pc Tray Holder	Tray holder for 3× vial racks
3 pc Rack for 10/20 mL vials	15 pos. 20 mL vial racks
1 pc SPME Arrow tool	Tool for SPME Arrow operation
1 pc Heatex Stirrer	Heated module for stirring during SPME extraction
1 pc Arrow conditioning module	Conditioning of SPME Arrows before/after analysis
1 pc Agitator	Sample incubation for up to 6× 20 mL vials
1 pc Mounting kit	Installs the x,y,z-robot GC instrument top
1 pc GC system	With split/splitless, or PTV type injector
1 pc SPME Arrow kit	Arrow inlet adapter for GC inlet
1 pc MS system	Single quadrupole analyzer, EI ion source

Analysis Parameter

The analytical parameters for the headspace extraction, GC separation, and MS detection are listed in Table 6.3, and in Table 6.4 for the selected ion monitoring (SIM) detection of the analytes involved.

Workflow

Initial Manual Steps

- 20 mL screw cap vials with 3 g of sodium chloride added were placed into the sample tray.

Automated Workflow

- Before extraction, the sample is transferred by the robot to the agitator and incubated at 60 °C for two minutes.
- Then, the sample is transferred by magnetic transport into the Heatex stirrer for extraction.
- The SPME Arrow penetrates the sample vial septum, and its sorbent phase is extended into the headspace above the liquid sample, while maintaining the vial temperature at 60 °C for a 30 minutes extraction.

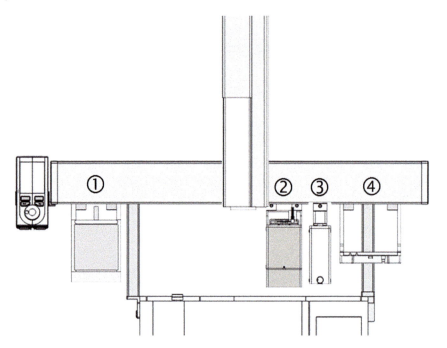

Figure 6.23 Recommended configuration for the SPME Arrow extraction workflow. 1 Agitator/Incubator; 2 Heatex stirrer; 3 Conditioning station; and 4 Sample tray.

Table 6.3 Analysis parameter for the T&O analysis by HS-SPME GC-MS.

Headspace extraction	
Sample volume	10 mL
Incubation temperature	60 °C
Incubation time	5 min
Agitator speed	1500 rpm
Extraction temperature, time	60 °C, 30 min
Conditioning temperature, time	250 °C, 8 min
GC	
Desorption temperature, time	250 °C, 1 min
Inlet mode	Splitless
Carrier gas	He, constant flow at 2 mL/min
Oven program	40 °C (1 min) → 25 °C/min → 240 °C → 10 °C/min → 280 °C (5 min)
Transfer line temperature	250 °C
MS	
System, Ionization	Single quadrupole, EI mode
Ion source temperature	250 °C
Scan mode	Full scan, SIM
Scan m/z	m/z 40–400 for retention times
SIM masses	See Table 6.4

Table 6.4 SIM masses used for T&O compound detection (in bold quantifier ions).

Compound	SIM masses [m/z]
IPMP	124, **137**, 152
IBMP	94, **124**, 151
MIB	**95**, 108
TCA	**195**, 197, 210
GSM	**112**, 125
BIN	135, **177**
TBA	331, **344**, 346

- For analysis, the SPME sorbent phase is retracted into the Arrow. The SPME Arrow depenetrates the vial and moves to the injection port and gets thermally desorbed.
- During the desorption time, the injector split is kept closed.
- The graphical representation of the automated workflow steps using an SPME Arrow is shown in Figure 6.39.

Quantitative Calibration

A series of standard dilutions with seven concentration levels in the range of 1–2000 ng/L were prepared and analyzed in triplicate. The calibration showed very good linearity, over a range of three orders of magnitude with a correlation factor $R^2 > 0.99$ for all compounds.

Recovery and Precision

Concerning the choice of sorbent materials, the DVB-CWR phase showed the lowest response. The highest response was achieved with the CWR-PDMS phase for IPMP, IBMP, and TCA. DVB-PDMS provided a higher response for MIB, GSM, BIN, and TBA. As the best performing compromise for all T&O compounds involved, the DVB-PDMS phase was selected.

Recoveries for regular tap water and spiked water samples were determined with 95–110% at concentrations of 10 and 20 ng/L. The precision of the method reported delivered RSD values <11.0% at these trace level concentrations. The limit of detection (LOD) values, based on a signal-to-noise ratio of three, were determined for all compounds in the range of 0.05 ng/L for TCA and 0.6 ng/L for TBA.

Sample Measurements

The salting-out effect showed a different effect for the individual analytes. The compounds IPMP, IBMP, MIB, GSM, and BIN, except of the both anisoles TCA and TBA, showed a significant improvement with an up to 5-fold increase in response. An extraction time of 30 minutes at 60 °C gave the maximum extraction efficiency. A shorter extraction time would even be practical for increased sample throughput.

Conclusion

Using the described automated workflow with a HS-SPME Arrow technique, T&O compounds can be quantified very sensitive in the low ng/L range below their odor threshold level. The DVB-PDMS phase was used as best performing for all T&O compounds.

6.4 Sulfur Compounds in Tropical Fruits

Tropical fruits become increasingly popular for consumers globally. Besides the different varieties, the degree of ripeness is most important for the aroma profile and finally the customer sensation, and price. The volatile sulfur compounds (VSCs) are often responsible for the juicy, fresh aroma of tropical fruits [23]. This poses a challenge for analytical chemists to identify these compounds as most often VSCs are found at low concentrations in most tropical fruits [24]. Many different headspace methods are proposed in the literature, of which the static HS allows the monitoring of the major compounds, and dynamic methods dive into the low levels of the characteristic aroma profiles [25, 26].

Durian, the "Queen of the Fruits" is the most popular fruit in south-east Asian countries available in different varieties. High demand, also due to significant export, leads to steadily increasing prices. Profiling of Durian varieties becomes an important analytical task, requesting fast and routine-proof analytical methods. The first analyses of VSCs in Durian were made from pulp distillates [27], later by vacuum distillation and extractions with dichloromethane (DCM) [28]. Also, GC-Olfactometry was used to characterize the VSCs [29].

The goal of the further described configuration is to provide an analytical solution for both, the analysis of the major components, and a practical way to analyze the characteristic low-level sulfur compound aroma profile in one automated workflow. The featured system setup allows the simultaneous extraction of samples by static and dynamic headspace techniques.

System Configuration

The minimum robotic system hardware requirements for the analysis of the high- and low-level VSCs as shown in Figure 6.24 are:

System – Tool – Module	Task
PAL RTC System	x,y,z-Robotic system with automated tool change, 850 mm rail length
1 pc Park Station	Park station for tools not in use during workflow
1 pc Tray Holder	Tray holder for 3× vial racks
3 pc Rack 15× 10/20 mL	15 pos. racks for 10/20 mL vials
1 pc Rack 60× 10/20 mL vial	60 pos. rack for 10/20 mL vials (optional)
1 pc Agitator	Sample incubation of up to 6× 20 mL vials
1 pc Headspace Tool	For a 2500 µL HS syringe operation
1 pc ITEX DHS Tool	For dynamic HS operation
1 pc ITEX DHS Syringe	With Tenax TA adsorbent
1 pc Heatex Stirrer	Optional for enhanced ITEX liquid extraction
1 pc Mounting kit	Installs the x,y,z-robot GC instrument top
1 pc GC system	With split/splitless, or PTV type injector, and SSL/MMI Arrow inlet adaptor
1 pc MS system	Single or triple quadrupole analyzer, EI ion source

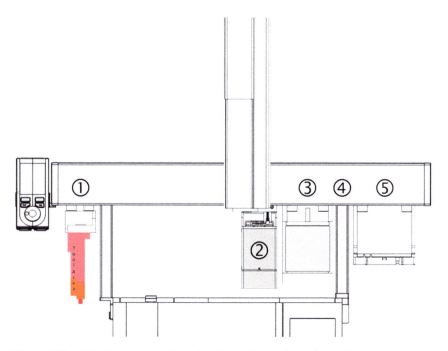

Figure 6.24 Robotic system configuration for static and dynamic headspace analyses. 1 Tool park station; 2 Heatex stirrer (optional); 3 Agitator/Incubator; 4 Instrument top mounting legs; and 5 Trayholder for sample vials.

Consumables

- 20 mL headspace screw-top clear vials with magnetic screw caps.
- GC column: RTX-VMS, 60 m × 0.32 mm ID × 1.8 μm.
- Headspace syringe, 2500 μL.

Samples for Testing

- The Mao Shan Wang, a premium durian variety, is originally cultivated in Malaysia, known for its bright yellow flesh, bitter and sweet taste, and creamy texture [30].

Initial Manual Sample Preparation

- Remove pulps from the Durian.
- Place the flesh of durian into beaker and mesh.
- Weigh 3 g (±0.1 g) of durian into 20 mL headspace vials.
- Add 1 mL of water into the sample vial with the Durian flesh.
- Close the headspace vials with magnetic screw caps.
- Place prepared vials into the trayholder racks.

Automation Workflow

After the Durian samples were prepared, the sample vials are separated into two batches. One batch is undergoing the automated static headspace extraction (Part A). The other batch of Durian vials is undergoing the dynamic headspace extraction (Part B).

Analysis Parameters

The important operational parameters of these fully automated static and dynamic headspace techniques are tabulated below in Table 6.5. The extraction parameters were separated into two sections, whereby Part A is focusing on the extraction of major volatile organic compounds from Durian. Part B is using the ITEX DHS tool for the dynamic headspace extraction of the VSC trace concentrations from the Durian sample. The extracted components by both techniques were analyzed under the same GC-MS full scan conditions.

Part A – Static Headspace Analysis Workflow

Static headspace is used to analyze the most abundant volatile compounds. This Part A of the automated workflow is used for the detection of major VSCs in tropical fruits, here as example Durian, by static HS GC-MS (Figure 6.25).

Part B – Dynamic Headspace Analysis Workflow

The dynamic headspace method is used to extract and concentrate the low-level VSC profile of the sample. This Part B of the automated method covers the detection of the trace volatiles in Durian by ITEX DHS GC-MS (Figure 6.26).

Table 6.5 Selected important analysis parameters by static and dynamic headspace techniques.

	Part A HS GC-MS	Part B ITEX-DHS GC-MS
	To extract major compounds	To extract trace compounds
Headspace extraction		
Incubation		
Incubation temperature	80 °C	80 °C
Incubation time	2 min	2 min
Agitator speed	700 rpm	700 rpm
Extraction		
Extraction strokes	n/a	20
Trap extract temperature	n/a	40 °C
Syringe temperature	70 °C	70 °C
Trap desorption		
Desorb temperature	n/a	250 °C
GC		
Column	RTX-VMS, 60 m × 0.32 mm × 1.8 µm	
Inlet temperature	240 °C	
Inlet mode	Split	
Split ratio	1 : 10	
Oven temperature	35 °C (10 min) → 3 °C/min → 60 °C → 5 °C/min → 130 °C → 30 °C/min → 240 °C (2 min)	
MS		
Interface temp	240 °C	
Scan m/z	20–400	
Scan time	0.3 s	
Source temperature	240 °C	

Part A and Part B – Combined

Both volatile extraction techniques can be combined in one comprehensive and automated workflow. After injection of the static headspace extract, the robotic system selects the dynamic headspace tool and proceeds with the collection of the VSCs in the prep-ahead mode. The injection into the same GC column follows after the first run is completed.

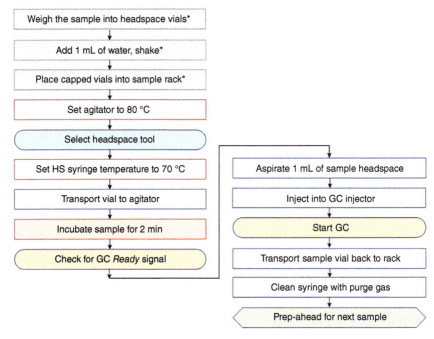

Figure 6.25 Part A – Automated workflow for static headspace analysis (*initial manual steps).

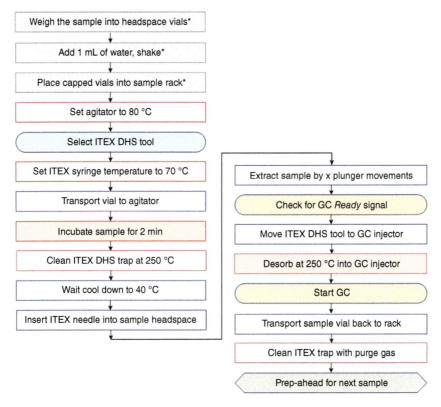

Figure 6.26 Part B – Automated workflow for dynamic headspace analysis (*initial manual steps).

Results

The distinctive fruity and onion-like odorous compounds of the Durian fruit were detected by both static and dynamic headspace technique in the combined automated workflow. In Part A with the very straightforward static headspace extraction technique, 13 major volatile organic compounds were detected and identified by GC-MS and NIST11 library search. The detected compounds were mainly from the thio-, alcohol, ester, and sulfide families. The ethanethiol peak #4 was detected with the highest intensity, as shown in Figure 6.27. In Part B, a total of 58 compounds were extracted by the dynamic headspace technique. Of those 45 additional compounds were detected by ITEX DHS GC-MS in addition to the major compounds detected in Part A by the headspace technique. The total ion chromatogram (TIC) of the dynamic HS extraction is shown in Figure 6.28.

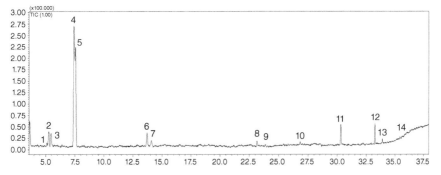

Figure 6.27 GC-MS full scan total ion chromatogram of the major VSCs by static headspace extraction. Source: Courtesy CTC Analytics AG.

Peak No.	Retention time [min]	Compound name
1	5.04	Methanethiol
2	5.20	Acetaldehyde
3	5.40	Methyl alcohol
4	7.34	Ethanethiol
5	7.57	Ethanol
6	13.64	Propyl mercaptan
7	14.03	1-Propanol
8	23.14	Propanoic acid, ethyl ester
9	23.88	Butanoic acid, 2-methyl, methyl ester
10	26.81	Butanoic acid, 2-methyl-, ethyl ester
11	30.41	Not identified by library search
12	33.26	Diethyl disulfide
13	33.91	Butanoic acid, 2-methyl-, propylester
14	35.42	Disulfide, ethyl 1-methylethyl

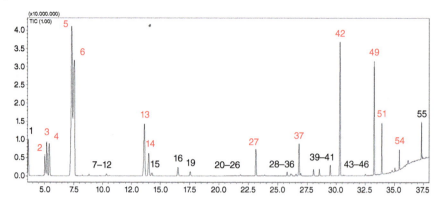

Figure 6.28 GC-MS full scan total ion chromatogram of the trace VSCs by dynamic headspace extraction. The major compounds also extracted by static HS technique are labeled in red. Source: Courtesy CTC Analytics AG.

Peak No.	Retention time (min)	Compound name
1	3.59	Hydrogen sulfide
2	5.04	Methanethiol
3	5.20	Acetaldehyde
4	5.40	Methyl alcohol
5	7.34	Ethanethiol
6	7.57	Ethanol
7	8.24	Dimethyl sulfide
8	8.83	Propanal
9	9.79	Acetone
10	10.36	Acetic acid, methyl ester
11	10.61	n-Hexane
12	12.99	Propanal, 2-methyl-
13	13.64	Propyl mercaptan
14	14.03	1-Propanol
15	14.29	Ethane, (methylthio)-
16	16.50	Ethyl acetate
17*	16.83	2-Butanone, 3-methyl-
18*	17.10	2-Butanone
19	17.57	Methyl propionate
20	19.32	1-Propanol, 2-methyl
21	21.37	Propanoic acid, 2-methyl-, methyl ester
22	21.61	1-Butanethiol
23	21.84	1-Butanol

Peak No.	Retention time (min)	Compound name
24	21.93	2-Propanal, 2-methyl
25	22.26	Methyl thiolacetate
26	22.69	1-Penten-3-ol
27	23.14	Propanoic acid, ethyl ester
28	23.54	2-Pentanone
29	23.64	n-Propyl acetate
30	23.88	Butanoic acid, methyl ester
31	24.52	Disulfide, dimethyl
32	25.29	Toluene
33	25.78	Propanoic acid, 2-methyl-, ethyl ester
34	26.11	1-Butanol, 3-methyl-
35	26.18	1-Butanol, 2-methyl
36	26.56	S-ethyl enthanethioate
37	26.81	Butanoic acid, 2-methyl, methyl ester
38*	26.96	Acetoin
39	28.06	Butanoic acid, ethyl ester
40	28.55	Propanoic acid, propyl ester
41	29.49	Methyl ethyl disulfide
42	30.33	Butanoic acid, 2-methyl-, ethyl ester
43	30.55	Butanoic acid, 3-methyl-, ethyl ester
44	30.63	Propanoic acid, 2-methyl-, propyl ester
45	30.73	2-Butenoic acid, ethyl ester, (E)-
46	31.04	S-Ethyl thiopropionate
47*	32.28	1-Hexanol
48*	32.52	Butanoic acid, propyl ester
49	33.26	Diethyl disulfide
50*	33.57	Disulfide, methyl propyl
51	33.91	Butanoic acid, 2-methyl-propylester
52*	34.80	Ethane, 1,1-bis(ethylthio)-
53*	35.05	Hexanoic acid, ethyl ester
54	35.42	Disulfide, ethyl 1-methylethyl
55	37.33	Trisulfide, diethyl
56[a]	37.39	1,2,4-Trithiolane, 3,5-dimethyl-
57[a]	37.47	1,2,4-Trithiolane, 3,5-dimethyl-
58	37.72	Octanoic acid, ethyl ester

a) Likely to be isomers and cannot be differentiated from mass spectra.
* Low intensity peak not labeled in Figure 6.28 full scan chromatogram

6.5 Ethanol Residues in Halal Food

Application

Ethanol in foods is not permitted to achieve the Halal certification. Halal product assurance has to ensure that all foods, beverages, drugs, cosmetics, chemical products, biological products, genetically engineered products, and similar goods are compulsory to obtain Halal certificate. Halal, means "permissible" in English, is regarded as the things or actions permitted by "Hukum Syara," such as the ingredients do not contain any parts or products of forbidden (non-halal) animals, najis (ritually unclean) and are safe for consumption.

Alcohol, by the means of Halal, is referred to as ethanol (CAS No. 64-17-5). In different countries, the permitted amount of ethanol varies, see Table 6.6. In general, ethanol <1% and produced by natural fermentation is considered as a preserving agent and Halal [31].

Scope and Principle of Operation

In this application, the ethanol content is screened and determined by using the static headspace extraction technique. During the described automated workflow, the vials are automatically incubated to release the containing ethanol from the food sample to the vapor phase. A heated syringe is used to penetrate the vials after a constant equilibration time and takes 1 mL of the gas phase and injects into the GC for analysis.

Table 6.6 Percent of alcohol permitted in Halal food.

Country	Regulating body	Alcohol max. [%]
Malaysia	JAKIM	0.01
Indonesia	MUI	1.0
Thailand	AOI	1.0
Singapore	MUIS	0.5
Brunei	BIRC	0
Europe	n.a.	<0.5
United Kingdom	n.a.	Not allowed
Canada	n.a.	Not allowed

AOI, Administration of Organizations of the Islamic Act; JAKIM, Department of Islamic Development Malaysia; MUI, Majelis Ulama Indonesia; MUIS, Majlis Ugama Islam Singapura; BIRC, Brunei Islamic Religious Council.
Source: Adapted from Alzeer and Hadeed [32].

With an overlapped sample heating, only 10 minutes cycle time are needed to complete one sample analysis. This short analysis time allows a high throughput ethanol screening.

Solvents and Chemicals

- Ethanol, CAS No. 64-17-5, >99.9%, for GC analysis.
- Water, CAS No. 7732-18-5, HPLC quality.
- Ethanol standard preparation:
  1. Add 1 mL of ethanol into a 100 mL volumetric flask.
  2. Fill the 100 mL volumetric flask for the most part with DI water and shake.
  3. Fill up to volume, close and shake vigorously.
  4. The solution will contain 1% (v/v) of ethanol.
  5. Transfer 1 mL of the 1% vol ethanol standard into a headspace vial.
  6. Close the headspace vial with a magnetic screw cap.
  7. Place the vial to the sample tray for analysis.

Consumables

- 20 mL headspace screw-top clear vials with magnetic screw caps.
- GC Column: RTX-VMS, 60 m × 0.32 mm ID × 1.8 µm, or equivalent.

System Configuration

The minimum robotic system hardware requirements for ethanol analysis by static headspace extraction, as shown in Figure 6.29, are:

System – Tool – Module	Task
PAL RSI System	x,y,z-Robotic system for HS analysis with manual tool change
1 pc Tray Holder	Tray holder for 3× vial racks
3 pc Rack VT15	15 pos. 20 mL vial racks
1 pc Agitator/Incubator	Sample incubation of up to 6× 20 mL vials
1 pc Headspace Tool	For a 2500 µL HS syringe operation
1 pc Headspace Syringe	2500 µL HS syringe
1 pc MHE Adapter	For optional multiple headspace quantitation
1 pc Mounting kit	Installs the x,y,z-robot GC instrument top
1 pc GC system	With split/splitless, or PTV type injector
1 pc MS system	Single quadrupole analyzer, EI ion source

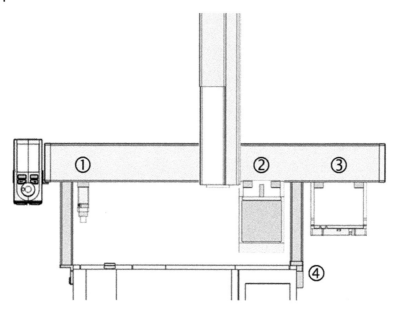

Figure 6.29 Minimum configuration for the ethanol HS-GC screening. 1 Multiple headspace adapter (MHE); 2 Agitator/incubator; 3 Sample tray; and 4 GC mounting legs.

Analysis Parameters

The parameters to automatically extract ethanol and analyze from foods are tabulated in Table 6.7.

Workflow

Initial Manual Pre-treatment

- Depending on the nature of the sample, about 1 g of homogenized solid material, or 1 mL of liquids is weighed exactly into HS sample vials.
- The vials are capped with magnetic caps and placed into the racks of the robotic sampler.

Automated Workflow

- The sample vials are transferred into the incubator in the overlapping mode of the robotic sampler so that all samples are equilibrated exactly the same time.
- Six sample vials are handled at the same time in a batch mode, scheduled according to the expected analysis cycle time of 10 minutes.
- After incubation, the HS workflow starts with the pressurization of the sample vials with inert or just carrier gas.
- Several filling strokes equilibrate the sample vapor in the heated HS syringe before injection.
- The automated workflow is graphically illustrated in Figure 6.30.

6.5 Ethanol Residues in Halal Food

Table 6.7 Selected analysis parameters for the HS GC-MS analysis of residual ethanol in food.

Headspace extraction	
Incubation temperature	80 °C
Incubation time	5 min
Agitator speed	700 rpm
Auxiliary gas	Nitrogen
Vial pressurization	2250 µL nitrogen
Syringe temperature	100 °C
GC	
Inlet temperature	200 °C
Inlet mode	Split
Injection volume	1.0 mL
Split ratio	10 : 1
Flow	1.50 mL/min
Pressure	58.8 kPa
Oven program	35 °C (1 min) → 20 °C/min → 200 °C (0.75 min)
MS	
Interface temperature	200 °C
Scan m/z	29–400
Scan time	0.3 s
Source temperature	200 °C

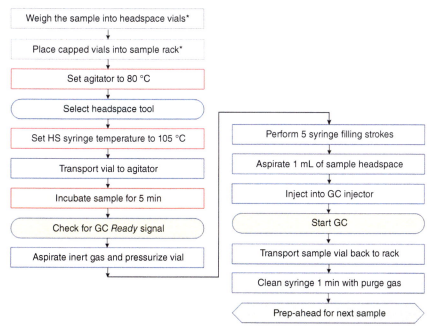

Figure 6.30 Workflow steps for the automated determination of ethanol residues (*initial manual steps).

Results

The precision of the described HS method was determined with 1.2% RSD based on the peak areas from 10 ethanol standard runs. Food samples were screened with reference to the 1% ethanol standard.

For practical reasons, the standard concentration of $1.0 \pm 0.2\%$ was used for reference and decision level as a one-point calibration, as of the country-specific regulation for Indonesia in Table 6.6. Samples with an ethanol content exceeding 1.2% were considered "fail," below 0.8%, the samples were considered "pass" (see Table 6.8). Samples with contents between 0.8% and 1.2% require a quantitative multi-point calibration for final decision. Samples showing an ethanol peak with an individually pre-determined signal-to-noise ratio, e.g. below S/N 10, would be considered as "not detected." A chromatogram of a soy sauce sample is shown in Figure 6.31.

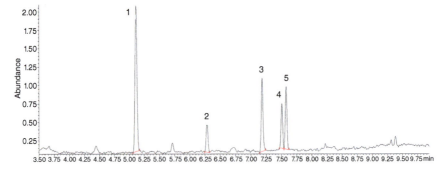

Figure 6.31 Analysis of the "Naturally Brewed Dark Soy Sauce" with a concentration of <0.8% ethanol. Source: Courtesy CTC Analytics AG.

Peak No.	Retention time [min]	Peak area [cts]	Compound name
1	5.11	470 497	Ethanol
2	6.28	83 937	2-Methyl-propanal
3	7.18	198 650	Ammonium acetate
4	7.51	112 796	3-Methyl-butanal
5	7.58	162 565	2-Methyl-butanal

Table 6.8 Ethanol screening results from different foods using the automated HS-GC-MS workflow.

Food type	Food sample	Ethanol (vol%)
Seasoning	Japanese Kyushu Soy Sauce	>1.2%
Seasoning	Naturally Brewed Dark Soya Sauce	<0.8%
Beverage	Red Bull	<0.8%
Beverage	Tiger Beer	>1.2%
Processed food	Sky Flakes Crackers	<0.8%
Fresh vegetable	Kang Kung (or Water Spinach)	Not detected

6.6 Volatile Organic Compounds in Drinking Water

Application

The classical standard method for the analysis of volatile organic compounds (VOCs) in drinking water is achieved by purge and trap extraction, concentration, and thermal desorption of the analytes to GC-MS. This application describes the comprehensive automation of VOC purge and trap analysis from laboratory samples and online water streams with the potential to be combined with other extraction methods for semivolatile organic compounds (SVOCs) like SPME.

Scope and Principle of Operation

The United States Environmental Protection Agency (US EPA) developed the method 524.3 for the "measurement of purgeable organic compounds in water by capillary column gas chromatography/mass spectrometry." The EPA 524.3 is a widely used standard method to determine more than 70 VOC compounds in drinking water for human consumption.

Coupling the purge and trap concentrator (P&T) with a robotic autosampler allows the unattended online and offline analysis of large sample series with integrated sample sequence control and transfer of the analytes for GC-MS analysis [33]. Laboratory samples collected from target sites can be readily processed not only from the standard 40 mL P&T vials but also from regular 20, 10, or 2 mL vials as shown in Figure 6.32. Flow cells can also be attached to the robotic sampling system for the unattended monitoring of online water streams, demonstrated in Figure 6.33 for the automated online sampling of two or more water streams (see also Section 4.1.6). Even the sampling from customized containers or bottles is possible with programmable x,y,z-robotic systems. The robotic sampler uses a multi-dilutor

Figure 6.32 Robotic autosampler sampling from different vial sizes connected to the U-tube sparger. Source: Courtesy CDS Analytical LLC. 1 U-tube sparger for water sample purging; 2 Inlet transfer tube from dilutor tool for automated sample transfer from vials or online; 3 Purge gas line; 4 Purge&Trap tool; 5 Sampling from standard 40 mL Purge & Trap vials; 6 Sampling from 20 mL vials; and 7 Sampling from 2 mL vials.

290 | 6 Solutions for Automated Analyses

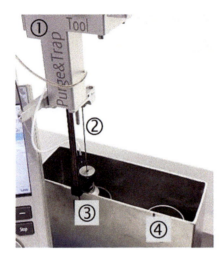

Figure 6.33 Online water sampling from two water streams using the Purge & Trap tool of the robotic autosampler. Source: Courtesy CTC Analytics AG. 1 Purge & Trap tool of the dilutor module; 2 Needle for sample aspiration; 3 Water stream #1 inflow*; and 4 Water stream #2 inflow* (*water overflows are collected at the bottom, a downward waste tube is not shown).

tool connected via a flexible tube to a purge and trap tool in the head of the unit for sampling either from vials or the online flow cell for the transfer of a sample volume to the P&T U-tube. A methanol and hot water reservoir is used for cleaning the tool and the sample lines. For priming the dilutor, a waste port is connected.

Solvents and Chemicals

- Internal standard mix, e.g. Supelco 861183, or similar.
- Surrogate standard mix, e.g. Supelco 861135, or similar.
- Volatiles calibration mix, e.g. Supelco 500607, or similar.
- Volatile organic compounds mix, e.g. Supelco 47408, or similar.
- Methanol, CAS No. 67-56-1, $\geq$99.9%, HPLC grade.
- Water, CAS No. 7732-18-5, HPLC grade.

Consumables

- 40, 20, 10, or 2 mL clear vials with screw caps.
- GC column: RTX-VMS, 30 m × 1.40 µm × 0.25 mm, or equivalent.
- Liquid syringe, 10 µL, for optional GC injection of liquid standards, needle gauge 22 or 23, point style conical.

System Configuration

The minimum hardware requirements of a robotic sampling system for automated off- and online purge and trap analysis as shown in Figure 6.34 are:

System – Tool – Module	Task
PAL RTC System	x,y,z-Robotic system with automated tool change for sample preparation and liquid injection, 850 mm rail length
1 pc Park Station	Park station for tools not in use during workflow

System – Tool – Module	Task
1 pc D7/57 Liquid Tool	Tool for a 10 µL GC injection syringe (optional for liquid injections)
1 pc Multi-Dilutor and Tool	For liquid handling with dispensing from solvent or reagent reservoirs, or transfer from and to sample vials
1 pc Tray Holder	Tray holder for 3× vial racks
3 pc Rack 12× 40 mL	12 pos. racks for 40 mL P&T vials
3 pc Rack 15× 10/20 mL	15 pos. racks for 10/20 mL vials (optional)
1 pc Standard Wash Module	Wash module for up to 5× syringe wash solvents
1 pc Purge & Trap unit	Sparger unit, with 5 mL U-tube for GC-MS analysis (25 mL U-tube for GC non-MS detection), built-in Tenax TA sorbent tube
1 pc ISTD addition module	Accessory of the P&T unit for automated dosing of an ISTD solution
1 pc Mounting kit	Installs the x,y,z-robot GC instrument top
1 pc GC system	With split/splitless, or PTV type injector
1 pc MS system	Single quadrupole analyzer, EI ion source

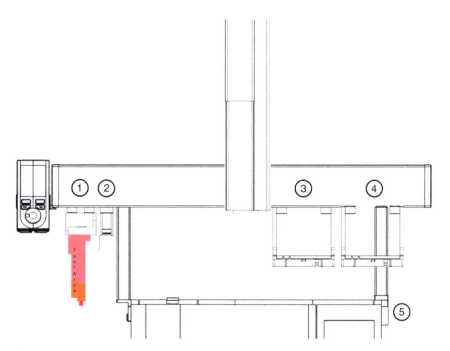

Figure 6.34 Robotic sampler configuration for automated purge and trap analysis. 1 Tool park station; 2 Standard wash station; 3 Trayholder for sample vials; 4 Trayholder extension, opt. for large sample series; and 5 GC mounting legs.

Table 6.9 Analysis parameter for the purge and trap concentrator and GC-MS analysis.

Purge & Trap concentrator	
Sample volume	5 mL
ISTD volume	2 µL
Purge gas, flow	N_2 or He, 40 mL/min
Purge temperature	40 °C
Adsorption trap temperature	35 °C
Dry purge time	0.5 min
Dry purge temperature	35 °C
Dry purge flow	100 mL/min
Pre-desorb temperature	230 °C
Desorb temperature	250 °C
Desorb time	2 min
Desorb flow	300 mL/min
Desorb gas	He (carrier gas), or N_2
Trap bake temperature	260 °C
Trap bake time	8 min
Trap bake flow	200 mL/min
Wet trap ready temperature	45 °C
Trap bake temperature	260 °C
Valve oven temperature	130 °C
Hot water rinse Temperature	70 °C
Transfer line temperature to GC	130 °C
GC	
Inlet temperature	135 °C
Inlet mode	Split
Injection volume	1.0 µL
Split ratio	30 : 1
Purge flow	3.0 mL/min
Carrier gas	He, constant flow at 1 mL/min
Oven program	35 °C (4 min) → 5 °C/min → 90 °C → 12 °C/min → 150 °C → 30 °C/min → 220 °C (2.7 min)
MS	
System, ionization	Single quadrupole, EI mode
Ion source temperature	200 °C
Interface temperature	220 °C
Scan mode	Full scan
Scan m/z	m/z 40–500
Scan time	0.3 s

Analysis Parameter

The typical analysis parameters for the P&T concentrator and GC-MS analysis are listed in Table 6.9.

Workflow

Initial Manual Steps

- 5 mL of an ISTD stock solution is prepared from a commercial standard with methanol for a concentration range of 50–100 µg/mL and transferred into the vial of the ISTD addition module attached to the P&T unit.

Automated Workflow (Figure 6.35)

- The dilutor tool is selected and primed, then aspirates the desired sample volume from the sample vial or online device stream, and transfers the sample volume to the U-tube.
- The ISTD solution, typically 2 µL, is added by the ISTD module to each sample, to produce a 25 µg/L concentration in the water sample.
- The purging process starts, trapping the purged volatiles. At the end of the purge process, the desorption of the trap takes place to transfer the VOCs to the GC injector, start of the GC-MS run.
- During the GC run, the U-tube, sample lines, and dilutor tool are washed with methanol, and optionally with hot water, for the next sample.
- The next sample processing starts with trapping the purged analytes in the prep-ahead mode during the GC run to be ready for the thermal desorption at the expected GC *Ready* time.

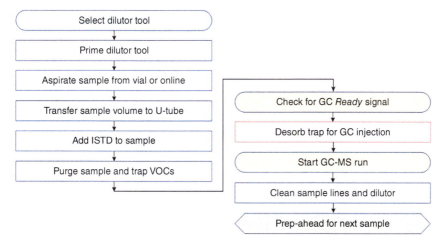

Figure 6.35 Workflow steps for the automated Purge & Trap analysis of VOCs in drinking water.

Quantitative Calibration

Using the robotic sampling system, the quantitative calibration can automatically be prepared in the same workflow within the CDS sample sequence. The commercially available calibration standard mix is prepared as the VOC working standard solution in a concentration of for instance 100 µg/L. From the working standard, the robotic system prepares and measures according to the workflow parameters and used customized dilutions of the calibration standard in the range of 1% to 100% of the employed standard solution. A typical calibration is shown for naphthalene in Figure 6.36 with an excellent precision of R^2 0.9964. This way a continuous system calibration can be achieved without a necessary operator interference.

Regulations

The described automated P&T system setup is compliant to DIN/ISO, U.S. EPA, China GB, and other international standards method protocols for VOC drinking water control.

Conclusion

Using a robotic autosampler system connected to a P&T sparger unit provides the fully automated operation for laboratory drinking water samples or online water monitoring. Different and customized vial formats are supported. A standard x,y,z-robotic platform offers the flexibility to maximize operational capacity with applications beyond P&T only, with additional static or dynamic headspace, SPME or liquid/liquid extraction (LLE) applications for higher concentrated or SVOC samples. Overlapping in the sample sequences reduces the total analysis time for large sample series significantly. This permits a sample throughput of about two water samples/hour.

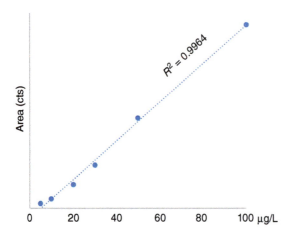

Figure 6.36 Linear calibration of naphthalene in tap water, automatically prepared for the range 1–100 µg/L. Source: Courtesy CTC Analytics AG.

6.7 Geosmin and 2-MIB

Application

Geosmin (*trans*-1,10-dimethyl-*trans*-9-decalol (GSM), Figure 6.37) and 2-MIB (Figure 6.38) are internationally strongly controlled for drinking water quality. Both compounds are responsible for a musty, muddy, or as earthy described unpleasant smell [34]. GSM also contributes besides TCA to the "cork taint," the earthy off-odor spoiling occasionally red and white wines. These compounds are also found in farmed and wild fish. Both compounds are produced by cyanobacteria in soil, lakes, reservoir and river waters, especially during an algae bloom, mainly leaching from dead bacteria, and finally get into drinking water from these sources. They break down in acid conditions. The human nose is extremely sensitive to GSM and 2-MIB. Depending on the individual, we can detect <5 ppt (=5 ng/L) in water, 60–65 ng/L in white wine, 80–90 ng/L in red wine [35]. Sensitive analytical methods are required to provide quantitative calibrations starting from 1 ng/L.

Scope and Principle of Operation

The described workflow uses HS-SPME for the determination of GSM and 2-MIB in drinking water and raw water samples. For extraction, the SPME Arrow technology is used. The chromatographic analysis of the analytes uses GC-MS for detection and quantification.

Solvents and Chemicals

- Standard solution of target analytes 2-MIB and GSM, each 100 mg/L in methanol, as certified standard solutions.
- ISTD 2-isobutyl-3-methoxy-pyrazine, 100 mg/L in methanol, as a certified standard solution, diluted to 10 ng/mL for use.
- NaCl, CAS No. 7647-14-5, analytical grade, baked at 450 °C for two hours.
- Methanol, CAS no. 67-56-1, chromatographic purity.
- Water, CAS No. 7732-18-5, HPLC quality.

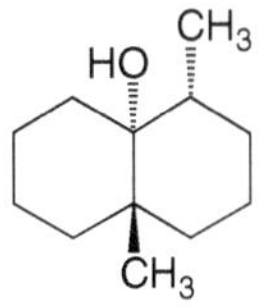

Figure 6.37 Geosmin (GSM).

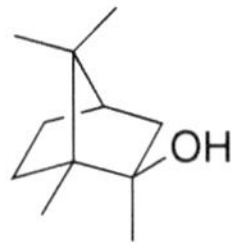

Figure 6.38 2-Methylisoborneol (2-MIB).

Consumables

- 20 mL headspace screw-top clear vials with magnetic screw caps.
- SPME Arrow, 1.1 mm OD, DVB/CAR/PDMS sorbent phase.
- GC inlet liner, 1.3 or 1.7 mm ID.
- GC column: DB-5MS UI, 30 m × 0.25 mm × 0.25 µm, or equivalent.

System Configuration

The minimum robotic system hardware requirements for the analysis of GSM and 3-MIB as shown in Figure 6.23 are:

System – Tool – Module	Task
PAL RTC System	x,y,z-Robotic system for liquid injection, 1200 mm rail length
1 pc Park Station	Park station for tools not in use during workflow
1 pc D7/57 Liquid Tool	Tool for an optional 10 µL GC injection syringe
1 pc Tray Holder	Tray holder for 3× vial racks
2 pc Rack VT15	15 pos. 20 mL vial racks
1 pc Rack VT54	54 pos. 2 mL vial rack
1 pc Standard Wash Module	Wash module for up to 5× syringe wash solvents
1 pc SPME Arrow tool	Tool for SPME Arrow operation
1 pc Heatex Stirrer	Heated module for stirring during extraction
1 pc Arrow conditioning module	Conditioning of SPME Arrows before/after analysis
1 pc Agitator/Incubator	Sample incubation of up to 6× 20 mL vials
1 pc Mounting kit	Installs the x,y,z-Robot GC instrument top
1 pc GC system	With split/splitless, or PTV type injector, and SSL/MMI Arrow inlet adaptor
1 pc MS system	Single or triple quadrupole analyzer, EI ion source

Analysis Parameter

Arrow Conditioning Module

- Conditioning temperature 270 °C.
- Pre- and post-desorption time five minutes.
- Before use, the SPME Arrow is conditioned as of manufacturer specifications.

Agitator

- Agitator temperature 60 °C.
- Before SPME extraction, the sample is incubated for two minutes to the extraction temperature and then moved to the Heatex stirrer.

Heatex Stirrer

- Stirrer speed 1500 rpm and temperature 60 °C.
- SPME Arrow extraction for 30 minutes.

GC-MS

- GC oven temperature program: 60 °C (2 min) → 10 °C/min → 270 °C (2 min).
- Injector temperature 250 °C, splitless two minutes.
- Carrier gas He, 5.0 quality, constant flow mode, 1.0 mL/min.
- Transfer line temperature 280 °C.
- Ion source temperature 250 °C.
- Acquisition mode SIM (bold quantification masses):
  2-MIB: m/z **95**, 107, 108
  GSM: m/z 111, **112**, 125
  ISTD: m/z **94**, 124

Workflow

Initial Manual Steps

- Place 1.5 g of NaCl in empty 20 mL screw cap vials.
- Add 5.0 mL of a water sample.
- Add 5.0 µL ISTD solution, 10 ng/mL.
- The capped vials are placed into the sample racks of the robotic system for analysis.

Automated Workflow

- The automated workflow is shown stepwise in Figure 6.39.
- For GC analysis, the SPME Arrow is moved to the GC injection port and thermally desorbed at 250 °C for five minutes.
- During the desorption time, the injector split is closed.
- The prep-ahead mode is used for increased sample throughput.

Quantitative Calibration

A series of standard dilutions with concentrations of 1, 10, 20, 50, and 100 ng/L are prepared automatically from the stock solution, see Section 6.1.2. A working solution of 1 mg/L is prepared by diluting the stock solution 100 mg/L with methanol. Then, the working solution is diluted with water step by step to prepare a series of solutions 0.5, 5, 10, 25, and 50 µg/L. Ten microliter each is transferred into a headspace vials with 5 mL NaCl solution (30%, w/w). The calibration curves of 2-MIB and GSM are obtained with good precision and linearity as shown in Figures 6.40 and 6.41. The calibration curves were generated by calculating the relative response to the ISTD. The calibration precision shown here is excellent in particular in the low sensitivity range. The linear correlation coefficient for the calibration range up to 100 ng/L was achieved with a correlation factor better than 0.999 for both compounds.

298 6 Solutions for Automated Analyses

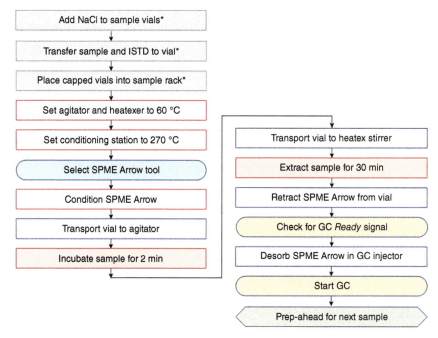

Figure 6.39 Workflow steps for the automated SPME Arrow analysis of GSM and 2-MIB (*initial manual pre-treatment).

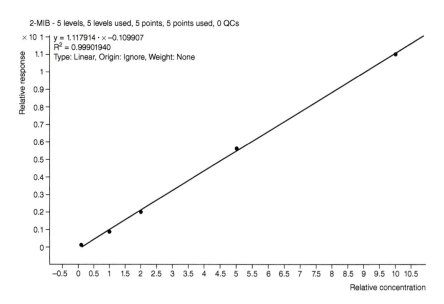

Figure 6.40 Calibration for 2-MIB in the low range up to 10 ng/L. Source: Courtesy CTC Analytics AG.

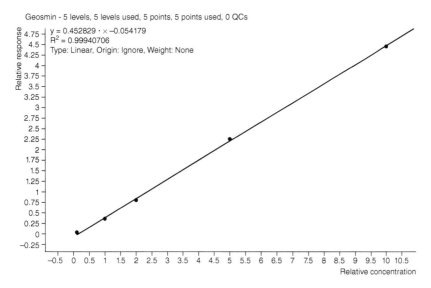

Figure 6.41 Calibration for geosmin in the low range up to 10 ng/L. Source: Courtesy CTC Analytics AG.

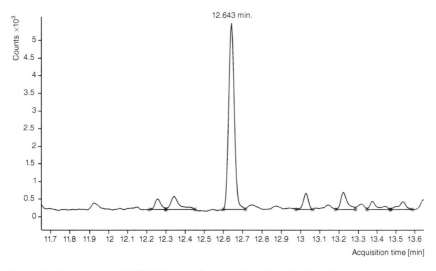

Figure 6.42 Geosmin HS-SPME Arrow signal at 10 ng/L spiked to drinking water (*m/z* 112.0). Source: Courtesy CTC Analytics AG.

Recovery and Precision

Regular tap water and spiked samples were tested for recoveries. The spiked standard was 10 ng/L with threefold repeated measurements. The recovery rates of 2-MIB and GSM achieved were 116.0% and 98.4%, respectively. The precision for 2-MIB and GSM in the repeated measurements of real-life samples was 3.8% and 2.8% RSD, respectively.

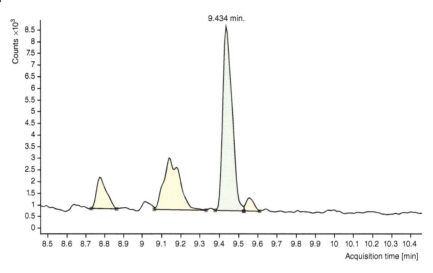

Figure 6.43 2-MIB HS-SPME Arrow signal at 10 ng/L spiked to drinking water (*m/z* 95.0). Source: Courtesy CTC Analytics AG.

Sample Measurements

Regular tap water was analyzed. The chromatograms in Figures 6.42 and 6.43 show the addition of a 10 ng/L spike to tap water. With the described HS-SPME Arrow method, a dominant peak of both target compounds can be achieved at the regulated decision level. Also, the online analyses from drinking water streams are described using the automated SPME workflow [36].

Regulations

The regulated limit for drinking water is set in most countries to 10 ng/L, e.g. Japan DWQS 2005 and China GB 5749-2006. In Europe, the Drinking Water Directive (Council Directive 98/83/EC of 3 November 1998 on the quality of water intended for human consumption) applies. In 2016, the China national standard test method GB/T 32470-2016 for GSM and 2-MIB testing in drinking water was established. According to this standard method, SPME is used as described to analyze GSM and 2-MIB [37].

Conclusion

Using the automated HS-SPME Arrow method, a large number of water samples can be screened at the relevant sensitivity level with excellent quantitative precision. For increased sample throughput, the prep-ahead mode with overlapped incubation in the agitator is implemented in the workflow.

6.8 Solvent Elution from Charcoal

Application

Charcoal-based sorbent materials (coconut shell-activated carbon) are in use, often for passive sampling of volatiles. Charcoal cannot be desorbed thermally and requires solvent desorption. Frequently hazardous solvents like CS_2 or DCM are involved in manual operation [38]. The charcoal sorbent material is transferred into autosampler vials and capped. The solvent extraction and analysis can be achieved fully automated by using a robotic autosampler integrated with GC or GC-MS.

Scope and Principle of Operation

This method can be used for the automated elution of charcoal material with different elution solvents, and succeeding online GC analysis. The charcoal sample is eluted with solvent supported by vortexing. Magnetic caps allow the vial transport to the Vortex mixer. The automated elution of sorbent materials allows the reduction of solvent volumes and a significantly reduced operator exposure compared to manual operations. ISTD can be added as required. After sedimentation of the charcoal powder, an aliquot of the extract is injected to GC.

Solvents and Chemicals

- Elution solvents as of application, e.g. CS_2, pentane, ethylenglycolmonophenylether [39].
- ISTD solution.
- Polar and non-polar syringe wash solvents.

Consumables

- 10 mL headspace screw-top clear vials with magnetic screw caps.
- GC column: DB-5MS UI, 30 m × 0.25 mm × 0.25 µm, or equivalent. For volatiles separation, a thick film column with e.g. 1.8 µm film is recommended.
- Liquid syringe of suitable volume for solvent dispensing, e.g. 1000 µL, needle length 57 mm, PTFE plunger, gauge 22, point style flat.
- Liquid syringe of suitable volume reserved for ISTD dispensing only, e.g. 500 µL, needle length 57 mm, PTFE plunger, gauge 22, point style flat (as an alternative).
- Liquid syringe for GC injection, e.g. 10 µL, needle length 57 mm, metal plunger, gauge 23s, point style conical.

System Configuration

The minimum hardware requirements of a robotic sampling system for automated charcoal elution and GC analysis as of Figure 6.44 are:

System – Tool – Module	Task
PAL RTC System	x,y,z-Robotic system with automated tool change for sample preparation and liquid injection, 850 mm rail length
1 pc Park Station	Park station for tools not in use during workflow
1 pc D7/57 Liquid Tool	Tools for a 10 µL GC injection syringe
2 pc D8/57 Liquid Tool	Tool for a 500 µL ISTD and 1000 µL sample preparation syringe
1 pc Tray Holder	Tray holder for 3× vial racks
3 pc Rack 54× 2 mL	54 pos. for 2 mL vials
1 pc Solvent Module	For 3× 100 mL solvent reservoirs
1 pc Standard Wash Module	Wash module for up to 5× syringe wash solvents, and ISTD vials
1 pc Adapter	Standard Wash Station Adapter to place 2 mL vials instead of one 10 mL wash vial
1 pc Vortex Mixer	Vortexing unit for intense sorbent/solvent mixing
1 pc Mounting kit	Installs the x,y,z-robot GC instrument top
1 pc GC system	With split/splitless, or PTV type injector
1 pc MS system	Single or triple quadrupole analyzer, EI ion source

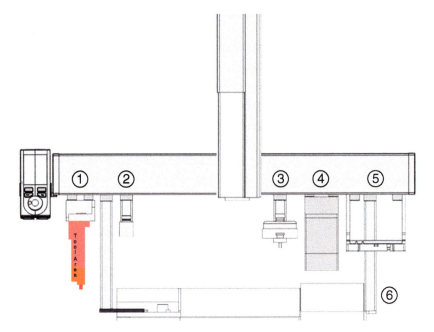

Figure 6.44 Robotic sampler configuration for automated charcoal elution. 1 Tool park station; 2 Standard wash station; 3 Solvent station for elution solvents; 4 Vortex mixer; 5 Sample tray with racks for 10 mL vials; and 6 Instrument top mounting legs.

Workflow

Initial Manual Steps

- The charcoal material is transferred from the sorbent tube into 2 mL vials. Alternatively, the use of 10 mL vials is possible with appropriate racks, solvent reservoirs, and liquid syringes.
- The capped vials are placed into the rack of the robotic sampler.

Automated Workflow

- The automated workflow for charcoal elution is graphically illustrated with stepwise operation in Figure 6.45.

Conclusion

Using the automated charcoal elution workflow with 2 mL vials, a significant reduction of potentially harmful solvents can be achieved. The amount of solvents required for elution is reduced to only microliter use. Due to the low elution volume, the analyte dilution in the used solvent is significantly reduced, with better detection sensitivity, compared to manual operations. The workflow can use the prep-ahead mode for an increased sample throughput.

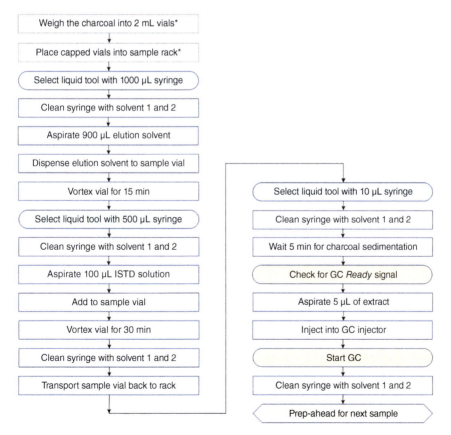

Figure 6.45 Workflow for automated charcoal solvent elution and GC injection (* initial manual steps).

6.9 Semivolatile Organic Compounds in Water

Application

Industrialization has created many new organic compounds that got distributed into the environment during production, transport, use, or disposal. The control and protection of natural resources is key in particular for a safe drinking water supply. Usually, the continuous online measurement of sum parameters like electric conductivity and pH value or sum parameters like TOC is standard. The targeted monitoring of VOCs and the less volatile dissolved SVOCs in the low µg/L range is the major analytical task in routine environmental analysis. The U.S. EPA provides a comprehensive overview of analytical methods, not limited to VOCs and SVOCs, within the "Selected Analytical Methods for Environmental Remediation and Recovery" (SAM 2017) program for analysis of solids, non-drinking water, drinking water, air and wipe samples [40].

The EPA Method 8270E is used "to determine the concentration of semivolatile organic compounds in extracts prepared from many types of solid waste matrices, soils, air sampling media and water samples" [41]. The method lists more than 250 compounds for GC-MS detection: "This method can be used to quantitate most neutral, acidic, and basic organic compounds that are soluble in methylene chloride (or other suitable solvents provided that the desired performance data can be generated) and are capable of being eluted, without derivatization, as sharp peaks from a gas chromatographic fused-silica capillary column coated with a slightly polar silicone. Such compounds include PAHs, chlorinated hydrocarbons, chlorinated pesticides, phthalate esters, organophosphate esters, nitrosamines, haloethers, aldehydes, ethers, ketones, anilines, pyridines, quinolines, aromatic nitro compounds, and phenols (including nitrophenols)." The EPA 8270 method suggests different appropriate extraction and concentration methods based in common on solvent extractions, i.e. the regular separation funnel (EPA 3510), continuous liquid extraction (EPA 3520), or Soxhlet extraction (EPA 3540/41), also the assisted methods using pressurized liquid extraction (PLE) (EPA 3545), or ultrasonic extraction (EPA 3550). Due to the different chemical nature of the analytes and the different extraction principles, some of the sample preparations cover the list of SVOC compounds with certain known exceptions.

Scope and Principle of Operation

The described automated LLE method is based on the EPA Method 8270E. It is applied for drinking water, and ground and surface waters for drinking water production.

Solvents and Chemicals

- Standard for water quality control, as of Table 6.10, for quantitative calibrations. The working standard concentration is 10 µg/L in water; the final concentration in the sample vial is 0.5 µg/L (ppb range).
- Phosphate buffer to adjust pH to 4.8–5.5.
- Na_2HPO_4, CAS No. 7558-79-4, 0.5 g.

- KH_2PO_4, CAS No. 7778-77-0, 49.5 g.
- NH_4Cl, CAS No. 12125-02-9, 0.3 g.
- Weigh 0.167 g of the salt mixture into the sample vial.
- NaCl p.a., CAS 7647-14-5, for the "salting out" effect.
- ISTD as of Table 6.12, 0.5 ppb each.
- *tert*-Butyl methyl ether p.a. (MTBE), CAS No. 1634-04-4, as extraction solvent.
- Water, CAS No. 7732-18-5, HPLC quality.

Consumables

- 20 mL headspace screw-top clear vials with magnetic screw caps.
- 2 mL micro-vials with ca. 300 µL volume (or, 2 mL vials with micro-inserts) with magnetic screw caps.
- GC column: DB-UI 8270D 30 m × 0.25 mm × 0.25 µm, or equivalent.

Table 6.10 SVOC standard for method development.

Compound Name	Concentration [µg/L]
Naphthalene	1.0
Dichlorvos	0.05
2,4,6-Trichlorophenol	0.05
2,6-Dinitrotoluene	0.1
2,4-Dinitrotoluene	0.1
Hexachlorobenzene	0.01
Dimethoate	0.05
Carbofuran	0.05
Atrazine	0.05
Pentachlorophenol	0.05
Benzo(a)pyrene	0.002
γ-HCH	0.01
Anthracene	1.000
Chlorothalonil	0.05
Methylparathion	0.05
Heptachlor	0.01
Malathion	0.05
Chlorpyrifos	0.05
Fluoranthene	1.0
DDTs	0.01
DEHP	3.0
PCBs	0.05
Benzo(b)fluoranthene	0.1
HCHs	0.01

- Liquid syringe, 25 µL, for GC injection, needle gauge 23, point style conical.
- Liquid syringe, 1000 µL, for liquid handling, needle gauge 22 or 23, point style 3/LC.
- Standard GC inlet liner with glass wool.

System Configuration

The minimum hardware requirements of a robotic sampling system for the SVOC analysis via LLE are (Figure 6.46):

System – Tool – Module	Task
PAL RTC System	x,y,z-Robotic system with automated tool exchange for sample preparation and GC injection, 850 mm rail length
1 pc Park Station	Park station for tools not in use during workflow
1 pc D7/57 Liquid Tool	Tool for a 10 µL GC injection syringe (optional for liquid injections)
1 pc D8/57 Liquid Tool	Tool for 1000 µL sample preparation syringe
2 pc Tray Holder	Tray holder for 3× vial racks, to be extended as required for high sample throughput
2 pc Rack 15× 10/20 mL	15 pos. racks for 10/20 mL vials
1 pc Rack 54× 2 mL	54 pos. rack for 2 mL and micro-vials
1 pc Solvent Module	For 3× 100 mL solvent reservoirs
1 pc Standard Wash Module	Wash module for up to 5× syringe wash solvents, and ISTD solution
1 pc Vortex Mixer	For intense vial mixing
1 pc Centrifuge	For solid or sediment samples (optional, requires 1200 mm rail length)
1 pc Mounting kit	Installs the x,y,z-robot GC instrument top
1 pc GC system	With SSL or PTV type injector for LVI
1 pc MS system	Single or triple quadrupole analyzer, EI ion source

Analysis Parameter

The analytical parameters for the LLE and GC-MS/MS analysis are provided with Tables 6.11 and 6.12.

6.9 Semivolatile Organic Compounds in Water

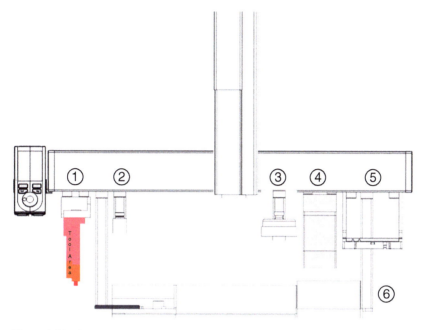

Figure 6.46 Robotic system configuration for the automated LLE extraction of water samples. 1 Tool park station; 2 Standard wash station, and position of ISTD; 3 Solvent storage module; 4 Vortex mixer; 5 Trayholder for 2 × 15 20 mL and 1 × 54 2 mL sample vials; and 6 Mounting legs for GC-MS unit.

Table 6.11 Selected analysis parameters for the GC-MS analysis of SVOCs.

LLE extraction	
Extraction temperature	Room temperature
Vortex mixer speed	1500 rpm
Vortexing time	100 s
GC	
Injector type	PTV, MMI or similar for LVI
Inlet liner	Standard liner with glass wool
Inlet temperature	40 °C (0.3 min) → 600 °C/min, 325 °C
Inlet mode	LVI concurrent solvent recondensation mode
Purge flow	60 mL/min, on at 2.85 min
Injection volume	20 µL
Carrier gas	He, constant flow at 1 mL/min
Pressure	58.8 kPa
Oven program	35 °C (2 min) → 40 °C/min → 170 °C → 10 °C/min → 310 °C (3 min)
Retention time locking	Phenanthrene-D10, RT 12.662 min
Transfer line temperature	300 °C
MS	
System, ionization	Triple quadrupole, EI mode
Quadrupole temperature	180 °C
Ion source temperature	300 °C
Interface temperature	280 °C
Scan mode	MRM, see Table 6.12

Table 6.12 SVOC compounds by LLE GC-MS analysis with linear range, MDL, and acquisition parameter (bold: MRM transitions for quantification).

No.	Compound name	RT [min]	Calibr. precision R^2	Linear range [µg/L]	MDL [µg/L]	Precursor ion [m/z]	Product ion [m/z]	Collision energy [V]
1	Dichlorvos	9.039	0.999	0.01–2.0	0.005	109.0	79.0	5
						184.9	**93.0**	**10**
2	2,4,6-Trichlorophenol	9.663	0.997	0.01–2.0	0.01	131.9	97.0	10
						196.0	**97.0**	**30**
3	2,6-Dinitrotoluene	10.205	0.995	0.01–2.0	0.01	165.0	90.1	15
						165.0	**63.0**	**25**
4	Acenaphthene-d10 (ISTD)	10.454				**162.1**	**160.1**	**20**
						164.1	162.1	15
5	2,4-Dinitrotoluene	10.676	0.992	0.02–2.0	0.02	**165.0**	**63.0**	**45**
						165.0	119.0	5
6	HCH, alpha-	11.920	0.999	0.01–2.0	0.005	216.9	181.0	5
						180.9	**145.0**	**15**
7	Hexachlorobenzene	12.001	0.999	0.01–2.0	0.002	182.9	147.0	15
						281.8	**211.9**	**30**
8	Dimethoate	12.076	0.998	0.01–2.0	0.01	**87.0**	**46.0**	**20**
						125.0	47.0	15
9	Carbofuran	12.134	0.996	0.05–2.0	0.05	164.2	149.1	10
						149.1	**121.1**	**5**
10	Atrazine	12.220	0.998	0.01–2.0	0.01	214.9	58.1	10
						200.0	**94.0**	**20**

11	HCH, beta-	12.278	0.998	0.01–2.0	0.01	**181.0**	**145.0**	**15**
						216.9	181.1	5
12	Pentachlorophenol	12.336	0.997	0.05–2.0	0.05	265.9	167.0	25
						267.9	**167.0**	**25**
13	HCH, gamma-(Lindane)	12.436	0.998	0.01–2.0	0.005	**216.9**	**181.0**	**5**
						181.0	145.0	15
14	Phenanthrene-d10 (ISTD)	12.670				**188.2**	**160.1**	**30**
						184.1	156.0	30
15	Chlorothalonil	12.670	0.998	0.01–2.0	0.01	265.9	230.9	20
						265.9	**133.0**	**45**
16	Anthracene	12.802	0.994	0.01–2.0	0.01	188.2	108.0	40
						178.1	**152.1**	**25**
17	HCH, delta-	12.854	0.998	0.01–2.0	0.01	**217.0**	**181.1**	**5**
						181.1	145.1	15
18	2,4,4′-Trichlorobiphenyl (BZ #28)	13.366	0.998	0.005–1.0	0.005	**256.0**	**186.0**	**25**
						258.0	186.0	25
19	Parathion-methyl	13.435	0.995	0.01–2.0	0.02	262.9	109.0	10
						125.0	**47.0**	**10**
20	Heptachlor	13.636	0.999	0.01–2.0	0.01	**271.7**	**236.9**	**15**
						273.7	238.9	15
21	2,2′,5,5′-Tetrachlorobiphenyl (BZ #52)	13.933	0.998	0.005–1.0	0.005	289.9	219.9	25
						291.9	**219.9**	**25**
22	Malathion	13.971	0.997	0.01–2.0	0.01	126.9	99.0	5
						157.8	**125.0**	**5**

(continued)

Table 6.12 (Continued)

No.	Compound name	RT [min]	Calibr. precision R^2	Linear range [µg/L]	MDL [µg/L]	Precursor ion [m/z]	Product ion [m/z]	Collision energy [V]
23	Chlorpyrifos	14.117	0.999	0.02–2.0	0.02	196.9	169.0	15
						313.8	**257.8**	**15**
24	Fluoranthene	15.079	0.994	0.05–2.0	0.005	201.1	200.1	15
						200.1	**174.0**	**25**
25	2,2′,4,5,5′-Pentachlorobiphenyl (BZ #101)	15.421	0.999	0.005–1.0	0.005	325.9	255.9	30
						327.9	**255.9**	**30**
26	DDE-p,p′	15.913	0.999	0.01–2.0	0.005	246.1	176.2	30
						317.8	**248.0**	**15**
27	2,3′,4,4′,5-Pentachlorobiphenyl (BZ #118)	16.518	0.996	0.005–1.0	0.005	325.9	255.9	30
						327.9	**255.9**	**30**
28	DDD-p,p′	16.674	0.999	0.01–2.0	0.005	237.0	165.1	25
						165.1	115.0	35
29	DDT-o,p′	16.729	0.999	0.01–2.0	0.005	235.0	165.2	20
						235.0	**199.1**	**15**
30	2,2′,4,4′,5,5′-Hexachlorobiphenyl (BZ#153)	16.886	0.998	0.005–1.0	0.005	359.9	289.9	25
						287.9	**217.9**	**40**
31	DDT-p,p′	17.353	0.991	0.01–2.0	0.005	**235.0**	**165.2**	**20**
						237.0	165.2	20

#	Compound	RT	R^2	Range		Ion 1	Ion 2	%
32	2,2′,3,4,4′,5′-Hexachlorobiphenyl (BZ #138)	17.362	0.998	0.005–1.0	0.005	**359.9**	**289.9**	**30**
						287.9	217.9	40
33	Triphenylphosphate (ISTD)	17.660				**214.9**	**168.1**	**15**
						326.0	325.0	5
34	Chrysene-d12 (ISTD)	18.300				**240.2**	**236.2**	**35**
						236.1	232.1	40
35	2,2′,3,4,4′,5,5′-Heptachlorobiphenyl (BZ #180)	18.538	0.999	0.005–1.0	0.005	393.8	323.8	30
						395.8	**323.8**	**30**
36	2,2′,3,3′,4,4′,5,5′-Octachlorobiphenyl (BZ #194)	20.121	0.998	0.005–1.0	0.005	427.8	357.8	30
						429.8	**359.8**	**30**
37	Benzo[b]fluoranthene	20.625	0.999	0.01–2.0	0.005	**252.1**	**250.1**	**35**
						126.0	113.1	10
38	2,2′,3,3′,4,4′,5,5′,6-Nonachlorobiphenyl (BZ #206)	20.723	0.998	0.005–1.0	0.005	461.8	391.7	30
						463.8	**393.7**	**30**
39	Benzo[a]pyrene	21.276	0.999	0.01–2.0	0.005	**252.1**	**250.1**	**35**
						125.0	124.1	10

Workflow

Initial Manual Steps

- A few initial manual steps prepare the 20 mL vials for the automated LLE process:
- Add 4 g of NaCl to each vial.
- Add 0.167 g phosphate buffer powder to adjust pH to 4.8–5.5 (NH_4Cl/phosphoric acid) to each vial.
- Add 10 mL of sample water and cap the vial. Alternatively, the sample can be added online from a flow cell using a pipette or dilutor tool.
- Place the vial into the rack of the robotic sampler.

Automated Workflow

- The automated workflow of the LLE extraction of SVOCs is illustrated graphically in Figure 6.47.

Quantitative Calibration

A series of spiked distilled water samples with standard concentrations of 0.01, 0.02, 0.05, 0.1, 0.2, 1.0, 2.0, and 4.0 µg/L, in which the polychlorinated biphenyl (PCB) concentrations are 0.005, 0.01, 0.025, 0.05, 0.1, 0.5, and 1.0 µg/L, were run for

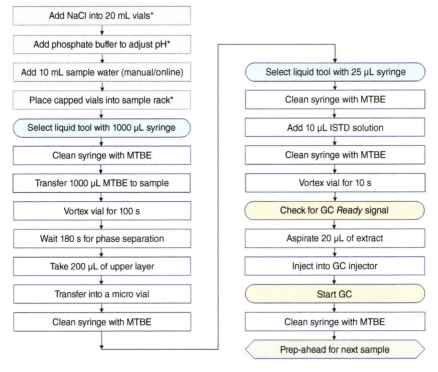

Figure 6.47 Automated workflow of the LLE extraction for SVOC analysis (*initial manual steps).

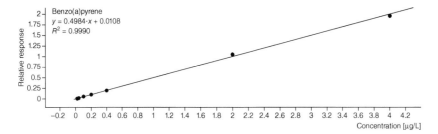

Figure 6.48 Quantitative relative response calibration of benzo(a)pyrene in the target concentration range: 0.01, 0.02, 0.05, 0.1, 0.2, 1.0, 2.0, and 4.0 μg/L, LLE extracted from spiked dest. water. Source: Courtesy CTC Analytics AG.

quantitative calibration. As a lead substance, the calibration for benzo(a)pyrene is shown in Figure 6.48. A very good linear calibration is achieved with a correlation factor R^2 of 0.9990.

Sample Measurements

The goal for the automated method is to avoid the time-consuming extract evaporation with the additional risk of the loss of analytes during extract blowdown. The final injection volume of the extract was determined for a reliable peak integration of the lead substance benzo(a)pyrene. The particular GC peaks of benzo(a)pyrene with the different injection volumes of 1, 10, 20, and 50 μL of the LLE extract 2.0 μg/L are given in Figure 6.49. The injection volume of 20 μL was chosen for sample analysis. Figure 6.50 shows the chromatogram of a spiked water sample with all the 35 SVOC compounds at a concentration level of 2.0 μg/L with 20 μL volume injected.

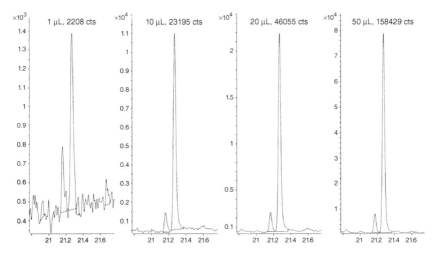

Figure 6.49 Benzo(a)pyrene LVI chromatograms of 1, 10, 20, and 50 μL of the LLE extract, 20 μL injection volume was selected for routine analysis. Source: Courtesy CTC Analytics AG.

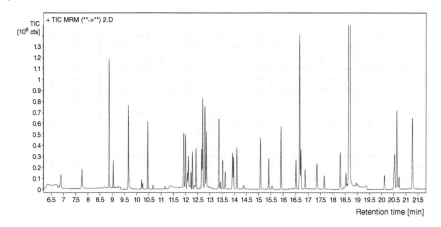

Figure 6.50 GC-MS chromatogram (MRM TIC) of 35 SVOC compounds and 4 ISTDs after LLE, each 2.0 μg/L. Compounds and retention times see Table 6.12. Source: Courtesy CTC Analytics AG.

Recovery

The recoveries for regular tap water and spiked sample were achieved in the range of 52–120%.

Regulations

The described automated LLE method complies with the China national standard test method GB/T 14848-2017 "Standard for Groundwater Quality" setting the maximum residue levels in the range of 0.01 μg/L for the organochlorine pesticides (OCPs) like HCB, HCHs or heptachlor, to 1 μg/L for the polyaromatic hydrocarbons (PAHs) naphthalene, anthracene, or fluoranthene. Benzo(*a*)pyrene is regulated with 0.002 μg/L. The analysis of the individual target compounds is regulated by a series of HJ standard methods for water quality control HJ 676-2013, HJ 478-2009, HJ 744, HJ 753-2015, HJ 715-2014, or HJ 676-2013. The described method comprises the requirements of the individual methods and allows the analysis in one automated LLE sample preparation.

Conclusion

The automated analysis method requires less organic solvent for extraction and a smaller sample volume than for the comparable manual procedure. This contributes to the operational safety by avoiding a solvent evaporation step and is a true green approach to environmental protection with less solvent waste. The automated workflow provides the required high sensitivity detection, in particular for the lead compound benzo(*a*)pyrene. The treatment of several samples in parallel for individual compounds as found in some regulations is replaced by one comprehensive multi-compound method. The large volume GC injection improves sensitivity, simplifies, and shortens the pretreatment process. The application of the triple quadrupole mass spectrometer in MRM mode with several mass transitions per analyte overcomes potential matrix interferences for a reliable qualitative and quantitative analysis. The seamless automation and integration of sample extraction and analysis increases the number of sample analyses per time.

6.10 Polyaromatic Hydrocarbons in Drinking Water

Application

PAHs are a class of organic compounds with two or more fused aromatic rings, containing only carbon and hydrogen. They are ubiquitous pollutants formed from combustion processes. PAHs are not usually found in water in notable concentrations. Their presence in surface water or groundwater is an indication of a source of pollution [42]. Some PAHs are known carcinogens and mutagens, in particular those having four or more aromatic rings forming a "bay region," such as benzo[a]pyrene, the best-studied cancerogenic PAH, shown in Figure 6.51 [43, 44].

Scope and Principle of Operation

The conjugated electron ring system of PAHs allows several sensitive analysis methods, of which most common are GC-MS, LC-UV, or FD detection [45]. The regulated low levels in drinking water require a high recovery extraction and selective detection method. PAHs are soluble in many organic solvents, but relatively insoluble in water. The compounds are susceptible to photooxidation; hence, amber vials should be used.

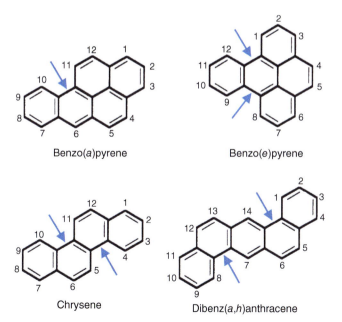

Figure 6.51 "Bay regions" of selected cancerogenic PAH compounds (Source: Adapted from Agency for Toxic Substances and Disease Registry (ATSDR) [44]).

The described method employs the SPME immersion technique for a sorbent-based extraction directly from the water sample, followed by GC-MS analysis in SIM mode for quantification. It was reported that the SPME extraction of PAHs from water became challenging in particular for hard and chlorinated waters with high dissolved natural minerals, or a high chlorine content due to disinfection processes [46]. Sodium thiosulfate is added for compensating such influences [47].

Solvents and Chemicals

- PAH mix: 16 components in cyclohexane (here 100.8 pg/mL for each PAH).
- ISTD: 4 deuterated PAH compounds (here 136.4 pg/mL for each deuterated PAH).
- Benzo[a]pyrene-d12 in cyclohexane, CAS No. 63466-71-7.
- Sodium thiosulfate, CAS No. 10102-17-7, aqueous solution 1.8%.
- Acetone, reagent grade, CAS No. 67-64-1.
- Water, CAS No. 7732-18-5, HPLC quality.

Consumables

- 20 mL headspace screw-top amber vials with magnetic screw caps.
- SPME Arrow, 1.1 mm OD, 100 µm PDMS sorbent phase.
- GC inlet liner, 1.3 or 1.7 mm ID.
- GC column: Rtxi-5MS, 30 m × 0.25 mm × 0.25 µm, or equivalent.
- Liquid syringe, 100 µL, for liquid handling, needle gauge 22 or 23, point style 3/LC.

System Configuration

The minimum hardware requirements of a robotic sampling system for the analysis of PAHs in drinking water as shown in Figure 6.52 are:

System – Tool – Module	Task
PAL RTC System	x,y,z-Robotic system with automated tool change, 1200 mm rail length
1 pc Park Station	Park station for tools not in use during workflow
1 pc D7/57 Liquid Tool	Tool for a 100 µL liquid syringe
1 pc SPME Arrow tool	Tool for SPME Arrow operation
1 pc Tray Holder	Tray holder for 3× vial racks
3 pc Rack 15× 10/20 mL	15 pos. racks for 10/20 mL vials
1 pc Rack 60× 10/20 mL vial	60 pos. rack for 10/20 mL vials (optional)
1 pc Standard Wash Module	Wash module for up to 5× wash solvents
1 pc Fast Wash Module	Wash module for 2× active wash solvents
1 pc Vortex Mixer	For intense vial mixing
1 pc Heatex Stirrer	Heated module for stirring during SPME Arrow extraction

6.10 Polyaromatic Hydrocarbons in Drinking Water

System – Tool – Module	Task
1 pc SPME conditioning module	Conditioning of SPME Fibers/Arrows before/after analysis
1 pc Agitator/Incubator	Sample incubation of up to 6× 20 mL vials
1 pc Mounting kit	Installs the x,y,z-robot GC instrument top
1 pc GC system	With split/splitless, or PTV type injector
1 pc SPME Arrow kit	Arrow inlet adaptor
1 pc MS system	Single or triple quadrupole analyzer, EI ion source

Analysis Parameter

The analysis parameters for the SPME immersion extraction of PAHs from water are listed in Table 6.13 and for GC-MS analysis in SIM mode in Table 6.14.

Workflow

Initial Manual Steps

- Prepare a 1.8% aqueous sodium thiosulfate solution in a 10 mL vial as a reducing agent for water samples.
- Place the $Na_2S_2O_3$ vial into a dedicated position of the standard wash station.
- Prepare 15 mL water samples in 20 mL vials and place them into the rack of the trayholder.

Automated Workflow

- The steps of the automated PAH extraction and analysis are graphically illustrated in Figure 6.53.

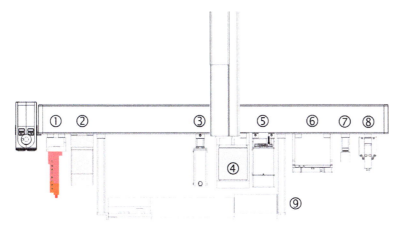

Figure 6.52 Robotic system configuration for SPME extraction of PAHs from water samples. 1 Tool park station; 2 Vortexing unit; 3 SPME conditioning station; 4 Agitator/incubator; 5 Heatex stirrer; 6 Sample trayholder; 7 Standard wash station; 8 Fast wash station; and 9 Mounting legs.

6 Solutions for Automated Analyses

Table 6.13 Selected analysis parameters for the SPME GC-MS analysis of PAHs in water.

SPME extraction	
Sample water volume	15 mL (10 mL)
Agitator incubation temperature, time	35 °C, 10 min
Agitator speed	700 rpm
Pre-conditioning temperature, time	250 °C, 10 min
Extraction temperature, time	35 °C, 30 min
Heatex stirrer	1500 rpm (1000 rpm for 10 mL)
Vial penetration depth	50 mm (55 mm for 10 mL)
SPME cleaning	2 min, water, fast wash station
Post-conditioning temperature, time	250 °C, 5 min
GC	
Inlet temperature	280 °C
Desorption time	5 min
Inlet mode	Splitless
Carrier gas	He, constant flow at 1 mL/min
Oven program	35 °C (5 min) → 40 °C/min → 150 °C → 20 °C/min → 250 °C → 10 °C/min → 305 °C (22 min)
Transfer line temperature	310 °C
MS	
System and ionization	Single quadrupole, EI mode
Ion source temperature	250 °C
Scan mode	SIM (as of Table 6.14)

Sample Measurements

The chromatogram in Figure 6.54 shows the GC-MS analysis of the SPME Arrow extraction of a spiked water sample at the level of 0.1 µg/L of the PAH standard. The quantitative calibration in the range of 0.01–0.20 µg/L was run in a regular drinking water matrix in duplicate runs. A very good precision with a determination coefficient R^2 of 0.9947 was achieved (Figure 6.55).

It should be noted that the wash steps of the workflow are of particular importance with "hard" waters of high mineral content. Any potential mineral deposit after desorption needs to be washed off for extended use of the SPME sorbent material for large sample series.

Regulations

Benzo(a)pyrene is generally used as a marker for the occurrence of carcinogenic PAHs in food [48]. In 2015 in Canada, a maximum acceptable concentration of

Table 6.14 PAH analytes GC-MS acquisition parameter.

No.	Compound name	Abbrev.	Target/ISTD	SIM group	Ret. time [min]	Quan. ion [m/z]
1	Naphthalene	N	Target	1	9.2410	128
2	Acenaphthylene	AY	Target	1	10.688	152
3	Acenaphthene-d10	AY-d10	ISTD	1	10.832	164
4	Acenaphthene	AE	Target	1	10.863	153
5	Fluorene	F	Target	1	11.418	166
6	Phenanthrene-d10	P-d10	ISTD	2	12.482	188
7	Phenanthrene	P	Target	2	12.508	178
8	Anthracene	A	Target	2	12.569	178
9	Fluoranthrene	FL	Target	3	14.004	202
10	Pyrene	PY	Target	3	14.326	202
11	Benz(a)anthracene	BaA	Target	3	16.258	228
12	Chrysene-d12	C-d12	ISTD	3	16.275	240
13	Chrysene	C	Target	3	16.320	228
14	Benzo(b)fluoranthene	BbF	Target	4	18.204	252
15	Benzo(k)fluoranthene	BkF	Target	4	18.252	252
16	Benzo(a)pyrene-d12	BaP-d12	ISTD	4	18.762	264
17	Benzo(a)pyrene	BaP	Target	4	18.803	252
18	Indeno(1,2,3-cd)pyrene	IP	Target	4	21.349	276
19	Dibenz(a,h)anthracene	DA	Target	4	21.436	278
20	Benzo(g,h,i)perylene	BghiP	Target	4	22.043	276

0.04 µg/L for benzo[a]pyrene (BaP) in drinking water was proposed [49]. The European regulation limits benzo(a)pyrene at 0.010 µg/L, and other polycyclic aromatic hydrocarbons at 0.10 µg/L as the sum of concentrations of benzo(b)fluoranthene, benzo(k)fluoranthene, benzo(ghi)perylene, and indeno(1,2,3-cd)pyrene [50]. China specifies a maximum benzo(a)pyrene concentration of 0.01 µg/L, and total PAHs concentration of 2 µg/L in the regulation GB 5749-2006 [51].

Conclusion

The SPME Arrow technology with immersion sampling provides a fast, robust, and automated method for the determination of PAH residues in drinking water. The method is compliant with the current regulations to control the set MRL levels in drinking water. The automated workflow and conditioning procedure for the SPME Arrow ensures good reproducible results for large sample series.

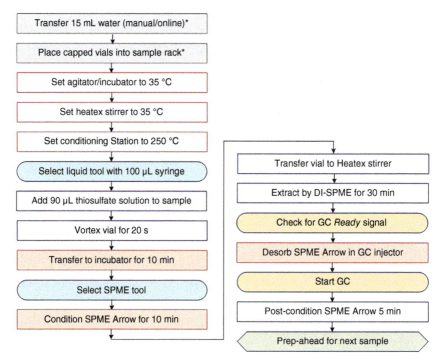

Figure 6.53 Automated workflow steps for the DI-SPME analysis of PAHs in drinking water (*initial manual steps).

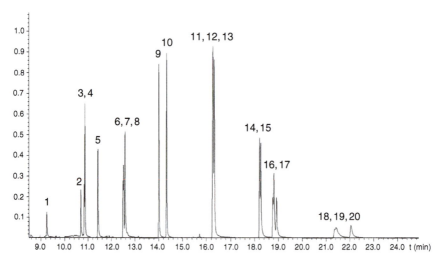

Figure 6.54 GC-MS total ion chromatogram of the PAH standard after DI-SPME Arrow extraction (peak # as of Table 6.14, benzo(*a*)pyrene **#17**). Source: Courtesy CTC Analytics AG.

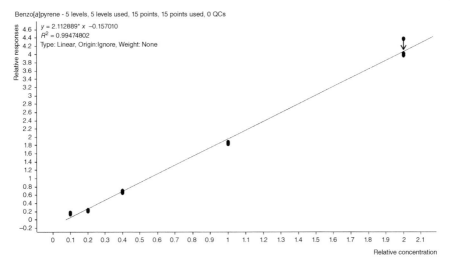

Figure 6.55 Quantitative calibration for Benzo(*a*)pyrene in the range of 0.01–0.20 μg/L. Source: Courtesy CTC Analytics AG.

6.11 Fatty Acid Methylester

6.11.1 Application

The analysis of the fatty acid profile or the quantification of total fat and individual fatty acids are basic analytical requirements in food, nutrition, and medical sciences. Fatty acid profiling is commonly performed using gas chromatography (GC) to separate the compounds followed by quantification with a flame ionization detector (FID). Fats in food can be present as free fatty acids (FFA), but are primarily bound in mono-, di- and triacylglycerides (MAG, DAG, TAG), and other types of lipids, e.g. the phospholipids. The boiling points of these lipids are high and complicate the direct analysis by GC with the problematic of a discrimination-free injection of the high boilers [52], require the use of high-temperature GC column material [53], and reveal complex chromatograms due to the complexity of the fatty acid glycerol substitution pattern [54]. TAG analysis by GC can be applied inter alia for authenticity control [55, 56]. For the fatty acid profiles or individual fatty acid quantification, the samples require preparation steps to release the fatty acids from the bound form and make them GC-amenable by derivatization.

Scope and Principle of Operation

The classical sample preparation method involves the hydrolysis of the fatty acids from lipids to the FFA followed by methylation to form the fatty acid methyl esters (FAMEs) [57]. As such, the method determines the FAMEs as a sum parameter and profile from all fatty acid carrying lipids in a sample. The FAMEs formed have lower boiling points than their FA/triglyceride counterparts and hence are easily amenable for a compound-specific separation by GC. The generated FAMEs are isolated by LLE into an organic solvent and then analyzed by GC-FID or GC-MS. It needs to be noted that the response factors of FAMEs may individually differ from GC-FID to GC-MS affecting results in 100% methods. The analytical benefit of the FID is the wide dynamic range, a response proportional to the number of carbon atoms, and its high sensitivity for FAMEs, reported to be more than 13 times more sensitive compared to triacylglycerides (TAGs) or FFAs [58].

Official methods have been established on gas chromatographic methods by the American Organization of Analytical Chemists (AOAC), the American Oil Chemist Society (AOCS), and the International Standards Organization (ISO), for instance with the AOAC standard methods 996.01 and 996.06 for fats in food [59, 60], the AOCS method Ce 1i-07 for polyunsaturated fatty acids (PUFA) in fish oils [61], or the ISO/EN 12966 method for animal and vegetable fats and oils [62], besides several similar methods for matrix-specific applications. These methods all have in common time-consuming manual sample handling steps typically prone to unavoidable individual variations, also can expose the analyst to hazardous chemicals. The automation of the FAMEs sample preparation and measurement using automated workflows is highly desirable for comparable results, operator safety, and increased sample throughput.

The described FAME method is an automated workflow for the analysis of fat in foods. The preparation of FAMEs takes place directly from vegetable oils or animal fat. Oil and fat samples can be applied directly without further preparation. Fat-containing food needs a prior fat extraction with a diethyl ether/petroleum ether mixture.

Depending on the choice of GC separation, the described sample preparation workflow can be used for the chromatographic determination for different analytical tasks including the determination of the total fat contents, registration of fatty acid profiles, or the quantification of cis/trans or the PUFA concentrations. The described workflow and default parameter set were developed for the AOCS method 996.01. Analogous FAME methods may require parameter adaptations. The AOAC method 996.01 is prepared for cereal products containing 0.5–13% total fat. The AOAC method 996.06 works analogous, but is prepared for general foods with modifications for dairy products and cheese.

Method Description

Oils and pure fats and fatty acids can be applied directly without prior extraction. Other food commodities need to be extracted before. The fat extraction is a manual step before the extracted fat extract can be applied to the described FAME workflow. The automated workflow includes saponification, methylation, and optional, the online GC analysis.

The described FAME workflow for AOAC 996.01 is an automated sample preparation method for the determination of FAMEs from fat samples by alkaline esterification using methanolic sodium hydroxide, generating the FFA from triglycerides and other lipids. The subsequent methylation using methanolic boron trifluoride

(BF_3) catalyst generates the FAME derivatives from the FFA. An LLE with hexane or heptane isolates the FAMEs from the reaction mixture for GC analysis.

The liquid handling steps and derivatization are executed on an x,y,z-robotic sampler using a liquid tool with a 1000 µL preparative syringe. Multiple liquid transfers are executed when a transfer volume is greater than the syringe volume. In the offline method, no tool change is required. The GC injection is optional and requires the automated tool change to the GC injection syringe. The transport of the reaction vials is executed with magnetic transport and requires magnetic vial screw/crimp caps. Finally, the prepared FAME derivative vials can be transferred and distributed to separate GC or GC-MS systems for chromatographic analysis, or injected to GC in the online mode.

Derivatization Method

The principle of the preparation method is the basic hydrolysis of the sample triglycerides (saponification) followed by the methylation to the FAMEs using BF_3 as a catalyst. The liquid/liquid solvent extraction using a Vortex mixer isolates the generated FAMEs. The hexane layer with the generated FAMEs is transferred into vials for the following separate analyses, or used for direct injection of a microliter aliquot of the organic phase to GC-FID or GC-MS analysis. The described operation procedure in this application follows the AOAC Official Method 996.01 for the determination of total, saturated, and unsaturated fat.

Limitations

The described FAME workflow for AOAC 996.01 performs the sample preparation in standalone mode (no GC required) without injecting the prepared FAMEs into the GC inlet. Samples are prepared in parallel with batches of six vials, corresponding to the capacity of the incubator, with a maximum of 54 samples being prepared per sample sequence.

Solvents and Chemicals

For AOAC 996.01 fat extraction:

- Ethanol, CAS No. 64-17-5, analytical grade.
- Hydrochloric acid, CAS No. 64-17-5, 8M (25 + 11 v/v).
- Diethyl ether, CAS No. 60-29-7, peroxide-free.
- Petroleum ether, CAS No. 8032-32-4, peroxide-free.

For AOAC 996.01 automated saponification and methylation:

- Methanol p.a., CAS No. 67-56-1.
- Trichloromethane p.a., CAS No. 67-66-3.
- Boron trifluoride catalyst, CAS No. 7637-07-2, 14% BF_3 solution in methanol.
- Methanolic sodium hydroxide, CAS No. 1310-73-2, solution, 0.5M NaOH in methanol.

- Sodium chloride, CAS No. 7647-14-5, brine, saturated aqueous solution, ca. 26% w/v.
- n-Heptane, p.a., CAS No. 142-82-5.
- Triglyceride ISTD solution
  - C13:0 tritridecanoin (glyceryl tritridecanoate), CAS No. 26536-12-9, 5.00 mg/mL in $CHCl_3$.
  - The ISTD solution is stable for one week if stored in amber vials <10 °C.
- Acetone, p.a., CAS No. 67-64-1.

For AOAC 996.06 fat extraction:

- Pyrogallol, CAS No. 87-66-1.
- Ethanol, p.a., CAS No. 64-17-5.
- Hydrochloric acid, CAS No. 64-17-5, 12M and 8M (25 + 11 v/v).
- Ammonium hydroxide, CAS No. 1336-21-6, 58% (w/w).
- Phenolphthalein, CAS No. 77-09-8.
- Diethyl ether, p.a., CAS No. 60-29-7, peroxide-free.
- Petroleum ether, CAS No. 8032-32-4, peroxide-free.

For AOAC 996.06 automated saponification and methylation:

- Trichloromethane, p.a., CAS No. 67-66-3.
- Diethyl ether, p.a., CAS No. 60-29-7, peroxide-free.
- Boron trifluoride catalyst, CAS No. 7637-07-2, 14% BF_3 solution in methanol.
- Toluene, p.a., CAS No. 108-88-3.
- Water, CAS No. 7732-18-5, HPLC quality.
- n-Hexane, p.a., CAS No. 110-54-3.
- Sodium sulfate, p.a., CAS No. 7757-82-6.
- Triglyceride ISTD solution:
  - C11:0 triundecanoin (glyceryl triundecanoate), CAS No. 13552-80-2, 5.00 mg/mL in $CHCl_3$.
  - Accurately weigh 2.50 g C11:0-triundecanoin into 500 mL volumetric flask. Add c. 400 mL $CHCl_3$ and mix until dissolved. Stable for one month <10 °C.
- Acetone, p.a., CAS No. 67-64-1.

Health Risk

Caution: Boron trifluoride (CAS No. 7637-07-2) is toxic and corrosive, may be fatal if inhaled. Refer to the safety data sheet (SDS) for precautions and safe handling.

Consumables

For the automated saponification and methylation workflow:

- 10 mL vials, clear, magnetic screw or crimp caps.
- 2 mL autosampler vials, non-magnetic screw caps.
- Syringe 10 µL, for GC injection, needle gauge 23s, point style AS/conical (optional).
- Syringe 1000 µL, with needle gauge G22 or G23, point style 3/LC.

System Configuration

The minimum hardware requirements of the robotic sampling system for the FAME preparation as shown in Figure 6.56 are:

System – Tool – Module	Task
PAL RSI System	x,y,z-Robotic system with manual tool change for liquid injection, 1200 mm rail length (automated RTC tool exchange optional)
1 pc Park Station	(Optional park station only for RTC system for tools not in use during for automated tool change)
1 pc D8/57 Liquid Tool	Tool for 1000 µL sample preparation syringe
1 pc MHE Module	For overpressure compensation in the capped vials after liquid additions (optional)
1 pc D7/57 Liquid Tool	Tool for an optional 10 µL GC injection syringe (optional)
2 pc Tray Holder	Tray holder for 3× vial racks
1 pc Rack 60× 10/20 mL vial	60 pos. 10 or 20 mL vial rack
3 pc Rack 54× 2 mL	54 pos. 2 mL vial rack
1 pc Fast Wash Module	Wash module for 2× active wash solvents
1 pc Solvent Module	For 3× 100 mL solvent/reagent reservoirs.
1 pc Agitator/Incubator	Sample incubation of up to 6× 20 mL vials
6 pc Vial adaptors	Agitator slot insert for the use of 10 mL vials
1 pc Vortex Mixer	For vial vortexing
1 pc Mounting kit	For benchtop installation (instrument top GC installation optional)
1 pc GC system	With split/splitless, or PTV type injector, and FID detector

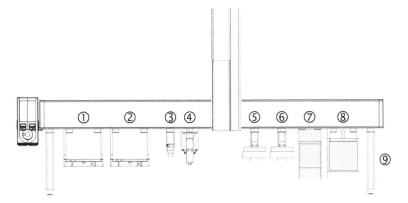

Figure 6.56 Configuration for the BF_3-catalyzed FAMEs derivatization workflow in benchtop installation. 1 Tray holder 1; 2 Tray holder 2; 3 MHE adapter; 4 Fast wash station; 5 Solvent module 1; 6 Solvent module 2; 7 Vortex mixer; 8 Agitator/Incubator; 9 Mounting legs (benchtop or instrument top).

The tray holders, solvent modules, and fast wash station are equipped with vials, reagents, and solvents as follows:

- Tray Holder 1
  - 60 pos. rack raw, sample vials with fat samples for derivatization.
  - The raw sample vials (10 mL) must be equipped with magnetic caps.
- Tray Holder 2
  - Slot 1: 54 pos. FAME derivative vials (2 mL) for subsequent GC injection.
  - The FAME derivative vials must be equipped with non-magnetic caps.
- Solvent Module 1
  - Pos 1: NaOH in MeOH.
  - Pos 2: BF_3 in MeOH.
  - Pos 3: NaCl solution.
- Solvent Module 2
  - Pos 1: ISTD solution.
  - (Alternatively, the ISTD solution can also be placed in a 10 mL vial in the 60 pos. rack on Tray Holder 1).
  - Pos 2: Heptane.
  - Pos 3: not used.
- Fast Wash
  - Pos 1: Water.
  - Pos 2: Acetone.
  - Pos 3: Waste port with drain tube.

Analysis Parameter

The finally prepared FAMEs are analyzed by GC with FID detection. The choice of analytical column and oven temperature program depends on the purpose of analysis [63]:

- Fat content – uses short nonpolar GC columns, e.g. 3 m × 0.25 mm ID × 0.3 µm film, 100% methyl polysiloxane.
- Fatty acid profiles – uses polar Carbowax columns (polyethylene glycol phase), typically of 30 m length × 0.25 mm ID × 0.25 µm film, e.g. DB-Wax, TraceGOLD TG-WaxMS, Rtx-Wax or similar.
- Nutritional values – requires a polar film, long columns for separation, e.g. 60 m length × 0.25 mm ID × 0.2 µm film DB-23, TraceGOLD TG-POLAR, BPX-90, or similar.
- Cis/trans fatty acid analysis – requires long polar phase GC columns with 70–95% cyanopropyl polysilphenylene-siloxane phase, 100–120 m length × 0.25 mm ID × 0.20–0.25 µm film, e.g. HP-88, TraceGOLD, TG-POLAR, TRACE TR-FAME, BPX70, or similar.

Workflow

Initial Manual Steps

- Fat extraction from food as necessary.

- Weigh approx. 50 mg of the fat sample exact into 20 mL vials.
- Cap the vials with magnetic screw or crimp caps.
- The fat sample vials are placed into the tray holder of the robotic sampler.

Automated Derivatization Workflow

- The automated workflow of the FAME sample preparation for GC analysis is illustrated stepwise in Figure 6.57.
- The online GC analysis of the prepared FAME derivatives can be integrated optionally.

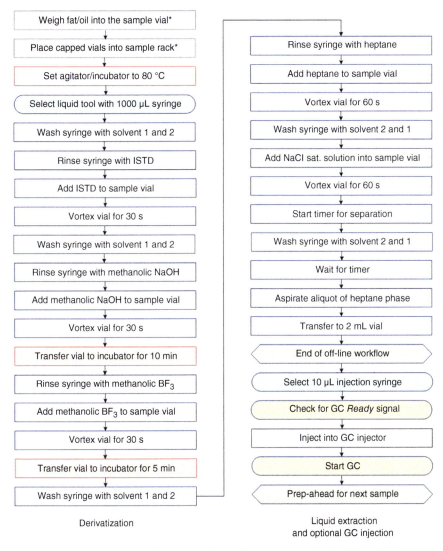

Figure 6.57 Automated workflow steps of the FAMEs preparation for GC analysis (*initial manual steps).

Sample Measurements

- With the default parameter set in this workflow, a maximum number of 40 samples can be processed in one sample sequence.
- With the default parameter set of this method, the runtime to prepare a batch of six samples is 1.50 hours.
- The runtime to prepare one single sample is 44 minutes.
- The total number of samples in one sample sequence is automatically split into batches of maximum six samples (the maximum number of positions in the agitator).
- Maximum 54 samples can be processed in one batch of samples.
- For the optional online GC analysis, the use of the prep-ahead mode is recommended.

Conclusion

The described automated method for triglyceride saponification and catalyzed methylation can be used for a wide spectrum of official analytical methods from different food commodities combined with GC-FID or GC-MS analysis. Unless oils are provided as samples for analysis, the prior extraction of fat from food is required.

6.12 MCPD and Glycidol in Vegetable Oils

3-Monochloropropane-1,2-diol (3-MCPD, Figure 6.58), 2-monochloropropane-1,3-diol (2-MCPD, Figure 6.59) and glycidol (GE, Figure 6.60) are known and regulated processing contaminants in foodstuff. 3-MCPD, GE, and its esters are formed unintentionally mainly from diacylglycerols (DAG) and monoacylglycerols (MAG) under the high-temperature conditions of the oil refining processes [64]. The routine control of the processing conditions is required to reduce their formation and to lower their levels in the final products [65].

3-MCPD and related substances are found in some processed foods and refined edible oils [66]. A high potential of 3-MCPD contamination is reported from walnut, hazelnut, olive pomace, grape seed, canola, fish, and palm oils [67]. MCPDs and glycidol were found as fatty acid-bound species not only limited to raffinated vegetable

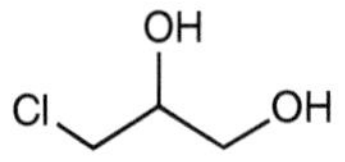

Figure 6.58 3-MCPD (3-monochloropropane-1,2-diol).

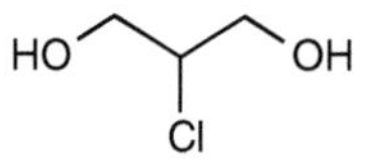

Figure 6.59 2-MCPD (2-monochloropropane-1,3-diol).

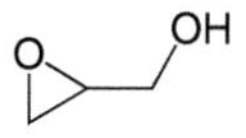

Figure 6.60 Glycidol (2,3-epoxy-1-propanol).

and fish oils, but also in margarine, bread and rolls, or even preserved meat. Only a small portion of 3-MCPD is present in its free form, but bound in some processed foods [68]. The compounds are generated at elevated temperature in the presence of acylglycerides and chloride-containing salts during processing. Already in 1978, free MCPD was found in soy sauce and hydrolyzed vegetable protein [69, 70]. It was shown that the major part of 3-MCPD and glycidol contaminations are bound to fatty acids as mono- and di-esters. The occurrence of bound MCPDs was reported as early as 2004 [71].

3-MCPD and GE are classified as potentially carcinogenic compounds. The maximum tolerated limit in food has been established at both national and international levels. In March 2016, the European Food Safety Authority (EFSA) declared a reduced value for the tolerable daily uptake of 0.8 µg/kg body weight 3-MCPD. The EFSA regulations also include maximum values for infant formula [72].

The available analysis methods can be distinguished into direct and indirect methods. Direct methods analyze MCPD and glycidol esters using LC-MS. A large number of individual compounds need to be monitored, according to the variety of bound fatty acids and their permutation. In contrast, indirect methods deliver a sum value for the total MCPD and glycidol content in a sample, independent of the bound states. Free and bound MCPD and glycidol can be differentiated with some methods. The sum value of total MCPD and glycidol in a sample is the basis for the current maximum residue limits (MRL). The methods for indirect analysis of MCPDs and glycidol hence are the basis of the international regulations setting the focus for the following method descriptions.

For analysis of MCPDs and glycidol, different methods are described as official methods. The analytical methods use common principles with the following steps:

- Addition of ISTD, usually as 3-MCPD-d5 or 3-MCPD-d5-fatty ester and glycidol-d5 or glycidol-d5-esters, depending on the method.
- Transesterification (hydrolysis) to the free form (glycidol and 2-/3-MCPD).
- Conversion of glycidol-esters to 3-monobrompropanediol-esters (3-MBPD).
- Extraction of the non-saponifiable phase and FAMEs with hexane (not analyzed, discarded).
- Derivatization of the free MCPD/MBPD with phenylboronic acid (PBA).
- Quantification by GC-MS, preferably GC-MS/MS for lowest detection limits.

Scope and Principle of Operation

The differences of the available sample preparation methods for MCPDs and glycidol are by hydrolysis (alkaline or acidic) and derivatization (in the aqueous or organic phase) as listed in Table 6.15. It was found that free and bound glycidol reacts under acidic conditions with chloride present to 3-MCPD at a fast rate and therefore leading to a potential overestimation of the 3-MCPD value. Hence, further method optimization was considered in the AOCS method 29c-13. The glycidol content transforms fully to 3-MCPD due to initial chloride presence, resulting in a total MCPD value. The true final sample 3-MCDP and glycidol values are calculated from the difference to a second reaction without the presence of chloride. The same reaction using

Table 6.15 Overview of the current MCPD and glycidol analytical methods.

AOCS method	AOCS Cd 29a-13	AOCS Cd 29b-13	AOCS Cd 29c-13	AOCS Cd 29 'd'-20	ISO 18363-4:2021
"Common Name"	"Unilever"	"3-in-1"	"DGF Fast & Clean"	"Zwagerman"	Zwagerman "Fast and Clean"
Sample size	100–110 mg	100 mg	100 mg	100–120 mg	100–120 mg
Hydrolysis	Acidic	Alkaline	Alkaline	Alkaline	Alkaline
Reaction time/temperature	16 h at 40 °C	16 h at −22 °C	3 min at RT	12 min at 10 °C	5 min at 10 °C
Workflow	Convert glycidyl esters into 3-MBPD esters before	Hydrolysis first, then convert free glycidol to 3-MBPD	Transesterification first, then convert free glycidol to 3-MBPD. Two consecutive analyses executed[a]	Transesterification first, then convert free glycidol to 3-MBPD	Transesterification first, then convert free glycidol to 3-MBPD
Derivatization	Phenylboronic acid 5 min at RT	Phenylboronic acid at RT	Phenylboronic acid at RT	Phenylboronic acid 10 s at RT	Phenylboronic acid 10 s at RT
Quantified analytes	3-MCPD, 2-MCPD, GE	3-MCPD, 2-MCPD, GE	3-MCPD, 2-MCPD, GE (calculated)	3-MCPD, 2-MCPD, GE	3-MCPD, 2-MCPD, GE

Quantitation		Calibration curve, ISTD: 3-MCPD-d5-ester, 2-MCPD-d5-ester and glycidol-d5-ester	Single point calibration, ISTD: 3-MCPD-d5-ester, glycidol-d5-ester, 2-MCPD-d5-ester	Single point calibration or calibration curve for 3-MCPD, ISTD: 3-MCPD-d5-ester	Single point calibration for 3-MCPD and GE; Calibration curve for 2-MCPD, ISTD: ^{13}C3-3-MCPD-ester, glycidol-dt5-ester	Single point calibration for 3-MCPD and GE; Calibration curve for 2-MCPD, ISTD: ^{13}C3-3-MCPD-ester, glycidol-dt5-ester
Samples per 24 h	Single head	10	26	30	32	>40
	Dual head[b]	10	30	36	37	>50
Special modules		Heated tray @ 40 °C	Cooled tray @ -22 °C	None	Cooled tray @ 10 °C	Cooled tray @ 10 °C
First results	After	18 h	17 h	48 min	46 min	36 min
Comments		• Batch approach • Potential overestimate by reactive matrix	• Batch approach • Expensive module • Reference method	• One-piece-workflow • At room temperature	• One-piece-Workflow • GC-MS/MS required • ^{13}C-labelled standard	• One-piece-workflow • GC-MS/MS required • ^{13}C-labelled standard

For "Zwagerman" methods see [73, 74].

a) Bound 3-MCPD and bound glycidol is determined as free 3-MCPD (Assay A), bound 3-MCPD only (Assay B), total glycidol by difference.
b) x,y,z-Robotic system with two parallel operating heads.

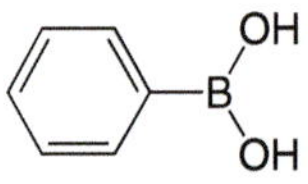

Figure 6.61 Phenylboronic acid (PBA).

bromide replacing chloride prevents the generation of additional 3-MCPD from glycidol. A conversion factor t is determined with standards. The AOCS method 29c-13 is fast, even though there are duplicate sample runs necessary, and precise, hence it became the basis of automated workflows for high productivity.

For GC-MS detection, 3-MCPD is derivatized with phenylboronic acid, Figure 6.61. After injection, an automated backflush prevents the analytical GC column from excess derivatization agent for quick cycle times.

Automated Analytical Methods

The AOCS Cd 29c-13 or DGF C-VI 18(10) is used as the most efficient method for automation of routine MCPD analyses. The principle of the automated workflow with the entire sample preparation including GC-MS analyses is shown in Figure 6.62. The evaporation step of the manual method is not necessary and optional due to the reduced sample scale, hence omitted for time-saving. A clean-up step of the sample extract uses a PBA scavenger after derivatization. Also, for

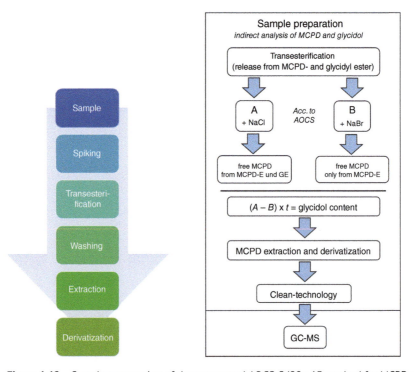

Figure 6.62 Sample preparation of the automated AOCS Cd29c-13 method for MCPD and glycidol analysis. Source: Courtesy Axel Semrau GmbH & Co KG.

quick cycle times, the protection of the analytical column from excess PBA reagent is achieved by a programmed GC injector backflush. The method is suitable for determination of 3-MCPD, 2-MCPD, and glycidol contents. One sample is being analyzed in <50 minutes. The results comply well with the conventional AOCS and DGF methods (AOCS Cd 29c-13, or DGF C-VI 18(10)).

The modular x,y,z-robotic system also allows the automation of the other MCPD methods AOCS 29a-13 and 29b-13 (see Table 6.15). For example, the AOCS method Cd 29b-13 is automated by using an additional and customized cooled tray for the required $-22\,°C$ treatment, and for the method AOCS Cd 29a-13, an evaporation unit and a centrifuge are integrated additionally.

Solvents and Chemicals for the Automated AOCS Cd29c-13 Method

- ISTD solutions of
  - 3-MCPD-d5 ester, CAS No. 342611-01-2.
  - 2-MCPD-d5 ester, CAS No. 1216764-05-4.
  - Glycidol-d5-ester, CAS No. 1346598-19-3.
- rac-1,2-Bis-palmitoyl-3-chloropropanediol, CAS No. 51930-97-3.
- rac-1,3-Distearoyl-2-chloropropanediol, CAS No. 26787-56-4.
- Glycidyl stearate, CAS No. 7460-84-6.
- MTBE, CAS No. 1634-04-4.
- Toluene, CAS No. 108-88-3.
- Methanol, CAS No. 67-56-1.
- Methanolic NaOH, CAS No. 1310-73-2.
- Acidic NaCl solution, CAS No. 7647-14-5.
- Iso-Octane, CAS No. 540-84-1.
- n-Hexane, CAS No. 110-54-3, or n-Heptane, CAS No. 142-82-5.
- Diethyl ether, CAS No. 60-29-7/Ethylacetate, CAS No. 141-78-6, mixture 3:2 (v/v).
- Phenylboronic acid, purum, CAS No. 98-80-6.
- Ethylene glycol, for analysis, CAS No. 107-21-1.

Consumables for the Automated AOCS Cd29c-13 Method

- 2 mL screw-top clear vials with magnetic screw caps.
- Syringe 1000 µL, for sample preparation, needle gauge G22 or G23, point style 3/LC.
- Syringe 250 µL, for sample preparation, needle gauge G22 or G23, point style 3/LC.
- Syringe 100 µL, for sample preparation, needle gauge G22 or G23, point style 3/LC.
- Syringe 10 µL, for GC injection, needle gauge 23s, point style AS/conical.
- GC split/splitless inlet liner, standard.
- GC pre-column: Rxi-5MS, 15 m × 0.25 mm × 0.25 µm, or equivalent.
- GC analytical column: Rxi-5MS, 15 m × 0.25 mm × 0.25 µm, or equivalent.

System Configuration

The instrument setup of the AOCS 29c method is represented in Figure 6.63. For the automation of the MCPD sample preparation, a 160 cm wide system rail with dual head operation is employed. This robotic system offers enough space for all required modules and can be additionally adapted to other methods or applications. A dilutor unit is mounted on the backside with solvent reservoirs for the extracting solvents.

The robot is mounted on top of the GC of a single or triple quadrupole GC-MS unit. This allows the immediate injection of the sample after preparation. Long residence times of the derivatization reagent PBA from the analytical column are avoided using the injector backflush option. The backflush provides shorter analysis times and the extended lifetime of the GC column due to a then reduced column bake-out temperature and time. The mass spectrometer ion source requires less ion source maintenance.

System – Tool – Module	Task
PAL DHR RSI/RTC System	x,y,z-Robotic system with dual head operation, automated tool change for sample preparation and GC injection, 1600 mm rail length
2 pc Park Station	Park stations for tools not in use during workflow
2 pc D7/57 Liquid Tool	Tool for a 10 µL GC injection syringe (optional for liquid injections)
2 pc D8/57 Liquid Tool	Tool for 1000 µL sample preparation syringe
1 pc Multi-Dilutor and Tool	For solvents dispensing
1 pc Tray Holder	Tray holder for 3× vial racks
3 pc Rack 54× 2 mL	54 pos. for 2 mL vials
2 pc Solvent Module	For 3× 100 mL solvent/reagent reservoirs
1 pc Fast Wash Module	Wash module for 2× active wash solvents
1 pc Standard Wash Module	For ISTD and derivatization agent
1 pc Agitator/Incubator	Sample incubation of up to 6× 20 mL vials
6 pc Adapter	Adapter for the insertion of 2 mL vials into the Agitator/Incubator
1 pc Vortex Mixer	For efficient liquid/liquid extraction
1 pc Mounting kit	Installs the x,y,z-robot GC instrument top
1 pc GC system	With PTV type injector and backflush
1 pc MS system	Single or triple quadrupole analyzer, EI ion source

Analysis Parameter

The standard analysis parameters for the derivatization step, GC analysis, and MS detection are listed in Tables 6.16 and 6.17.

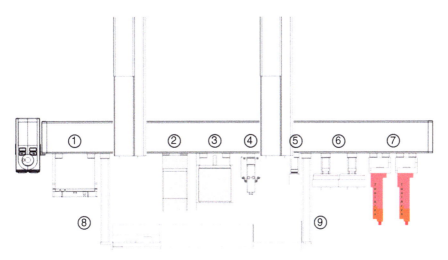

Figure 6.63 Typical configuration for an MCPD workstation installed on a GC-MS unit.
1 Tray holder for sample vials; 2 Vortex mixer; 3 Agitator/Incubator; 4 Fast wash station;
5 Standard wash station; 6 Solvent/reagent reservoirs (2×); 7 Tool park stations (2×);
8 Dilutor (installed backside, not shown); and 9 GC mounting legs.

Table 6.16 Selected analysis parameter for MCPD and glycidol analysis.

Derivatization	
Sample amount	100 mg
Reaction temperature	70 °C
Incubation time	30 s
Agitator speed	2500 rpm
GC	
Injector	250 °C
Inlet mode	Splitless
Backflush on after	8 min
Injection volume	1.0–3.0 µL
Split ratio, time	30 : 1, after 1 min
Carrier gas	He, constant flow at 1.5 mL/min
Oven program	70 °C (1 min) → 20 °C/min → 200 °C (0 min) → 40 °C/min → 300 °C (4.0 min)
Transfer line temperature	250 °C
MS	
System, ionization	Triple quadrupole, EI mode
Ion source temperature	250 °C
Scan mode	MRM, see Table 6.17

Table 6.17 MRM mass transitions for MCPDs (bold the quantifier ions).

Compound	Collision energy [V]	Transition [m/z]
2-MCPD	15	198.0 > 104.0
	15	**196.0 > 104.0**
2-MCPD-d5	10	**203.0 > 201.0**
	10	201.0 > 93.0
3-MCPD	10	**196.0 > 147.0**
	10	198.0 > 149.0
3-MCPD-d5	10	**201.0 > 150.0**
	10	201.0 > 93.0
3-MBPD[a]	8	**240.0 > 147.0**
	8	242.0 > 147.0
3-MBPD-d5[a]	8	**245.0 > 150.0**
	8	247.0 > 150.0

a) The glycidol reaction products.

Workflow

Initial Manual Steps

- Weigh 100 mg oil for analysis exact into 2 mL vials.
- Cap the vials with a septum and magnetic screw cap, and place into the tray holder rack 1.
- Prepare additional 2 mL vials with 100 mg Na_2SO_4 for the drying step, place into tray holder rack 2.

Automated Workflow

- Syringe cleaning steps are required repeatedly after syringe selection and after dispensing.
- The workflow steps are illustrated in detail in Figure 6.64.
- The cleaning steps with solvents are omitted in the graphics for improved clarity of the workflow.

Sample Measurements

3-MCPD and 2-MCPD and the deuterated standards were typically detected using GC-MS/MS triple quadrupole instruments for achieving the required selectivity and sensitivity. For each compound, two ion transitions are selected. One transition serves for quantification, the other one for qualification by the ion ratio. The used ion transitions for the GC-MS/MS analysis are shown in Table 6.16.

For validation, the 2- and 3-MCPD-esters rac-1,2-bis-palmitoyl-3-chloropropanediol and rac-1,3-distearoyl-2-chloropropanediol are used. Virgin olive oil is

6.12 MCPD and Glycidol in Vegetable Oils

Figure 6.64 Automated workflow of the AOCS Cd 29c-13 method for MCPD analysis (*initial manual steps, [1]Part A uses NaCl (Assay A) and Part B uses NaBr (Assay B), as of Figure 6.62).

suitable as a blank matrix as it is pressed at low temperatures only and does not undergo any refining process, hence suitable for blank runs. It should not contain any MCPD esters. Defined amounts of the two validation compounds are added to the olive oil sample and prepared using the robotic system as described.

Sensitivity, Recovery, and Precision

The 3-MCPD recoveries in the validation experiments were achieved with 91.6% for Part A, 101.9% for Part B, and 116.2% for 2-MCPD in Part B. The measurements gave a LOD of 0.026 mg/kg sample with a LOQ of 0.041 mg/kg sample for 3-MCPD.

338 | 6 Solutions for Automated Analyses

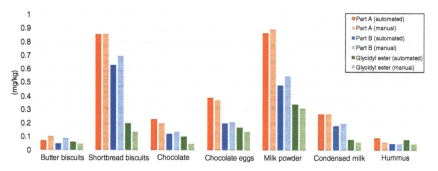

Figure 6.65 Comparison of automated workflow data with manual measurements of real-life samples. Source: Courtesy Axel Semrau GmbH & Co KG.

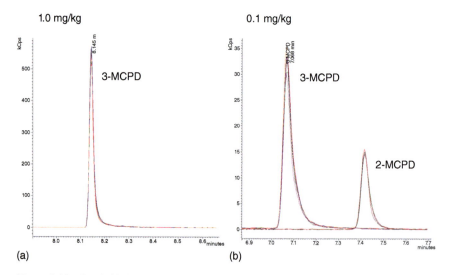

Figure 6.66 Overlaid chromatograms of six repeat sample runs of spiked virgin olive oil. Source: Courtesy Axel Semrau GmbH & Co KG.

The reproducibility with measurements on four consecutive days was in the range of 7.7–8.9%. The blank value measurements confirmed the assumption with a value <LOD for 3-MCPD and 2-MCPD.

For further validation of the automated workflow, sample results were compared with the manual procedure using a reference sample. The results are presented in Figure 6.65. The chromatograms in Figure 6.66 demonstrate the reproducibility of the automated measurements of 2- and 3-MCPD in the concentrations of 1.0 and 0.1 mg/kg.

Conclusion

The overlapped processes of the automated workflow provide high productivity for routine production control of MCPD with a capacity of 36 sample measurements

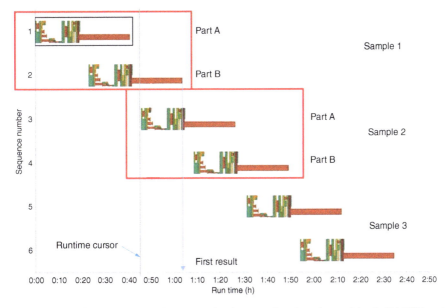

Figure 6.67 Overlapped sample preparation in prep-ahead mode provides 2-/3-MCPD and glycidol results of the first sample already after 1:05 hours. Part A and Part B of the method are shown for Sample 1 and Sample 2. Screenshot from CHRONOS sequence control software. Source: Courtesy Axel Semrau GmbH & Co KG.

in 24 hours. The first results are available after finishing the Part B of a sample at approx. 95 minutes, as shown in Figure 6.67. While the Part A of a sample is analyzed by GC-MS, the preparation of Part B is executed in prep-ahead mode. The sample preparation of the next sample starts overlapped with the GC run of Part B of the previous sample.

6.13 Mineral Oil Hydrocarbons MOSH/MOAH

Application

The contamination of food with mineral oil components is known since long [75, 76]. The risk assessment, as well as short- and long-term measures with conducting analyses of food and food contact materials, was addressed already in 2009 by the German Federal Institute for Risk Assessment (BfR), an independent institution within the German Federal Ministry of Food and Agriculture [77]. In 2012, revised in 2013, the EFSA Panel on Contaminants in the Food published their comprehensive "Scientific Opinion on Mineral Oil Hydrocarbons in Food". This document provides the current status on the definition and chemistry of the mineral oil hydrocarbon contamination of food, risk assessment, and toxicity and gave directions for sampling, analysis, and reporting [78]. In 2019, the EFSA reported about the contamination of infant formula and follow-on formula by mineral oil aromatic hydrocarbons (MOAH) [79].

Scope and Principle of Operation

Mineral oil hydrocarbons (MOH) or mineral oil products considered in the EFSA document are hydrocarbons containing 10 to about 50 carbon atoms, which clearly defines the challenge of analysis from low to very high molecular weight compounds. MOH comprises complex mixtures, principally of straight and branched open-chain alkanes (paraffins), largely alkylated cycloalkanes (naphthenes), collectively classified and categorized as the mineral oil saturated hydrocarbons (MOSH), and mineral oil aromatic hydrocarbons (MOAH). The European Commission Joint Research Center provided the "Guidance on sampling, analysis and data reporting for the monitoring of mineral oil hydrocarbons in food and food contact materials" [80]. Here, the minimum performance requirements of the analytical methods for MOSH/MOAH monitoring are defined.

Due to the different toxicological characterization of the MOSH and MOAH components, and their differently regulated maximum residue levels, both MOH groups need to be separated during analysis. The performance of the total MOH determination by infrared spectroscopy is not sufficient anymore today. And, GC alone does not separate the saturated from the unsaturated fraction. The separation of the saturated from the aromatic compounds is achieved on normal phase LC with hexane or pentane as mobile phase. Two fractions are separated, which are transferred online to a GC column each by LC-GC coupling (see Section 5.3). Each of the MOSH and MOAH fractions elute as a broad hump from the respective GC column. Individual compound separation and identification requires the comprehensive GC × GC column separation preferably with HR/AM-MS detection. For the quantitative MOSH and MOAH determination, the GC separation with FID detection is the standard method. The quantification is based on peak areas of calibrated retention time windows for defined C-number intervals according to the EFSA guidelines [80].

The analytical benefits of the LC-GC method are manifold: LC has a high sample capacity and allows the direct injection of edible oils after dilution, is not limited in molecular weight, and provides an efficient clean-up for matrix separations. If in one LLE step a separation plate number is 1, the LC column achieves a separation power of more than 2000 plates. GC also offers a very high separation efficiency and detectors with a very high dynamic range and sensitivity like the FID, and the mass selective detection for identification. The normal phase separation allows a direct transfer of calibrated LC fractions to a GC column using hexane as a GC compatible solvent. An auxiliary ultraviolet (UV) detector for LC is used for retention time calibration with marker substances like cholestane for the end of the MOSH elution, 1,3,5-tri-*tert*-butyl benzene (TBB) for the beginning and perylene for the end of the MOAH fraction. The LC-GC "injection" mode avoids any discrimination in the required wide molecular range from C10 up to C50 and beyond. A solvent split technique as known from programed temperature vaporizer (PTV)-type GC injectors is not suitable for this application due to severe discrimination effects of the lower boiling analytes. The LC fraction transfer uses a GC pre-column with a timed solvent vapor exit and backflush. The GC transfer of LC fractions can be reliably automated by retention time calibration for both of the required fractions MOSH and MOAH with a transfer in separate GC columns for a parallel FID detection.

Figure 6.68 meta-Chloroperoxybenzoic acid (mCPBA) for olefin epoxidation.

One limitation of the LC separation of certain oil or food matrices is a potential co-elution of biogenic olefins like squalene or carotenoids in the MOAH elution region. Such polyunsaturates would interfere with the MOAH quantification, e.g. with olive or palm oil, and lead to an overestimate of the result. The epoxidation of such interfering compounds with meta-chloroperoxybenzoic acid (mCPBA), Figure 6.68, is an optional purification step as part of the workflow prior to LC injection [81]. The increased polarity of the reaction products allows the LC separation from the MOAH elution region. The epoxidation also removes certain olefins co-eluting with the MOSH fraction and hence may also be employed as a purification step for the MOSH fraction as well [80]. Sodium thiosulfate is used for quenching the reaction after a preset reaction time.

The LC fractions eluting the MOSH and MOAH compounds are transferred at the marker calibrated retention times into the dedicated GC pre-column by switching the injection valve from the robotic system. The MOSH fraction elutes first. The second MOAH fraction is directed into the second pre-column by the column selection valve (see Figure 6.69, items 4 and 5). The evaporation of the solvent is achieved at moderate GC oven temperature through the built-in and at that time open solvent vapor exit. After the second fraction is transferred and the majority of the solvent

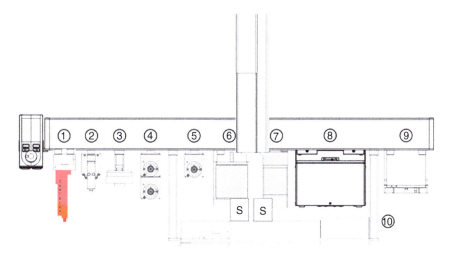

Figure 6.69 Suggested robot configuration for the automated MOSH/MOAH analysis. 1 Tool park station; 2 Fast wash station; 3 Solvent station; 4 Injection and fraction switch valves; 5 Selection valve for MOSH and MOAH GC column; 6 Agitator/Incubator; 7 Vortex module; 8 Centrifuge; 9 Trayholder for 2 and 10 mL vials; 10 GC top installation legs; and S Solvent vapor exits of the GC.

vapor evaporated, the solvent exit valves close, and the regular chromatographic separation starts by ramping the GC oven and FID detection. In parallel, the LC column gets backflushed and conditioned for the next sample.

In the sample pre-treatment, an additional purification step with activated aluminum oxide might become necessary. In some samples, odd-numbered plant n-alkanes in the range of n-C21 to n-C35 can dominate the chromatogram, tend to overload the column, and overlap the mineral oil fraction [80]. Activated aluminum oxide, heated to 350–400 °C, retains long-chain n-alkanes with more than about 20 carbon atoms, while the branched iso-alkanes of the MOH contamination pass the column unretained [82–84]. An additional second online purification step is proposed with an activated aluminum column to separate the n-alkanes from the MOH iso-alkanes prior to LC-GC transfer [83].

Solvents and Chemicals

- Retention time and ISTD mix [85], consisting of:
  - *n*-Undecane (C11), ≥99%, CAS No. 1121-21-04.
  - *n*-Tridecane (C13), ≥99%, CAS No. 629-50-5.
  - Cyclohexylcyclohexane (bicyclohexyl, Cycy), 99%, CAS No. 92-51-2.
  - Pentylbenzene (5B), analytical standard, CAS No. 538-68-1.
  - 1-Methylnaphthalene (MN), analytical standard, CAS No. 90-12-0.
  - 2-Methylnaphthalene (MN), analytical standard, CAS No. 91-57-6.
  - 5α-Cholestane, ≥97% for HPLC, CAS No. 481-21-0.
  - 1,3,5-*tri*-tert-Butyl benzene (TBB), ≥97%, CAS No. 1460-02-2.
  - 1,4-Di(2-ethylhexyl) benzene (DEHB), analytical standard, CAS No. 87117-22-4 (can replace TBB).
  - Perylene, analytical standard, CAS No. 198-55-0.
  - Solution 100 ng/µL in *n*-hexane.
- mCPBA, CAS No. 937-14-4, 200 mg/mL in ethanol.
- $Na_2S_2O_3$, CAS No. 7772-98-7, 100 mg/mL aqueous solution to remove excess mCPBA and induce phase separation.
- Na_2SO_4, ≥99%, CAS No. 7757-82-6, dried.
- Ethanol, CAS No. 64-17-5, HPLC grade.
- *n*-Hexane, CAS No. 110-54-3, HPLC grade.
- Dichloromethane, CAS No. 75-09-2, HPLC grade.

Consumables

- 10 mL screw-top clear vials with magnetic screw caps.
- 2 mL screw-top clear vials with magnetic screw caps.
- LC column: silica column 2.1 mm × 250 mm, particle size 5 µm, spherical, pore size 60 Å, Restek Allure, or similar.
- Pre-column: MXT stainless steel column, 10 m × 0.53 mm, uncoated, Siltek deactivated.
- GC column: Rtxi-5MS Column 15 m × 0.25 mm × 0.25 µm, or equivalent.

- Liquid syringe, 100 µL, for liquid handling, needle gauge 22 or 23, point style 3/LC.
- Liquid syringe, 1000 µL, for liquid handling, needle gauge 22 or 23, point style 3/LC.

System Configuration

The suggested hardware configuration of a robotic system for MOSH/MOAH analysis with optional epoxidation are:

System – Tool – Module	Task
PAL RTC System	x,y,z-Robotic system with automated tool change for sample preparation and liquid injection, 1200 mm rail length
1 pc Park Station	Park station for tools not in use during workflow
1 pc D7/57 Liquid Tool	Tool for a 100 µL LC injection syringe
1 pc D8/57 Liquid Tool	Tool for the 1000 µL sample preparation syringe
2 pc Tray Holder	Tray holder for 3× vial racks
3 pc Rack 54× 2 mL	54 pos. racks for 2 mL vials (samples)
3 pc Rack 15× 10/20 mL	15 pos. racks for 10 mL vials (epoxidation)
1 pc Solvent Module	For 3× 100 mL solvent/reagent reservoirs
1 pc Fast Wash Station	Active syringe wash station for 2 solvents
1 pc Vortex Mixer	For intense vial mixing
1 pc Agitator/Incubator	Sample incubation of up to 6× 20 mL vials
1 set Vial Inserts 10 mL	Inserts for the use of 10 mL vials in the agitator
3 pc Valve Drive	For LC injection and switching valves
1 pc LC injection Valve	For the LC injection
2 pc 6-port Switching Valve	For LC column backflush and selector valve for waste, MOSH or MOAH
1 pc Centrifuge	For the epoxidation step
1 pc Mounting kit	Installs the x,y,z-robot GC instrument top
1 pc GC system	With 2× FID detector, backflush electronic pressure control (EPC)
2 pc Solvent Exit Valve	With restrictor, GC rooftop-mounted
1 pc CHRONECT LC-GC control unit	For control of gas flows and valve timings
1 pc Clarity data system	For MOSH and MOAH integration and reporting
1 pc LC system	With binary pump and UV detector

Analysis Parameter

The standard analysis parameters for the MOSH/MOAH analysis by epoxidation purification, LC fractionation, LC-GC coupling, and FID detection are listed in Table 6.18.

Table 6.18 Analytical parameter for MOSH/MOAH analysis by LC-GC coupling.

LC separation	
Column temperature	Room temperature
Mobile phase	n-Hexane/DCM gradient
Flow	300 µL/min for separation and transfer
Gradient	Injection 100% n-Hexane
	1.5 min 70% n-Hexane/30% DCM
	6.0 min 100% DCM backflush (500 µL/min)
	15.0 min 100% n-Hexane (500 µL/min)
	30.0 min end of cycle, ready for injection
UV detection	230 nm for RT marker compounds
LC fraction transfer volume	2× 450 µL
GC	
Inlet mode	LC-GC transfer
Carrier gas	H_2, constant flow
Pressure	65 kPa for transfer, 150 kPa for analysis
Solvent vapor exit	Open 0.5 min before transfer, closed 0.3 min after transfer
Oven program	60 °C (4 min) → 30 °C/min → 400 °C (4.0 min)
FID	
Base temperature	410 °C
Airflow	280 mL/min
Hydrogen flow	40 mL/min
Nitrogen flow	25 mL/min

Workflow

Initial manual sample pre-treatment steps for different type of samples:

Paperboard:
- Cut into small pieces e.g. by a shredder, get pieces homogenized.
- 1.5 g is immersed in 50 mL hexane/ethanol (1:1) for two hours at room temperature.
- 4 mL of extract get mixed with 6 mL of water and centrifuged.
- The hexane phase is transferred to 2 mL vials and analyzed using the LC-GC workflow.

Vegetable oil or fat [86]:
- Weigh ca. 300 mg exact into a 10 mL autosampler vial.
- Place the capped vials into the system sample rack for analysis.
- Chose the epoxidation purification of the workflow if required.

Figure 6.70 Automated workflow of the MOSH/MOAH analysis with the optional epoxidation step (*initial manual steps).

346 6 Solutions for Automated Analyses

Cosmetic Products (Lip Balm) [87]:
- Cosmetic samples are extracted with *n*-hexane by sonication and centrifuged.
- The hexane phase is transferred to 2 mL vials and analyzed using the LC-GC workflow.

Automated Workflow
- The automated workflow for the MOSH/MOAH analysis is illustrated with the detailed steps graphically in Figure 6.70.

Quantitative Calibration

A series of standard dilutions from reference material in the concentration range of interest are prepared. A quantitative calibration between 0.5 and 7.0 ng is shown with good precision in Figure 6.71 with a correlation factor R^2 of 0.9979. Bicyclohexyl (Cycy) is used as the ISTD for the MOSH, 1-MN, and 2-MN for the MOAH compounds. The quantification performed by the CDS determines the total MOSH and MOAH values and also certain C-number sub-fractions in the chromatograms as defined by the elution of n-alkanes C10–C16, >C16–C20, >C20–C25, >C25–C35, >C35–C40, >C40–C50 for MOSH and C10–C16, >C16–C25, >C25–C35, and >C35–C50 for MOAH [80]. The alkane fractions are shown as vertical dotted lines in the MOAH hump of Figure 6.72.

Sample Measurements

The overlaid chromatograms in Figure 6.72 show both of the FID traces for the MOSH and MOAH channel. The humps around the retention time of ca. 15 min represent the saturated and aromatic hydrocarbon contamination. Detection limits of <5 mg/kg for MOSH and MOAH were reported for vegetable oils. For low-fat

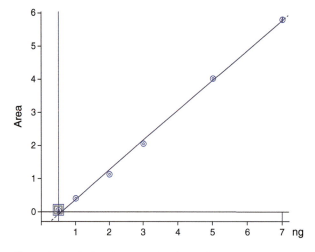

Figure 6.71 Calibration curve for C40 MOSH with a correlation coefficient of R^2 0.9979. Source: Courtesy Axel Semrau GmbH & Co KG.

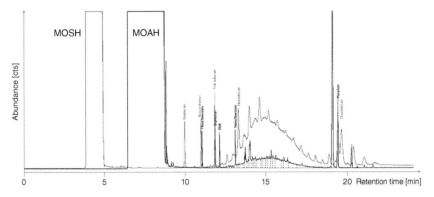

Figure 6.72 Parallel MOSH and MOAH LC-GC-FID analysis (chromatogram overlay, RT marker labeled, green MOSH, black MOAH trace with C-fraction integration lines). Source: Courtesy Axel Semrau GmbH & Co KG.

containing matrices, such as rice or pasta, detection limits as low as 0.5 mg/kg are feasible.

Conclusion

The described analysis method for MOSH/MOAH contamination using a hyphenation of normal-phase LC with GC-FID is applicable for the analysis of mineral oil hydrocarbon contaminations in vegetable oils, foodstuffs, cosmetics, and food contact materials at concentrations in the range of <1–100 mg/kg. The workflow is compliant with the standard method DIN EN 16995:2017-08. The automated workflow using a robotic x,y,z-sampling system is completed with first results in about 30 minutes. A high sample throughput of 48 samples/day can be achieved with a total capacity of up to 162 samples in unattended operation.

6.14 Pesticides Analysis – QuEChERS Extract Clean-Up

Application

The pesticide analysis using the QuEChERS extraction [88] is internationally standardized with EN 15662 [89] and AOAC 2007.01 [90] methods and in general use today even beyond pesticide analysis. In contrast to the standardized extraction with a citrate buffer in EN 15662 or acetate pH buffer in AOAC 2007.01, the clean-up method of the raw extract is not standardized and gets individually optimized for different food matrices. Many different clean-up procedures are published addressing the particular challenges of the diverse food matrices. Mostly, the dispersive solid-phase extraction (dSPE) approach with varying sorbent materials and amounts is used in manual operation with the consequence of countless optimized clean-up procedures published for particular food matrices. Far more than 1000 publications every year are dealing with modified and further improved clean-up procedures for various food commodities [91].

The automated workflow for QuEChERS extract clean-up uses micro-solid-phase extraction (µSPE) cartridges. This workflow describes the automated clean-up of the acetonitrile raw extracts from the QuEChERS extraction with the online injection to GC-MS and LC-MS. The µSPE cartridges applied for the automated extract clean-up use a specifically prepared mixture of sorbent materials optimized for high recovery and a fit-to-purpose clean-up efficiency [92]. Using the automated µSPE clean-up of QuEChERS extracts a food matrix-dependent clean-up is no longer necessary. For the analysis of a wide variety of food commodities, even such with higher fat content like fish or meat, only one cartridge type is used [93, 94]. A slightly different sorbent mix is used for pesticides determined by GC with anh. $MgSO_4$ for extract drying, and another type without $MgSO_4$ for the more polar compounds determined by liquid chromatography [95].

Scope and Principle of Operation

After QuEChERS extraction and centrifugation, the crude sample extracts in MeCN are transferred into 2 mL autosampler vials, capped with septa, and inserted into the robotic system sample tray.

The µSPE cartridges for GC-MS or LC-MS analysis used for the clean-up process are provided in a dedicated cartridge rack. The µSPE clean-up is carried out in the chemical filtration mode. Unwanted sample matrix is retained by the cartridge sorbent material, and the pesticide analytes are recovered in a sharp elution profile, see also section 4.5.2.3. As an important analytical advantage, an eluate concentration by evaporation is not required. The prepared eluates are sufficiently concentrated for analysis and significantly cleaner compared to dSPE [92].

The µSPE cartridge is operated like a micro-LC column, loaded, and eluted with a defined low flow for maximum separation of analytes from matrix. The syringe operates here as the solvent pump with the needle tip placed directly on top of the sorbent bed with only low dead volume. An optimum flow rate of 2 µL/s is required, only achievable and well reproducible by using a liquid syringe driven from a programmable robotic sampler. Increasing the sample volume or additional elution solvent volume beyond the recommended elution volume of the pesticides fraction will not increase the assay "sensitivity" but will elute retained matrix, making the desired clean-up effect increasingly void with increased elution volume. The cartridges are sealed with a crimped septum, which allows the transport between the racks while hanging on the syringe needle.

Optionally, APs for enhanced pesticide peak elution profiles and improved peak integration can be added as of user's choice [96, 97]. The AP addition requires and uses an additional liquid syringe typically of 25 or 100 µL volume. The mixing of added solvents is performed by "syringe mixing" with repeated aspiration and dispensing. Also, the sandwich injection can be used for adding APs for GC injection (see Section 5.2.1). For online injection to GC, a standard 10 µL GC injection syringe is used, a 25 µL syringe for LC injection. The cleaned extract is taken from the eluate collection vial in the desired volume for injection to analysis.

Table 6.19 µSPE clean-up workflow published by Morris and Schriner [92].

Step	Description
1	Select the µSPE tool with 1000 µL syringe
2	Wash the 1000 µL syringe with polar and non-polar solvents
3	**Condition** the cartridge with **150 µL of elution solvent**
4	Pull up 150 µL raw extract from the Sample Rack into the 1000 µL syringe
5	Transfer the µSPE cartridge by syringe needle transport to the Elution Rack
6	**Load 150 µL extract** onto the cartridge (2 µL/s) and collect cleaned extract in the Eluate tray
7	**Elute with 150 µL elution solvent** (2 µL/s) and collect cleaned extract in the same Eluate vial
8	**Discard the cartridge** into a waste bin (optional, or place back to the cartridge rack)
9	Wash the 1000 µL syringe with MeCN/MeOH/water (1:1:1 by volume)
10	Wash the 1000 µL syringe with MeCN
11	For addition of APs, switch to 25 or 100 µL syringe and wash with MeCN (optional), or use the Sandwich injection technique
12	Add 25 µL AP + QC solution to the collection vial in Eluate tray (optional)
13	Mix the solvents in the collection vial with syringe pull-up/dispense
14	Wash the 100 µL syringe with MeCN/MeOH/water (1:1:1 by volume)
15	Wash the 100 µL syringe with MeCN
16a	For GC injection, switch to the 10 µL GC syringe and wash with MeCN
16b	For LC injection, switch to the 25 µL LC syringe and wash with MeCN
17	Pull up 2–10 µL of the cleaned extract from the Eluate tray
18	Wait for GC/LC *Ready* signal, and inject, wash after injection with MeCN
19	Select the µSPE tool with 1000 µL syringe and move to home position
20	Prepare next sample as of optional prep-ahead mode

Source: Adapted from Morris et al. [92].

The described workflow follows the automated QuEChERS clean-up method published by Bruce D. Morris and Richard B. Schriner [92]. This procedure includes the conditioning of the cartridge before the sample load and elutes the cleaned extract with additional elution solvent, see Table 6.19.

The cartridge conditioning and additional elution steps can be skipped in the described workflow for a reduced preparation time and increased sample throughput. The workflow published in 2016 by Steven J. Lehotay et al. directly loads 300 µL of the raw extract in one step [95], as outlined with its details in Table 6.20. This creates a short workflow with excellent QuEChERS extract clean-up and good recovery for highly sensitive GC-MS analysis. The time axis for extract clean-up fits well the fast low-pressure gas chromatography (LPGC)-MS and UPLC-MS analysis for an optimized sample throughput in a routine laboratory. "The automated mini-SPE cleanup coupled with LPGC-MS/MS analysis not only achieved

Table 6.20 Fast μSPE workflow published by Lehotay, Han and Sapozhnikova [95].

Step	Description
1	Select the μSPE tool with 1000 μL syringe
2	Wash the 1 mL syringe with MeCN (2× 500 μL)
3	Pull up **300 μL raw extract** from the Sample Rack into the 1000 μL syringe
4	Transfer the μSPE cartridge by syringe needle transport to the Elution Rack
5	**Elute extract** through the μSPE cartridge at 2 μL/s
6	**Discard cartridge** into waste receptacle
7	Wash the 1000 μL syringe with 1:1:1 MeCN/MeOH/water (2× 500 μL)
8	Wash the 1000 μL syringe with MeCN (4× 500 μL)
9	Switch to 100 μL syringe and wash with MeCN (2× 50 μL)
10	Add 25 μL MeCN to eluate collection vial (with glass insert) in rack 2
11	**Add 25 μL AP + QC solution** to eluate collection vial (with glass insert) in rack 2
12	Mix the solvents in the collection vial with syringe aspiration/dispense
13	Wash the 100 μL syringe with 1:1:1 MeCN/MeOH/water (5× 50 μL)
14	Wash the 100 μL syringe with MeCN (3× 50 μL)
15	Switch to the 1000 μL syringe to prepare the next sample

Source: Based on Lehotay et al. [95].

high-quality results for diverse type of analytes and foods, the approach also enabled reliable, high-throughput operations without much labour or instrument maintenance" [95]. It is shown that due to the matrix scavenging operation of the μSPE cartridges, a total load volume of 300 μL raw extract (similar to the 150 μL raw extract plus 150 μL elution solvent in Table 6.19) shows a good analyte recovery and optimum performance for routine operation of even fatty matrices. Increasing the total applied liquid volume, raw sample plus an elution solvent volume, bears the risk of additional unwanted matrix elution. An ISTD can be used to correct for small dead volume fluctuations of the μSPE cartridge.

Fast LPGC for Short Sample Preparation Cycles

Within the concept of automated sample preparation, it is worth noting at this point that the applied low-pressure GC (LPGC) provides particular advantages not limited to but in particular for automated pesticide analysis. It is a distinctive example of how instrumental analysis methods follow the improved performance of automated workflows on a synchronized short time axis.

First published by Calvin Giddings as "vacuum chromatography," this particular GC setup found rare attention [98] until Jaap de Zeeuw patented and published in 2000 a setup to speed up GC separations compatible with any commercial GC-MS instrument [99]. A short uncoated narrow-bore restrictor column (pre-column) connects a regular injector with a wide-bore capillary column. The analytical column is installed as usual with GC-MS systems directly leading to the mass spectrometer ion source. Vacuum conditions apply to the column outlet and reaches far into the column. The vacuum condition shifts the Van Deemter carrier gas velocity optimum to significantly higher flows [100]. The GC injector can be of regular SSL or PTV type and is operated as usual with a regular head pressure and split and purge flows. Also, the GC oven programming is standard, even can be accelerated by faster heating rates. Column dimensions typically used for LPGC are 3–5 m in length with 0.15–0.18 mm ID for the pre-column, deactivated, no film, and 10–15 m with 0.53 mm ID, and a 1 µm film thickness for the analytical column. The column film type usually chosen is the known 5% phenyl type, but can be selected as it is required by the particular application. Pre- and analytical columns are tightly connected in the GC oven with a zero dead volume connector.

While the sample injection is performed as regular, also large volume injections are practical, the column separation is effected at sub-ambient pressure [101, 102]. The low pressure inside of the analytical column leads to a significantly lower gas viscosity and higher carrier gas velocity. It is the same consideration why in pressurized GC separations hydrogen is preferred for reduced analysis times.

Several analytical advantages of LPGC complement ideally the automated QuEChERS extract clean-up for pesticides [103]. The main aspect is the fast analysis and runtime compatibility to the short sample preparation. The optimum carrier gas velocity of 0.53 mm ID columns in LPGC is with 90–100 cm/s for helium about 10 times higher than with conventional GC separations. Optimum carrier gas flows are in the range of up to 2 mL/min being well compatible with the direct MS coupling and the standard ion sources. Peaks are fast eluting with typical sharp peak widths of about two seconds, which allows MS/MS instruments reasonable dwell times for switching the many compound selective MS/MS ion transitions for an optimum signal-to-noise ratio (S/N). The elution time saving results in more than 70%, increasing the sample throughput significantly (Figure 6.73). The GC cycle time is down to a similar time scale as the automated µSPE clean-up leading to an optimum duty cycle of the GC-MS system [104]. Compared to a traditional micro-bore column, about 10–20% less peak capacity in LPGC is tolerated with the advantage of higher speed, increased column load capacity, and improved sensitivity, expressed by an increased S/N ratio. Due to the short column and high carrier gas velocity, the compounds elute earlier at lower elution temperatures. The lower thermal stress is in particular advantageous for pesticides, but in general for many thermally labile compounds. Also, the analytes elute in a region of lower column bleed with reduced background noise and hence better detectability at trace concentrations. In practice, the less background-affected mass chromatograms facilitate significantly the automated peak integration for more reliable results on the lowest trace levels [105]. Less column bleed also leads to less MS ion source contamination by the silylated column film material and increases the ion source uptime for considerably reduced maintenance [104].

The LPGC setup is compatible with all modern GC instruments and any online automated sample preparation. In particular for pesticides analysis, the analytical time axis for sample preparation, automated clean-up, the UPLC separation for LC-MS and LPGC for GCMS, both with pesticide runtimes below 10 minutes,

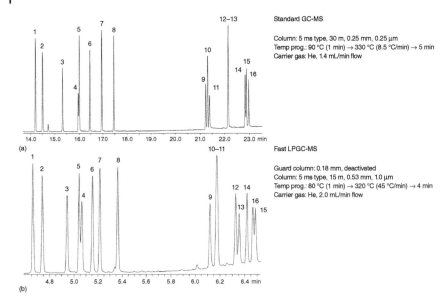

Figure 6.73 Comparison of GC elution times of a pesticide standard acquired with standard GC-MS (a) and LPGC-MS (b). Source: Redrawn from S. Lehotay, private communication. Analysis conditions: full scan data acquisition m/z 50-550, 1 µL splitless injection of a 2 ng/µL pesticide standard. 1 Diazinon; 2 Isazophos; 3 Chlorpyrifos-methyl; 4 Fenitrothion; 5 Pirimiphos-methyl; 6 Chlorpyrifos; 7 Pirimiphos-ethyl; 8 Quinalphos; 9 Pyridaphenthion; 10 Phosmet; 11 EPN; 12 Phosalone; 13 Azinphos-methyl; 14 Pyrazophos; 15 Azinphos-ethyl; and 16 Pyraclofos.

become compatible for the short automated extract clean-up in the prep-ahead mode. Long GC separations are any longer the determining factor for sample throughput.

Solvents and Chemicals

The following solvents for cartridge conditioning, syringe wash, and standards showed optimum performance in the automated QuEChERS extract clean-up process [92]. Analyte protectants (APs) and ISTDs are optionally in use [92, 96].

- Conditioning and elution solvent:
  - 100 mM formate buffer in MeCN/MeOH (1:1).
  - Formate buffer (100 mM) is prepared in MeOH/MeCN (1:1) by dissolving ammonium formate 70 mM, CAS No. 540-69-2, and adding 30 mM formic acid, CAS No. 64-18-6.
  - Lower concentration of formate buffer (50 and 10 mM) is prepared by dilution of the 100 mM solution with MeOH/MeCN (1:1).
- Wash solvents:
  - MeOH/MeCN/water (1:1:1, v:v:v).

- MeCN, HPLC quality.
- Lower concentration of formate buffer (50 and 10 mM) is prepared by dilution of the 100 mM solution with MeOH/MeCN (1:1).
- Internal standards:
  - 20 mg/L solution of triphenylphosphate in MeOH, CAS No. 115-86-6.
  - 2 mg/L solution of atrazine-d5 in MeCN, CAS No. 163165-75-1.
- Pesticides standards:
  - Standards can be applied for spiking experiments and diluted in 10 mM formate buffer (aqueous, pH 4)/MeOH/MeCN (13:2:1 by volume) to give calibration standards 0.02, 0.05, 0.1, 0.2, 0.4, 1, 2, 5, 10, 20, and 40 µg/L.
- Analyte protectants (APs) containing:
  - 25 mg/mL ethylglycerol (3-ethoxy-1,2-propanediol), CAS No. 1874-62-0.
  - 2.5 mg/mL gulonolactone, CAS No. 3327-64-8.
  - 2.5 mg/mL d-sorbitol, CAS No. 50-70-4.
  - 1.25 mg/mL shikimic acid, CAS No. 138-59-0.
  - Prepared in MeCN/water (3:2, v:v) containing 1.1% formic acid to enhance the pesticide stability of the final extract.

Consumables

- µSPE cartridges for GC analysis, e.g. CTC Analytics p/n Cart-uSPE-GC-QUE-0.3 mL, consumables for GC QuEChERS, including cartridges, 0.3 mL polypropylene (PP) vials with conical insert, and caps.
- µSPE cartridges for LC analysis, e.g. CTC Analytics p/n Cart-uSPE-LC-QUE-0.3 mL, consumables for LC QuEChERS, including cartridges, 0.3 mL PP vials with conical insert, and caps.
- Syringes
  - Syringe 1000 µL, needle length 57 mm, gauge 22, PST 3.
  - Syringe 25 µL or 100 µL for optional AP and ISTD addition, needle length 57 mm, PTFE plunger, gauge 22s, point style flat.
  - Syringe 25 µL for LC injection, needle length 57 mm, PTFE plunger, gauge 22s, point style flat.
  - Syringe 10 µL for GC injection, needle length 57 mm, gauge 23s, point style conical.
- Vials
  - 2 mL clear glass with 300 µL micro-insert without spring, or 300 µL micro-vials.
  - 2 mL amber glass, used for the extracts in case they are not analyzed online or/and need to be stored.
- Caps and Septa
  - Non-metallic screw caps with pre-cut silicon/PTFE septa for eluate collection.
  - Non-metallic screw caps with silicon/PTFE septa for general use.

System Configuration

Minimum hardware requirement for the robotic µSPE extract clean-up system as shown in Figure 6.74:

System – Tool – Module	Task
PAL RTC System	x,y,z-Robotic system for liquid injection (850 or 1200 mm rail length)
1 pc Park Station	Park station for tools not in use during workflow
2 pc Tray Holder	Tray holder for 3× vial racks, a second trayholder is optional for liquid injection samples
2 pc Rack 2 mL	54 pos. 2 mL vials for sample and eluate rack (see Figure 6.75)
1 pc µSPE Elution Tray	Vial Lock, 54 position aluminum (see Figure 6.77)
1 pc µSPE Cartridge Tray	54 pos. as cartridge reservoir
1 pc Waste Receptacle	With drain tube (see Figure 6.78)
1 pc Solvent Station	For 3× 100 mL solvent bottles
1 pc Standard Wash Station	For 4× 10 mL solvent vials, 1× 10 mL waste vial or waste tube adapter
2 pc Adapter for Standard Wash Station	Allows to place 2 mL vials instead of the 10 mL wash vials
1 pc Fast Wash Module	For syringe wash steps, 2 solvents
1 pc D8/57 µSPE Tool	Tool for a 1000 µL syringe for clean-up process with needle guide for µSPE cartridges
2 pc D7/57 Liquid Tool	Tool for a 10 and 25 µL syringe for GC and LC injection
1 pc D7/57 Liquid Tool	Tool for a 25 µL syringe for optional AP and ISTD addition (alternatively 100 µL syringe)
1 pc Mounting kit	Installs the x,y,z-robot GC instrument top, or as a standalone benchtop installation for LC only
GC-MS and LC-MS analysis	
1 pc GC system	With split/splitless, or PTV type injector
1 pc MS system	Single or triple quadrupole analyzer, ESI ion source
1 pc LC system	Binary LC gradient system, column oven
1 pc MS system	Single or triple quadrupole analyzer, ESI ion source

System Setup

The robotic system comprises a special trayholder setup for the µSPE workflow with three dedicated racks:

- All racks are designed for 54 cartridges and vial positions, see Figure 6.75.
- The SAMPLE RACK receives the vials with raw extract, see Figure 6.75.

6.14 Pesticides Analysis – QuEChERS Extract Clean-Up | 355

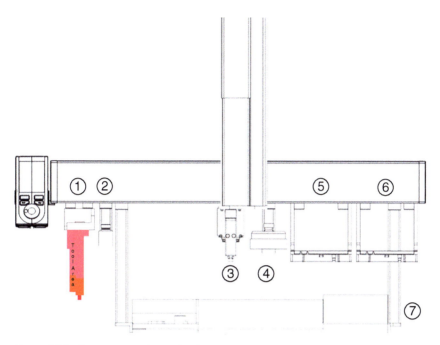

Figure 6.74 Suggested configuration for the µSPE clean-up workflow for QuEChERS extracts. 1 Tool park station; 2 Standard wash station; 3 Fast wash station; 4 Solvent module; 5 µSPE clean-up tray holder; 6 Tray holder for additional sample vials (optional); 7 Mounting kit for GC; and 8 LC injection valve with loop (not shown).

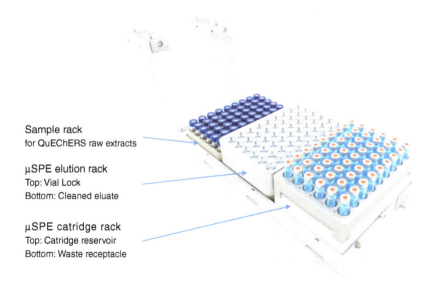

Sample rack
for QuEChERS raw extracts

µSPE elution rack
Top: Vial Lock
Bottom: Cleaned eluate

µSPE catridge rack
Top: Catridge reservoir
Bottom: Waste receptacle

Figure 6.75 Configuration and setup of the µSPE trayholder. Source: Courtesy CTC Analytics AG.

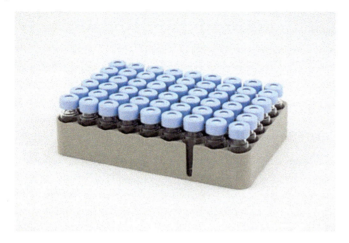

Figure 6.76 Eluate rack, 54× 2 mL vials with pre-cut septa. Source: Courtesy CTC Analytics AG.

Figure 6.77 μSPE vial lock installed on the eluate rack. Source: Courtesy CTC Analytics AG.

- The ELUATE RACK is a dedicated rack with empty vials with split septa for the cleaned extract collection. It is covered by an aluminum top (Vial Lock) as a cartridge guide, which keeps the cartridges vertically in place during the load and elution steps (extraction rack), see Figures 6.76 and 6.77.
- The CARTRIDGE RACK provides the required number of μSPE cartridges for the clean-up workflow. It is also used for an optional cartridge conditioning, see Figure 6.78.

The workflow for the pesticide extract clean-up is optimized for a liquid syringe tool with a 1000 μL syringe (larger volume syringes are optional for other μSPE

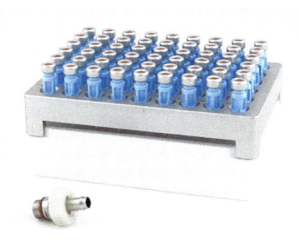

Figure 6.78 54 Position μSPE cartridge rack on top of the waste receptacle with tube adapter. Source: Courtesy CTC Analytics AG.

applications). The addition of APs, ethyl acetate, or ISTD solutions is optional and requires an additional liquid syringe tool for 25 or 100 μL syringes (to be added to the minimum hardware requirements shown in the system configuration). For the online GC injection, a liquid tool with 10 μL syringe, for LC injection for a 25 μL syringe, is required.

The workflow is designed this way that the sample vial index, the vial position in the raw extract rack, is used throughout the processing. The Sample Rack position 1 uses the cartridge in position 1 of the Conditioning Rack and elutes into the empty vial in position 1 of the Elution Rack.

Workflow

Initial Manual Steps

- After QuEChERS extraction and centrifugation approx. 1 mL of the supernatant is transferred to 2 mL autosampler vials.
- The vials get closed with regular septa and non-magnetic caps.
- The vials with raw extract are placed into the sample rack.
- Empty vials capped with non-magnetic caps and split septa are placed into the eluate rack, covered by the vial lock.
- μSPE cartridges are supplied to the cartridge rack.
- Solvents for elution and syringe wash are provided.

Automated Workflow

- The automated workflow for the clean-up of the raw QuEChERS extracts with μSPE cartridges is illustrated graphically in Figure 6.79.

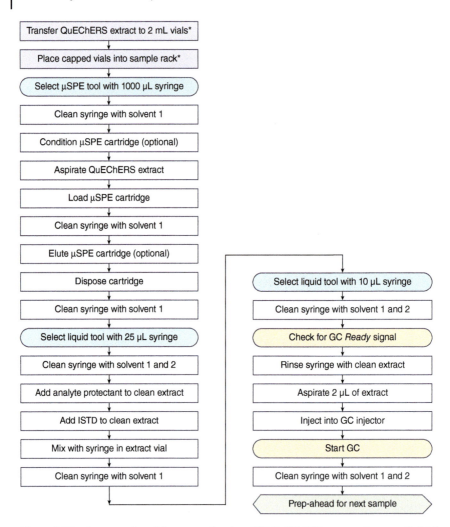

Figure 6.79 Automated workflow steps for the μSPE QuEChERS extract clean-up (*initial manual steps).

Sample Measurements

The use of the automated μSPE clean-up became standard in many pesticides laboratories. Besides the regular food commodities, the view is directed towards the critical matrices, often requiring special clean-up treatments with the dSPE regime. Figure 6.80 shows QuEChERS extracts with intense matrix from black tea and chlorophyll from green spinach. In both cases, the cleaned μSPE extracts, collected in micro-inserts, show already visibly a strong reduction of the unwanted tea and chlorophyll matrix. The chromatograms in Figure 6.81 show the GC-MS/MS analysis of the spiked black tea sample at the lowest calibration level of 1 ng/g (ppb)

Figure 6.80 Visual µSPE clean-up results for black tea (a) and spinach (b).

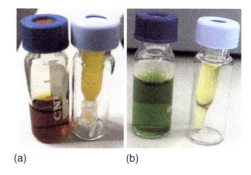

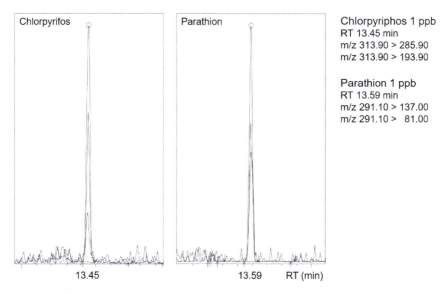

Figure 6.81 Black tea sample pesticides GC-MS/MS peaks at 1 ppb after µSPE clean-up.

of the pesticides chlorpyrifos and parathion both with a clean baseline and high signal response.

For the very complex matrix of spices, Arnab Goon et al. published about the clean-up of QuEChERS extracts for pesticides analysis [106]. The comparison to the manual dSPE clean-up in Figure 6.82 showed average recoveries of 50–90% for the set of about 70 pesticides. The recoveries when using µSPE were higher with an average of 70–100%. In particular, late-eluting compounds, e.g. cypermethrin, deltamethrin, etofenprox, and fenvalerate, improved using the µSPE workflow.

The µSPE clean-up, in general, extends the application of the QuEChERS extraction also to fatty matrices, which often require special pre-treatment with freezing out lipids for fat removal using the manual dSPE steps [107]. The chromatograms

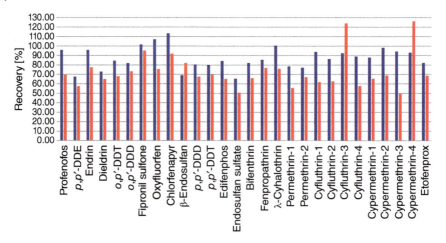

Figure 6.82 Comparison between μSPE and dSPE clean-up for spices using similar sorbent composition [106], blue bars μSPE, red dSPE. Source: Courtesy AOAC International.

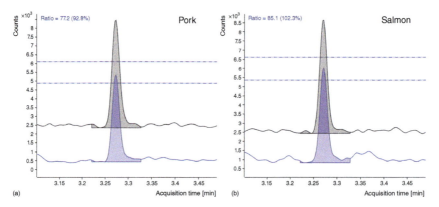

Figure 6.83 Results from fatty food matrices pork (a) and salmon (b) using the automated μSPE clean-up (acephate: m/z 136.0 > 42.0 (black), 136.0 > 94.0 (blue)). Source: Courtesy Chromatographia; Lehotay et al. [95].

for pork and salmon samples, both with a typical lipid content of approx. 30% or more, are shown in Figure 6.83. "Even the fatty salmon matrix showed very good peak shapes and perfectly linear calibration with the same high-quality results as the other matrices" [95]. The beneficial effect of the good clean-up effect of the automated μSPE workflow on the GC inlet system was demonstrated by Lehotay et al. with the remaining matrix deposit in the GC inlet liner after being reported 230 food sample injections, including those from fatty matrices fish and meat, demonstrated with the photo of this liner in Figure 6.84 [95]. The μSPE clean-up of difficult matrix extracts from rice, grapes, and black tea for LC-MSMS analysis showed high recoveries and a high method robustness with stable instrument response for 195 polar pesticides for a large series of sample analyses (Figures 6.85 and 6.86) [108]. The short separation of only 15 minutes cycle time with high instrument uptime allows high productivity in a fully automated workflow.

Figure 6.84 GC inlet liner and septum after 230 matrix samples of different food commodities. Source: Courtesy Chromatographia [95].

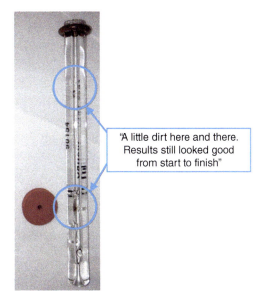

"A little dirt here and there. Results still looked good from start to finish"

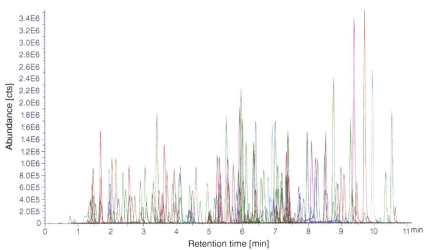

Figure 6.85 LC-MS/MS chromatogram of 195 pesticides compounds after µSPE clean-up. Source: Image used with permission of Thermo Fisher Scientific Inc.

Conclusion

The efficiency of the automated µSPE clean-up for pesticides extracts from food samples is demonstrated with critical matrices. In particular, the described µSPE clean-up extends the application of QuEChERS also to high lipid-containing samples as avocados, meat, and fish besides other food commodities like herbs and spices. A modification of the extract clean-up procedure for different food commodities is no longer required. A full automation of the pesticides analysis from clean-up to GC-MS and LC-MS analysis can be established for routine applications.

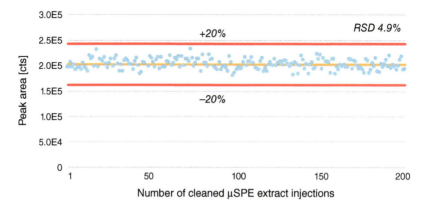

Figure 6.86 Method robustness with 200 injections of µSPE cleaned black tea extract gave a stable response of fenpyroximate at 50 µg/kg with 4.9% RSD. Source: Image used with permission of Thermo Fisher Scientific Inc.

6.15 Glyphosate, AMPA, and Glufosinate by Online SPE-LC-MS

Application

Glyphosate-based pesticides, first sold to market in 1974, are among the most used pesticide formulations worldwide. They are used as broad-spectrum systemic herbicides in agriculture primarily against weeds that compete with cultivated crops. In the discussion about the potential health risks of the suspected carcinogenic compound, and about the required maximum residue levels, the analysis of glyphosate and its metabolites are of particular focus. The EFSA revised its previous review of the existing MRLs in 2019 [109].

Scope and Principle of Operation

Residue levels are determined by analyzing glyphosate (Figure 6.86) and its metabolite aminomethylphosphonic acid (AMPA) (Figure 6.87), together with the contact herbicide glufosinate (Figure 6.88) due to its structural similarity. The compounds are highly soluble in water and insoluble in organic solvents. The use of classical LLE is ineffective. The high polarity, small molecular mass, and absence of a chromophore or fluorophore impose the use of derivatization techniques. The manual sample preparation for the analysis of glyphosate using cation exchange columns for clean-up of aqueous extracts from soil or food commodities, combined with the following derivatization with heptafluorobutanol and trifluoroacetic anhydride for GC-MS, requires significant effort in sample preparation [110]. The manual procedures are time-consuming, complex and can lead to increased

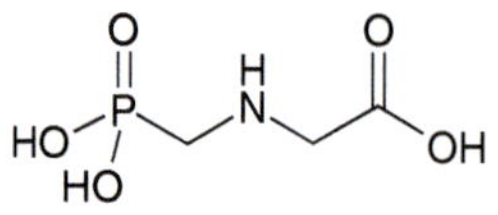

Figure 6.87 Glyphosate (PMG, N-(phosphonomethyl)glycine)

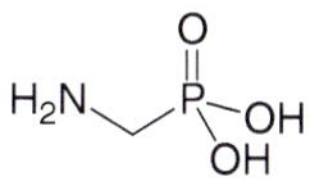

Figure 6.88 AMPA (aminomethylphosphonic acid).

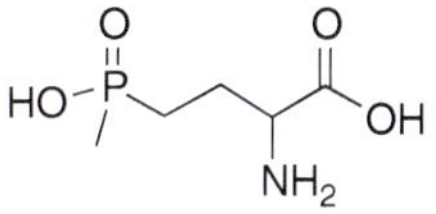

Figure 6.89 Glufosinate ((RS)-2-amino-4-(hydroxy(methyl)phosphonoyl)butanoic acid).

operator-induced errors. Analytical methods without derivatization are available with the "Quick Polar Pesticides" (QuPPe) method for these highly polar pesticides in foods for LC-MS [111], or the amperometric LC detection, but this one cannot provide mass spectra for additional compound confirmation [112]. The QuPPe method involves extraction with acidified methanol and direct LC-MS/MS determination without derivatization.

The here described fluorenylmethyloxycarbonyl derivatization (9-fluorenylmethyl chloroformate, FMOC-Cl) for primary and secondary amines became a standard sample preparation method for improved extraction and separation of glyphosate and related compounds for the analysis by LC-MS. The comprehensive automation of extraction, derivatization, and clean-up facilitates sample throughput and improves process safety. In a single workflow, the automated procedure described covers the FMOC derivatization in aqueous media (Figure 6.90), the analyte concentration, and sample clean-up using high-pressure online solid-phase extraction, and LC-MS/MS analysis for the identification and quantification of glyphosate, AMPA, and glufosinate for water and food samples. The online SPE connects directly to an LC system typically with triple quadrupole MRM detection [113]. The automated online SPE module uses a new clean-up cartridge for each sample, delivered by a built-in automated cartridge exchanger, eliminating potential carry-over effects between samples. The online SPE connects directly to an LC-MS/MS system, typically with triple quadrupole MRM detection. Using online SPE, the elution is quantitatively performed with a significantly smaller elution volume transferred to the LC system delivering a high concentration factor. The method is applicable for water and beverages, as well to complex food matrices e.g. cereals, pulses, and spices [114].

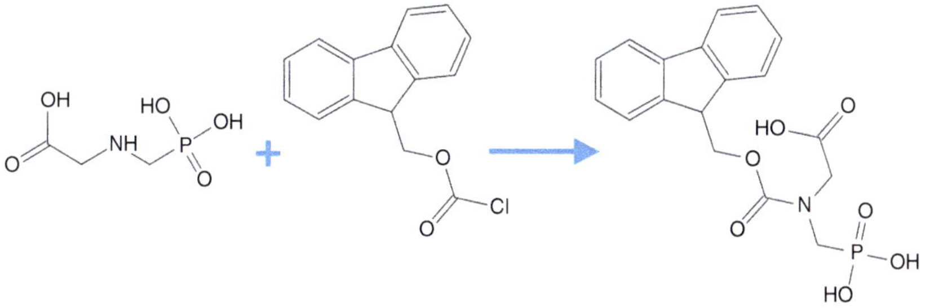

Glyphosate   FMOC-Cl   Glyphosate-FMOC

Figure 6.90 Derivatization reaction principle: glyphosate reacts with FMOC-Cl to form glyphosate-FMOC (AMPA, glufosinate analogue).

Solvents and Chemicals

- Glyphosate, CAS No. 1071-83-6.
- AMPA, CAS No. 1066-51-9.
- Glufosinate, CAS No. 51276-47-2.
- Borate buffer, pH = 9, CAS No. 10043-35-3.
- FMOC-Cl solution, 10 mM in acetonitrile, CAS No. 28920-43-6.
- H_3PO_4, 2% in water, CAS No. 7664-38-2.
- Formic acid, 100 mM in water, CAS No. 64-18-6.
- Ammonium acetate, 50 mM in water, CAS No. 631-61-8.
- Methanol, chromatographic purity, CAS No. 67-56-1.
- Acetonitrile, HPLC quality, CAS No. 75-05-8.
- Water, HPLC quality, CAS No. 7732-18-5.

Consumables

- C8EC-SE cartridge, 18.5 mg C8 sorbent material.
- 2 mL screw-top clear vials with magnetic screw caps.
- LC column: 3 µm C18, 150 × 2 mm, Phenomenex Gemini, or equivalent.
- Liquid syringe, 1000 µL, for liquid handling, needle gauge 22 or 23, point style conical.
- Liquid syringe, 5000 µL, for LC injection, needle gauge 22 or 23, point style 3/LC.

System Configuration

A dual-head robotic system is used for sample preparation, allowing parallel workflow operation, and avoiding a repeated syringe tool exchange. The minimum hardware requirements of a robotic sampling system for glyphosate, AMPA and glufosinate analysis as shown in Figure 6.91 are:

System – Tool – Module	Task
PAL DHR RSI/RSI System	x,y,z-Dual head robotic system with manual tool change for sample preparation and liquid injection, 1200 mm rail length
1 pc D8/57 Liquid Tool	Tool for 1000 µL sample preparation syringe
1 pc D18/57 Liquid Tool	Tool for 5000 µL sample preparation syringe, for LC injection
1 pc Tray Holder	Tray holder for vial racks
1 pc Rack 78× 4 mL	Dedicated 78 pos. rack for 4 mL vials
1 pc Solvent Module	For 3× 100 mL solvent/reagent reservoirs
1 pc Standard Wash Module	For 5× 10 mL reagent reservoirs
1 pc Fast Wash Module	Wash module for 2× active wash solvents, recommended for large syringe wash steps
1 pc Agitator/Incubator	6 Position agitator with inserts for 4 mL vials, for the derivatization step
1 pc Mounting kit	Installs the x,y,z-robot LC instrument top
1 pc online SPE system	With automated cartridge exchange
1 pc LC-MS system	Triple quadrupole analyzer, ESI ion source

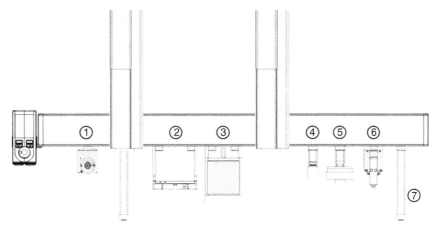

Figure 6.91 Robotic system dual head configuration for the automated glyphosate analysis. 1 Injection valve to online SPE clean-up; 2 Sample trayholder; 3 Heated agitator; 4 Standard wash station for reagents; 5 Solvent/reagent reservoirs; 6 Fast wash station; and 7 LC-MS instrument top mounting legs.

Analysis Parameter

The detailed LC-MS analysis parameters for the glyphosate, AMPA and glufosinate analysis by FMOC derivatization are listed in Table 6.21.

Workflow

Initial Manual Steps

- Water is analyzed without pre-treatment. Food and soil samples are extracted with acidified water, neutralized, and diluted with water to minimize possible matrix effects.
- 1 mL of a water sample, or filtered/centrifuged aqueous sample extract, is transferred to 2 mL vials and placed into the rack of the robotic sampler.

Automated Workflow

- The steps of the automated workflow as programmed are shown in Figure 6.92 using the GERSTEL Maestro control software [116].
- Automated Derivatization Procedure:
  1. Add 100 µL of borate buffer (pH 9) to 1 mL of sample.
  2. Add 200 µL of 10 mM FMOC-Cl solution.
  3. Agitate for 20 minutes at 50 °C.
  4. Cool to ambient temperature.
  5. Add 130 µL 2% H_3PO_4 to quench the reaction.
- Automated Online-SPE Clean-up Procedure:
  1. Condition a C8EC-SE cartridge with methanol, water with 100 mM formic acid.
  2. Load 1000 µL of the derivatized sample onto SPE.
  3. Wash with water and 100 mM formic acid.
  4. Wait for LC-MS *Ready* signal.
  5. Elute with LC pump gradient, and simultaneous start LC-MS run.

Table 6.21 Analysis parameter for the FMOC derivatization, online SPE clean-up, and LC-MS analysis.

Derivatization	
Borate buffer, pH 9	100 µL
FMOC-Cl solution, 10 mM	200 µL
Reaction time, temperature	20 min, 50 °C
H_3PO_4, 2%, to quench reaction	130 µL
Online SPE clean-up	
Cartridge	C8EC-SE, 18.5 mg
Conditioning	5000 µL, methanol
Rinsing	4000 µL, water with 100 mM formic acid
Sample load	1000 µL, 2 mL/min
Wash	2500 µL, water with 100 mM formic acid
Elution, online to LC	Mobile phase gradient
LC	
Mobile phase A	50 mM ammonium acetate, pH 9 adjusted
Mobile phase B	Acetonitrile
Gradient profile	See Table 6.22
LC column oven temperature	45 °C
MS	
System, ionization	Triple quadrupole, ESI negative mode
Ion source voltage	−4500 V
Source temperature	400 °C
Cone gas flow	345 kPa (50 psi)
Desolvation gas flow	483 kPa (70 psi)
Scan mode	MRM
Glyphosate-FMOC m/z, collision energy	390 > 168, −18 V
	390 > 150, −34 V
Glufosinate-FMOC m/z, collision energy	402 > 180, −16 V
	402 > 206, −20 V
AMPA-FMOC m/z, collision energy	322 > 110, −12 V
	322 > 136, −22 V

Source: Adapted from Helle and Chmelka [115].

Sensitivity

The reported LOQ and LOD of glyphosate and AMPA in wheat, water, tea leaves, and honey samples achieved by automated sample preparation and LC-MS/MS analysis are listed in Table 6.23.

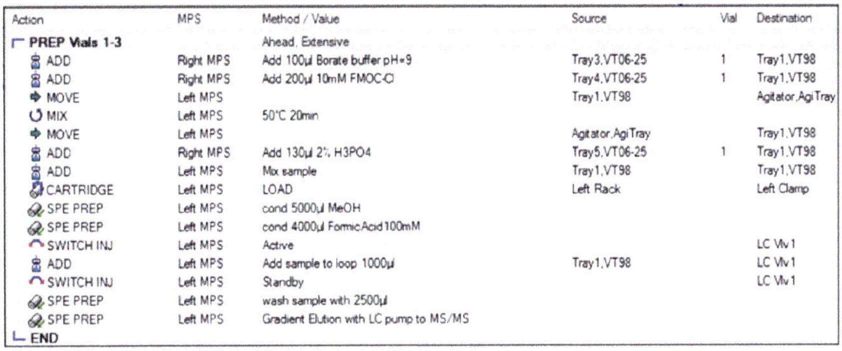

6.15 Glyphosate, AMPA, and Glufosinate by Online SPE-LC-MS

Action	MPS	Method / Value	Source	Vial	Destination
PREP Vials 1-3		Ahead, Extensive			
ADD	Right MPS	Add 100µl Borate buffer pH=9	Tray3,VT06-25	1	Tray1,VT98
ADD	Right MPS	Add 200µl 10mM FMOC-Cl	Tray4,VT06-25	1	Tray1,VT98
MOVE	Left MPS		Tray1,VT98		Agitator,AgiTray
MIX	Left MPS	50°C 20min			
MOVE	Left MPS		Agitator,AgiTray		Tray1,VT98
ADD	Right MPS	Add 130µl 2% H3PO4	Tray5,VT06-25	1	Tray1,VT98
ADD	Left MPS	Mix sample	Tray1,VT98		Tray1,VT98
CARTRIDGE	Left MPS	LOAD			Left Clamp
SPE PREP	Left MPS	cond 5000µl MeOH			
SPE PREP	Left MPS	cond 4000µl Formic Acid 100mM			
SWITCH INJ	Left MPS	Active			LC Vlv 1
ADD	Left MPS	Add sample to loop 1000µl	Tray1,VT98		LC Vlv 1
SWITCH INJ	Left MPS	Standby			LC Vlv 1
SPE PREP	Left MPS	wash sample with 2500µl			
SPE PREP	Left MPS	Gradient Elution with LC pump to MS/MS			
END					

Figure 6.92 Automated workflow for glyphosate analysis from derivatization, SPE clean-up, and LC-MS injection using Maestro PrepBuilder with point and click selection of individual steps [118]. Source: Courtesy GERSTEL GmbH & Co KG.

Table 6.22 Gradient profile for LC separation.

Time [min]	Flow [mL/min]	A [%]	B [%]
0	0.25	80	20
10.0	0.25	5	95
15.1	0.25	80	20
25.0	0.25	80	20

Source: Adapted from Helle and Chmelka [115].

Table 6.23 Limits of quantification (LOQ) and detection (LOD).

Sample	LOQ	LOD
Wheat	<1.0 µg/kg	<0.3 µg/kg
Water	<0.01 µg/L	<0.003 µg/L
Tea leaves	<1.0 µg/kg	<0.3 µg/kg
Honey	<1.0 µg/kg	<0.3 µg/kg

Source: Redrawn from Helle and Chmelka [115].

The MRLs for glyphosate in wheat are regulated in the EU with 10 mg/kg (in the US 30 mg/kg). EU regulations require glyphosate levels in water to be below 100 ng/L (in the US: 700 µg/L [117]). The MRL for glyphosate for tea leaves in the EU is regulated with 2 mg/kg (in the US: 1 mg/kg). The MRL in honey according to EU regulations is 0.05 mg/kg. LOQs achieved using the automated method are substantially lower [115].

Sample Measurements

The automated sample preparation process is completed in 25 minutes. Together with the LC-MS/MS cycle time of about 20 minutes, a total sample cycle time of

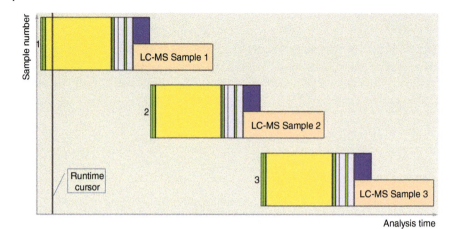

Figure 6.93 Analysis sequence in prep-ahead mode of the samples showing the individual sample prep steps and the actual runtime cursor indicating the ongoing preparation step and analysis status (Maestro screen display) (after). Source: Courtesy GERSTEL GmbH & Co KG.

<45 minutes is achieved. For maximum sample throughput using the automated system, the sample preparation and analysis are synchronized using the prep-ahead functionality (Figure 6.93). Samples are simultaneously prepared by the robotic sampler during the LC-MS/MS analysis of the previous sample.

Conclusion

The online SPE-LC-MS/MS method presented provides method detection limits that easily meet or exceed the requirements of regulations established in the US and EU for all three analytes in multiple matrices [113, 114, 120]. The online SPE requires significantly less solvent for clean-up and analyte elution. This increases the concentration factor for lower detection levels, avoids an additional evaporation step, and contributes with less environmental impact to a greener analytical method setup. The combination of robotic sampling, online sample preparation, clean-up, and LC-MS/MS analysis can be performed in one integrated system operated from one sequence table. The prep-ahead functionality delivers high efficiency and throughput for the complete system by multi-sample overlap.

6.16 Pesticides, PPCPs, and PAHs by Online-SPE Water Analysis

Application

The pre-concentration of water samples using online SPE replaces the manual handling of large water sample volumes. Typically, with manual SPE concentration, up to 1000 mL of water is concentrated 100–1000 times with the incurred consequences of a difficult transport, inefficiently long preparation times over hours, and the generation of large waste solvent volumes due to necessary high milliliter SPE elution and evaporation steps. Also, the one-time-use of SPE cartridges themselves are avoidable consumables and can reduce laboratory plastic waste for a greener analytical chemistry. Automated online SPE works with small sample and solvent volumes eliminating the time-consuming evaporation by using a built-in extraction column for analyte concentration, which can be conditioned for reuse for the next samples in sequence. The automated workflow runs unattended with small water volumes of typically a few milliliters only. The water samples are placed directly into the robotic sampler. A good example is reported with the EPA Method 543 for pesticides in drinking water, officially established in 2015 [121]. The concentration, elution, and separation are done by a full automation of the complete sample preparation process. Only 1–5 mL of sample volume is typically required. The analysis time per sample is short with <20 minutes allowing an unattended high throughput of water samples. For further increase of sample throughput, also a staggered parallel operation with two online SPE cartridges was realized [122].

Scope and Principle of Operation

The described online method is applied to various types of water matrices like drinking water, bottled water, fresh water, and even the influent and effluent from a wastewater treatment plant (WWTP) [123]. The technique is used as well for the determination of pharmaceuticals and personal care products (PPCP) in water along the EPA Method 1694 [124]. PAHs were determined in contaminated environmental samples, seawater, reclaimed water, and rainwater runoff samples [125]. The online method is also used for a fast drinking water control for the cyclic peptide algal toxins microcystin and nodularin in a water treatment plant [126]. After online clean-up and concentration, the eluates are measured by LC-MS/MS with positive and negative electrospray ionization (ESI) [127]. PAHs are detected most sensitive using dopant-assisted atmospheric pressure photoionization ionization (APPI).

Sample Preparation

Preservation reagents are added to each sample bottle before shipment to the field or prior to sample collection. Before analysis, the water samples are filtered and transferred into 10 mL amber vials and spiked with ISTD. Wastewater samples were diluted at a 1:100 ratio. As of EPA 1694, the sample pH is adjusted to 2 with acid and a second sample adjusted to pH 10 with a base.

Solvents and Chemicals

- Certified internal standard solutions of the target analytes.
- Aqueous solutions containing 5% to 20% acetonitrile (MeCN), adjusted to pH, were spiked with PPCPs at the low ng/L level.
- Acetonitrile, HPLC quality, CAS No. 75-05-8.
- Water, HPLC quality, CAS No. 7732-18-5.

- Methanol, HPLC quality, CAS No. 67-56-1.
- Preservation reagents as of EPA Method 543 [121]:
  - pH 7.0 buffer: blend of Tris (tris(hydroxy-methyl)aminomethane) and Tris HCL (tris(hydroxymethyl)aminomethane hydrochloride), e.g. Sigma-Aldrich Trizma #T-7193, or equivalent.
  - Ascorbic acid, CAS No. 50-81-7 – reduces free chlorine at the time of sample collection.
  - 2-Chloroacetamide, CAS No. 79-07-2 – inhibits microbial growth and analyte degradation.

Consumables

- 10 mL amber vials, crimp- or screw-capped.
- Online SPE cartridge e.g. Hypersil GOLD™, C18, 20 mm × 2.1 mm × 12 µm, Waters Oasis HLB with C18 blend, or similar.
- Analytical column, e.g. Hypersil GOLD™, C18, 50 mm × 2.1 mm × 3 µm, or similar.
- Syringe 2500 µL for liquid handling, needle length 57 mm, PTFE plunger, gauge 19, point style flat.
- Syringe 100 µL for ISTD addition, needle length 57 mm, PTFE plunger, gauge 19, point style flat.

System Configuration

The minimum hardware requirements of the robotic sampling system for automated online SPE analysis as of Figure 6.94 are:

System – Tool – Module	Task
PAL RTC System	x,y,z-Robotic system with automatic tool change for liquid injection, 850 mm rail length
1 pc Park Station	Park station for tools not in use during workflow
1 pc D18/57 Liquid Tool	Tool for a 2500 µL syringe for sample
1 pc D8/57 Liquid Tool	Tool for a 100 µL syringe for ISTD addition
1 pc Tray Holder	Tray holder for 3× vial racks
1 pc Cooled Tray	Often a cooled tray is required alternatively
3 pc Rack 15× 10/20 mL	15 pos. vial racks for 10 mL vials
1 pc Vortex Mixer	For sample mixing after ISTD addition
1 pc LC Injection Valve	6-Port injection valve with sample loop
1 pc 10 mL Sample Loop	Sample loop for 10 mL injections (5 mL optional)
1 pc LC Switching Valve	6-Port valve for online SPE cartridge load and backflush
1 pc Standard Wash Module	Wash module for up to 5× syringe wash solvents, resp. ISTD positions

System – Tool – Module	Task
1 pc Fast Wash Module	Wash module for 2× active wash solvents
1 pc Mounting kit	Installs the x,y,z-robot benchtop next to the LC unit
2 pc LC systems	Regular HPLC pump or sample load, a UPLC pump for analysis
1 pc MS system	Triple quadrupole analyzer, ESI ion source

The robotic sampling system configuration includes two 6-port LC valves as shown in Figure 6.94. The valve switching states are illustrated in Figure 6.95. One valve carries the sample loop with injection port served by the robotic sampler and is connected to the LC load pump. The second valve connects the sample loop with the online SPE clean-up column, analytical pump, separation column, and mass spectrometer.

Workflow

The first step in the workflow is the automated addition of the ISTD to the sample, followed by vortexing. The online SPE is performed using a set of two configured valves, the "loop valve" and the "SPE valve," mounted to the rail of the x,y,z-robotic sampler. The workflow steps are controlled by the robot [128]. The steps of the online clean-up and analysis workflow are illustrated in Figure 6.95a–d.

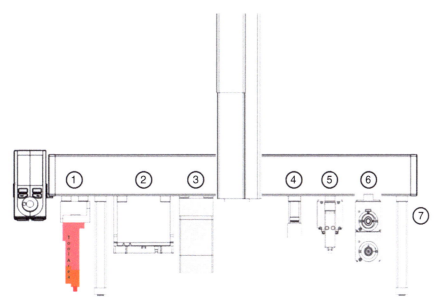

Figure 6.94 Robotic sampling system configuration for online SPE analysis. 1 Tool park station; 2 Trayholder for 10 mL vial racks (or cooled tray); 3 Vortex mixer; 4 Standard wash station for ISTD vials; 5 Fast wash station; 6 Injection and switching 6-way valves, sample loop; and 7 Benchtop mounting legs.

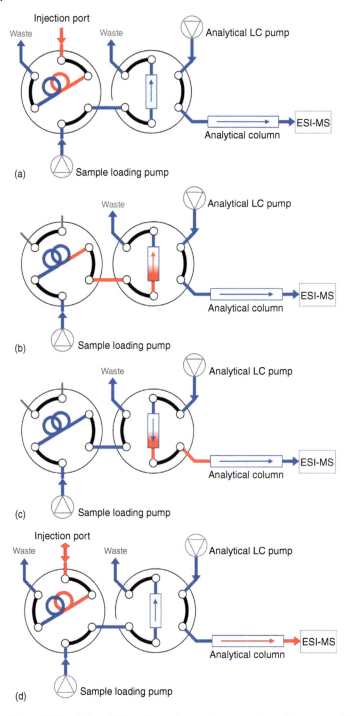

Figure 6.95 Online SPE configuration and automated workflow steps. The sample plug is shown in red. (a) Water sample injection into the sample loop, partial filling mode, column conditioning. (b) Load of water sample to online SPE cartridge. (c) Elution of sample to the analytical column, loop flushing. (d) Analyte separation and ESI-MS detection, SPE column conditioning, new sample load.

1) Sample injection:
   - The prepared samples, quality controls, and calibration solutions are loaded from the sample vials located on the trayholder to the 5 mL sample loop by the robotic sampler.
   - The typical injection volume is 2 mL with partial loop filling.
2) Load sample:
   - The loop valve switches and allows the load pump to send the sample plug in reverse direction to the online SPE column.
   - The SPE column is washed and prepared for elution.
3) Elution:
   - The SPE valve switches and connects the SPE column in-line with the analytical pump and separation column.
   - At the same time, the transfer by gradient elution of the cleaned sample plug to the analytical column starts in the reversed direction of the sample load.
4) Separation and next injection:
   - While the separation on the analytical column is ongoing, the loop valve switches back to prepare the sample loop for the next injection.
   - The robot injects solvents in sequence, e.g. methanol and water for cleaning.
   - A next sample is injected while the LC separation and detection of the previous sample is ongoing.
5) SPE column conditioning:
   - The SPE valve switches back, while the analysis is still ongoing.
   - The SPE column gets connected with the load pump for reconditioning by MeCN, then turning to aqueous conditions the SPE column for re-equilibration for the next sample.
   - An additional backflush valve can be used and activated here (not shown).
6) Next sample load:
   - After completion of the ongoing analysis, the analytical column is prepared to start conditions by the analytical pump unit.
   - Switching the loop valve starts the next analysis cycle by loading the next sample onto the SPE column.

Analysis Parameter

The parameters for online-SPE with LC-MS analysis are listed in Table 6.24, the gradient program for the loading and analysis HPLC pumps in Tables 6.25 and 6.26, respectively.

Conclusion

The online pre-concentration and clean-up with a reduced water sample volume of a few milliliters only allow for minimized sample manipulation and significant time savings over traditional SPE concentration methods. The routine method is robust with low maintenance requirements. The online SPE column can be used reportedly for 500–1000 drinking water analyses before a replacement becomes necessary.

Table 6.24 Selected analysis parameters for the online SPE LC-MS analysis.

LC	
Sample loading pump	HPLC pump type
Gradient for sample loading	See Table 6.25
Analytical LC pump	UPLC Pump type
Gradient analytical separation	See Table 6.26
Solvent A	Water with 0.1% formic acid
Solvent B	Acetonitrile with 0.1% formic acid
MS	
LC-MS System	Triple quadrupole LC-MS/MS system
Ion source	ESI, positive ion mode
Spray voltage	4500 V
Sheath gas	45 units (N_2)
Auxiliary gas	Not used
Transfer tube temperature	330 °C
Collision cell pressure	1.0 units (Ar)

Source: Adapted from Beck et al. [129].

Table 6.25 Gradient program for the loading pump.

Time	%A	%B	Flow rate [mL/min]
0	95	5	1.0
1.5	95	5	1.0
2.0	95	5	0[a]
12.5	95	5	0[a]
12.6	5	95[b]	1.0
14.5	5	95[b]	1.0
14.6	95	5	1.0
17.0	95	5	1.0

a) The flow is turned off from 2 to 12.5 min to conserve mobile phase.
b) The column is rinsed from 12.6 to 14.5 min with a high organic phase, before re-equilibrating to start conditions.

Source: Adapted from Beck et al. [129].

With the significant solvent savings, using the online SPE technology is a veritable contribution to green analytical chemistry. The detection and quantification of pesticides, PAH, and PPCP compounds is achieved at detection limits from sub to low nanogram per liter levels well compliant to the regulatory requirements.

Table 6.26 Gradient program for the analysis pump.

Time	%A	%B	Flow rate [mL/min]
0	95	5	0.2
1.5	95	5	0.2
10.0	0	100	0.2
12.0	0	100	0.2
12.1	95	5	0.2
17.0	95	5	0.2

Source: Adapted from Beck et al. [129].

6.17 Residual Solvents

Application

VOCs are not only regulated and monitored in drinking water and food packaging materials, but in particular, in pharmaceutical products. The volatiles to be analyzed in pharmaceuticals are defined "as organic volatile chemicals that are used or produced in the manufacturing of drug substances, excipients, or dietary ingredients, or in the preparation of drug products or dietary supplement products", as illustrated in Figure 6.96. The United States Pharmacopeia (USP) established the analytical method as "<467> Residual Solvents," known in short as "USP <467>", for the analysis of critical residues by GC headspace analysis with FID detection. Also, the USP <467> sets the residue limits for pharmaceutical drug and dietary supplement products, most current with a revision announcement of 1 November 2019, to become official 1 December 2020 [130].

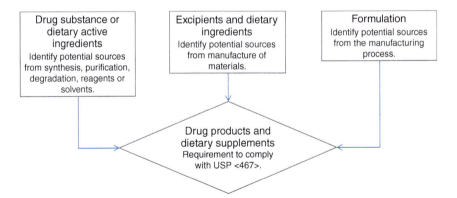

Figure 6.96 Potential sources of residual solvents in pharmaceutical drug products and dietary supplements (after, Source: Adapted from The United States Pharmacopeial Convention Inc. 2019 [130]).

Scope and Principle of Operation

The solvents are categorized in the USP <467> for their potential risk to human health with a maximum acceptable intake per day, and also due to their environmental impact. The compounds are classified into three groups [130]:

Class 1 (solvents to be avoided): Strongly suspected human carcinogens, solvents particularly known to have ozone-depleting properties, non-genotoxic animal carcinogens or possible causative agents of other irreversible toxicity, such as neurotoxicity or teratogenicity.

Class 2 (solvents to be limited): Solvents suspected of other significant but reversible toxicities.

Class 3 (solvents with low toxic potential): Solvents with low toxic potential to humans; no health-based exposure limit is needed.

The particular analytical challenge of this application is the chemical nature of different analytes with different volatility and solubility in water (k factor). The reference compounds are prepared in the standards with various concentration levels according to their permitted daily exposure (PDE) and vary widely with the GC detector response (FID). While the USP <467> does not state a specific precision requirement, a repeatability of better than 5% RSD ($n = 10$) is the generally preferred minimum quantitative performance.

Solvents and Chemicals

- Standard solutions of Class 1, 2, and 3 listed solvents, as of Tables 6.27–6.29.
- DMSO p.a., CAS No. 67-68-5, for standard mix dilution to a working standard.
- Water, HPLC grade, CAS No 7732-18-5, solvent for calibration dilution.

Consumables

- 20 mL headspace screw-top clear vials with magnetic screw caps.
- Headspace syringe, 2500 µL, PTFE plunger, needle length 65 mm, gauge 23, point style sideport.
- GC inlet liner: Straight glass liner with cup.
- GC column: Rxi-624Sil, 30 m × 0.32 mm × 1.8 µm, or equivalent.

Table 6.27 Class 1 residual solvents to be avoided.

Solvent	Concentration limit [ppm]
Benzene	2
Carbon tetrachloride	4
1,2-Dichloroethane	5
1,1-Dichloroethene	8
1,1,1-Trichloroethane	1500

Source: Adapted from The United States Pharmacopeial Convention Inc. 2019 [130].

Table 6.28 Class 2 residual solvents to be limited.

Solvent	Concentration limit [ppm]
Acetonitrile	410
Chlorobenzene	360
Chloroform	60
Cumene	70
Cyclohexane	3880
1,2-Dichloroethene	1870
1,2-Dimethoxyethane	100
N,N-Dimethylacetamide	1090
N,N-Dimethylformamide	880
1,4-Dioxane	380
2-Ethoxyethanol	160
Ethylene glycol	620
Formamide	220
Hexane	290
Methanol	3000
2-Methoxyethanol	50
Methylbutylketone	50
Methylcyclohexane	1180
Methylene chloride	600
Methylisobutylketone	4500[a]
N-Methylpyrrolidone	530
Nitromethane	0.5
Pyridine	2.0
Sulfolane	1.6
Tetrahydrofuran	7.2
Tetralin	1.0
Toluene	8.9
Trichloroethylene	0.8
Xylene	21.7

a) Official 1-Dec-2020.
Source: Adapted from The United States Pharmacopeial Convention Inc. 2019 [130].

Table 6.29 Class 3 residual solvents with low toxic potential[a].

Acetic acid	Isobutyl acetate
Acetone	Isopropyl acetate
Anisole	Methyl acetate
1-Butanol	3-Methyl-1-butanol
2-Butanol	Methylethylketone
Butyl acetate	2-Methyl-1-propanol
tert-Butylmethyl ether	Pentane
Dimethyl sulfoxide	1-Pentanol
Ethanol	1-Propanol
Ethyl acetate	2-Propanol
Ethyl ether	Propyl acetate
Ethyl formate	Triethylamine[b]
Formic acid	
Heptane	

a) 5000 ppm or 0.5% w/w considered acceptable.
b) Official 1-Dec-2020.
Source: Adapted from The United States Pharmacopeial Convention Inc. 2019 [130].

System Configuration

The minimum robotic system hardware requirements for residual solvents analysis as shown in Figure 6.97 are:

System – Tool – Module	Task
PAL RSI System	x,y,z-Robotic system for HS analysis with manual tool change
1 pc Tray Holder	Tray holder for 3× vial racks (extendable for high throughput analyses to 2 or more trays)
3 pc Rack VT15	15 pos. 20 mL vial racks
1 pc Agitator/Incubator	Sample incubation of up to 6× 20 mL vials
1 pc Headspace Tool	For a 2500 µL HS syringe operation
1 pc Mounting kit	Installs the x,y,z-robot GC instrument top
1 pc GC system	With split/splitless injector, FID detector
1 pc MS system	Single quadrupole MS, EI ionization (optional)

Analysis Parameter

The analytical parameters for the USP<467> headspace method are listed in Table 6.30.

6.17 Residual Solvents

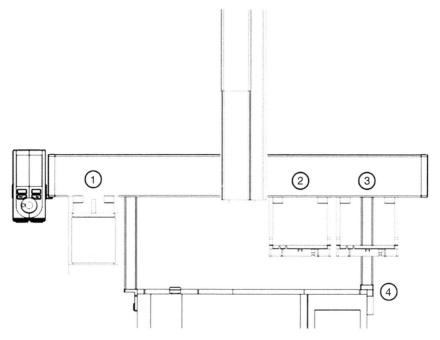

Figure 6.97 Robotic sampler configuration for USP <467> high throughput residual solvent analysis. 1 Agitator/Incubator; 2 Sample vial trayholder #1; 3 Sample vial trayholder #2; and 4 GC top mounting legs.

Workflow

Initial Manual Steps

- The test sample needs to be dissolved to release the residual solvent or is pulverized with minimized loss of volatile solvents [131].
- Transfer 1.0 mL of the standard stock solutions to headspace vials containing 5.0 mL of water.
- Transfer 5.0 mL of the sample solution to headspace vials, add 1.0 mL of water.
- Cap the vials and place into the trayholder of the robotic autosampler.

Automated Workflow

- Upon start of the workflow, the sample vials are successively transferred from the sample tray to the incubator unit.
- An overlapping function is activated using all six vial positions of the incubator.
- With the first sample, the cycle time of the overall measurement procedure is recorded by the robot.
- Next samples are transferred for incubation in time as calculated to achieve the completion of the incubation right at the time the *Ready* signal of the GC is expected.
- This schedule allows in the automated processing for all samples constant and reproducible incubation time. Each sample gets analyzed without any delay immediately after completion of the constant incubation time.
- The "prep-ahead" mode is used to incubate samples while the GC separation is ongoing.

Table 6.30 Selected analysis parameters for the USP <467> HS-GC analysis of residual solvents.

Headspace extraction	
Incubation temperature	80 °C
Incubation time	45 min
Agitator speed	250 rpm
Auxiliary gas	Nitrogen
Sample vial penetration depth	25 mm
Vial pressurization	1.6 bar
Syringe temperature	105 °C
Filling strokes	5
Stroke volume	1200 µL
GC	
Inlet temperature	200 °C
Inlet mode	Split
Injection volume	1200 µL
Split ratio	5:1
Carrier gas	He, constant flow at 2.2 mL/min
Linear velocity	35.6 cm/s
Oven program	40 °C (20 min) → 10 °C/min → 160 °C → 30 °C/min → 240 °C (5 min)
FID detector	280 °C

Sample Measurements

The chromatograms in Figure 6.98 show the separation of the compounds of the Class 2 calibration solution with low ppm concentration as of Table 6.31. The large difference in concentration and detector response can be seen with the elution of methanol/acetonitrile and 1,4-dioxane in separate windows, next to the dominant base peak of cyclohexane. The method reproducibility complies well with the general method expectations with a precision of 3–5% RSD ($n = 10$) for all compounds. The quantitative calibration shows excellent linearity even for low response analytes as acetonitrile in Figure 6.99.

Conclusion

The described automated workflow for USP <467> residual solvent analysis allows the quantification of the regulated compounds with high sample throughput, high sensitivity, and reproducibility compliant with the current performance regulations.

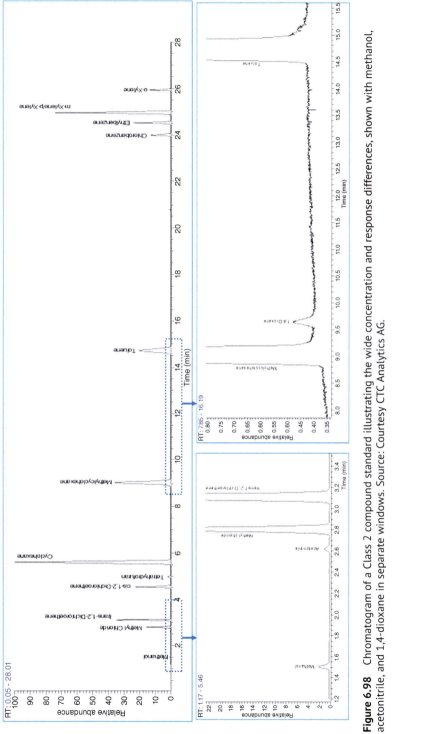

Figure 6.98 Chromatogram of a Class 2 compound standard illustrating the wide concentration and response differences, shown with methanol, acetonitrile, and 1,4-dioxane in separate windows. Source: Courtesy CTC Analytics AG.

Table 6.31 Concentrations of the Class 2 calibration test solution.

Compound	Concentration [ppm]
Methanol	8.34
Acetonitrile	1.14
Methyl chloride	1.67
trans-1,2-Dichloroethene	2.61
cis-1,2-Dichloroethene	2.62
Tetrahydrofuran	1.92
Cyclohexane	10.79
Methylcyclohexane	3.28
1,4-Dioxane	1.06
Toluene	2.47
Chlorobenzene	1.00
Ethylbenzene	1.03
m-Xylene	3.62
p-Xylene	0.85
o-Xylene	0.54

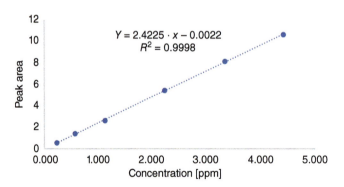

Figure 6.99 Quantitative calibration for the low-level compound acetonitrile with excellent linearity R^2 0.9998 in the range of 0.22–4.4 ppm. Source: Courtesy CTC Analytics AG.

6.18 Chemical Warfare Agents in Water and Soil

Application

The Chemical Weapons Convention (CWC) outlaws the production, stockpiling, and use of chemical weapons and their precursors. Also within this treaty, the destruction and potential chemical weapons production facilities are monitored [132]. The

Finnish Institute for Verification of the Chemical Weapons Convention (VERIFIN) started as a research project in 1973 for carrying out several roles in support of the chemical weapons disarmament and published the "Recommended operating procedures for analysis in the verification of chemical disarmament" in the so-called "Blue Book," in its most recent 2017 Edition [133]. The institute is the National Authority for Finland under the Chemical Weapons Convention, carrying out many of the tasks required by the treaty such as preparing the Finnish Government declarations to the Organization for the Prohibition of Chemical Weapons (OPCW) of inventories of controlled substances and escorting OPCW inspections of Finnish chemical production facilities. It is also one of the 18 OPCW-designated laboratories worldwide for performing chemical weapons verification tests [134].

Automated analysis of the poisonous and deadly chemical warfare agents is a highly desired feature to exclude human risk in particular during sample preparation. An automated workflow for the sample preparation of different types of chemical warfare agents namely VX, VG, and HD from four different matrices was first time described by Marc André Althoff et al. [135]. The optimized workflow for automated LLE extraction and µSPE clean-up is the subject of this application.

Scope and Principle of Operation

The described method is used for the automated extraction, clean-up, and chromatographic analysis of water, soil, and wipe samples. The samples are placed pH-adjusted in 2 mL regular or 10 mL conical vials into the robotic sampler. The LLE step is achieved after adding the extraction solvent DCM by the dilutor tool of the sampler and using a vortexing unit. After phase separation, which can be supported for soil samples by centrifugation, the organic phase is taken for a µSPE clean-up, followed by analysis with injection to GC-MS.

Solvents and Chemicals

- Standard solutions of the chemical warfare agents VX, VX-Disulfide, HD, and VG (Amiton) were prepared in DCM, made available from the Chemical Defense, Safety and Environmental Protection School Sonthofen, Germany.
- Dichloromethane p.a., CAS No. 75-09-2, as extraction solvent.
- Na_2SO_4, CAS No. 7757-82-6, anhydrous.
- Methanol, CAS No. 67-56-1, HPLC quality.
- Water, CAS No. 7732-18-5, HPLC quality.
- Hexane, CAS No. 110-54-3, GC purity $\geq$97%.
- Ethylacetate, CAS No. 141-78-6, GC purity $\geq$99.5%.

Consumables

- µSPE cartridges with sorbent materials:
  - 10 mg UCT C18 EC.
  - 10 mg UCT Silica.

- 2 mL screw-top clear vials with magnetic screw caps.
- 10 mL screw-top clear vials with magnetic screw caps.
  - Dedicated dispersive liquid/liquid micro-extraction (DLLME) vials of Billimex® type [136].
- GC column CP-Sil 8 CB Low Bleed/MS, 30 m × 0.25 mm × 0.25 µm, or equivalent.
- GC inlet liner 130 mm × 2.0 mm ID, straight empty glass liner.
- Liquid syringe 100 µL for LLE, needle gauge 22 or 23, point style 3/LC.
- Liquid syringe 1000 µL for µSPE clean-up, needle gauge 22 or 23, point style 3/LC.
- Liquid syringe 10 µL for GC injection, needle gauge 23s, point style AS/conical.

System Configuration

The minimum hardware requirements of a robotic sampling system for the automated analysis of warfare agents as shown in Figure 6.100 are:

System – Tool – Module	Task
PAL RTC System	x,y,z-Robotic system with automated tool change for sample preparation and GC injection, 1200 mm rail length
1 pc Park Station	Park station for tools not in use during the workflow
2 pc D7/57 Liquid Tool	Tool for a 10 µL GC injection and 100 µL syringe
1 pc D8/57 Liquid Tool	Tool for a 1000 µL sample preparation syringe
1 pc Dilutor and Tool	For liquid handling with dispensing from solvent or reagent reservoirs
2 pc Tray Holder	Tray holder for 3× vial racks, used for LLE and µSPE
1 pc µSPE kit	Incl. cartridge tray, waste receptacle, vial lock
3 pc Rack 15× 10/20 mL	15 pos. racks for 10 mL vials
2 pc Rack 54× 2 mL	54 pos. racks for 2 mL vials
1 pc Solvent Module	For 3× 100 mL solvent reservoirs
1 pc Standard Wash Module	Wash module for up to 5× syringe wash solvents
1 pc Fast Wash Module	Wash module for 2× active wash solvents, recommended for large syringe wash steps
1 pc Vortex Mixer	For intense LLE vial mixing
1 pc Agitator/Incubator	Sample incubation of up to 6× 20 mL vials. 10 mL vials with adaptor
1 pc Centrifuge	Recommended for LLE phase separation of soil samples, for 2 mL and 10 mL vials
1 pc Mounting kit	Installs the x,y,z-robot GC instrument top
1 pc GC system	With split/splitless, or PTV type injector
1 pc MS system	Single or triple quadrupole analyzer, EI ion source

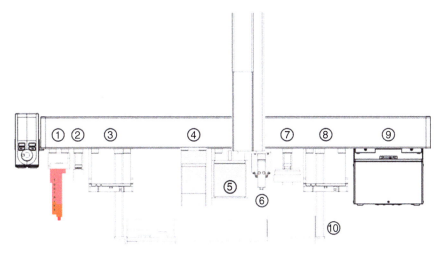

Figure 6.100 Robotic sampler in the configuration for the LLE and μSPE clean-up workflows. 1 Tool park station; 2 Standard wash station; 3 Trayholder for μSPE clean-up; 4 Vortexing unit; 5 Incubator/agitator; 6 Fast wash station; 7 Solvents reservoir for μSPE clean-up; 8 Trayholder for LLE samples; 9 Centrifuge; and 10 Instrument top mounting legs.

Analysis Parameter

The analysis parameters for the LLE extraction step, clean-up, and GC-MS analysis are provided with Table 6.32.

Workflow

The described workflows follow the recommended operating procedures defined in the so-called "Blue Book" [133]. As of the goal for the automated procedure, the volumes required are adjusted to the available instrument vial capacities of 2–10 mL.

Initial Manual Steps

- Water samples of 5 mL were adjusted to pH 7 and pH 11, respectively.
- 200 mg of homogenized (manual mortar grinding) soil samples were placed in 2 mL vials containing 100 mg of Na_2SO_4 and taken up in 600 μL of DCM.
- The wipe samples (cotton bud tips) were placed into 10 mL vials to which 2.5 mL of DCM were added. Extraction was done by using the vortexing unit.
- 2 mL vials for extract drying are prepared by adding 100 mg of dried Na_2SO_4 and placed into the rack of the autosampler.

Automated Workflow

- The automated LLE workflow is depicted in Figure 6.101.
- The centrifugation step to support the phase separation is configured optionally in the programmed workflow, in particular recommended for soil samples.

- An additional centrifugation step after vortexing with drying agent is not required as the sodium sulfate precipitates readily during the transport of the vial back to the rack position.
- Soil samples and contaminated water samples are subjected to an additional clean-up step before GC injection. Potential hydrocarbon background is removed using a silica µSPE cartridge. The clean-up with µSPE shown in Figure 6.101 can be integrated optionally into the described workflow.

Table 6.32 Analysis parameter for the extraction of CWAs from water and soil.

LLE extraction step	
Temperature	Room temperature
Vial size	2 mL, or 10 mL
Sample, extraction solvent volume	500 µL, 250 µL for 2 mL vials
	5.0 mL, 2.5 mL for 10 mL vials
Vortex mixing speed, time	1200 rpm, 5 s
Centrifugation	5000 g, 100 s for 2 mL vials
	2000 g, 100 s for 10 mL vials
µSPE clean-up step	
Conditioning	1000 µL methanol
	2000 µL water
Sample load volume	500 µL
Cartridge wash	2000 µL hexane
Elution volume	1000 µL ethylacetate
Flow rate for load and elution	12 µL/min
GC	
Inlet type, temperature	PTV, constant temperature, 250 °C
Inlet mode	Splitless
Injection volume	1.0 µL
Purge flow	1.2 mL/min
Carrier gas	He, constant flow at 1 mL/min
Oven program	70 °C (1 min) → 30 °C/min → 290 °C (1 min)
Transfer line temperature	280 °C
MS	
System, ionization	Triple quadrupole, EI mode
Ion source temperature	280 °C
Scan mode	SIM
Scan time	0.20 s

Source: Adapted from Althoff [135].

Figure 6.101 Automated LLE extraction workflow with optional μSPE clean-up step for soil and contaminated water samples, and online GC injection (*initial manual steps, ** μSPE clean-up optional for soil and contaminated samples).

Table 6.33 Recovery results of the automated LLE of chemical warfare agents VX, VG, and HD ($n = 3$).

Matrix[a]	Recovery rate ± RSD [%]		
	VX	VG	HD
Tap water (pH 7)	64.1 ± 9.6	79.8 ± 8.3	47.3 ± 3.8
Tap water (pH 11)	88.0 ± 3.4	85.9 ± 5.1	22.1 ± 2.7
Surface water (pH 7)	96.0 ± 3.4	99.9 ± 3.6	51.5 ± 4.0
Surface water (pH 11)	98.7 ± 1.4	99.9 ± 1.8	50.7 ± 1.4
Soil	66.2 ± 2.7	88.2 ± 4.0	46.8 ± 4.4
Wipe	65.6 ± 3.1	58.4 ± 2.3	51.2 ± 1.4

a) Aqueous samples were adjusted to the given pH-value prior to the extraction.
Source: Adapted from Althoff [135].

Table 6.34 Recovery results of chemical warfare agents from contaminated aqueous samples after µSPE clean-up, ($n = 3$).

Matrix	Cartridge	Recovery ± RSD [%]			
		VX	VX-disulfide	VG	HD
Water	C18-EC	56.5 ± 0.5	57.9 ± 2.2	65.8 ± 0.1	64.8 ± 1.9
Water samples	a)	24 ± 10	b)	b)	32 ± 7
Diesel fuel	silica	76.3 ± 1.3	77.4 ± 0.2	98.2 ± 1.5	c)

a) As of [119].
b) No values available.
c) Analyte not retained by the cartridge.
Source: Adapted from Althoff [135].

Recovery

Regular tap water and soil samples were spiked with the analytes and tested for recoveries. Table 6.33 shows very good recoveries of the analytes in the chosen matrix samples. The recovery rates depend on the pH value. The more acidic the matrix, the faster a degradation process for VX and VG can be observed. HD degrades rapidly with high pH values.

The recoveries of the complete workflow including the µSPE clean-up step of matrix contaminated samples in neutral condition are shown in Table 6.34.

Sample Measurements

The described automated workflow for the simultaneous extraction of different types of chemical warfare agents is used for different matrices like tap and surface waters, soil, and wipe materials.

For method development, samples were spiked with 10 ppm of the respective analytes. Matrix test samples prepared in tap water were processed with C18 µSPE cartridges, also spiked with a strong hydrocarbon background of 10 mg/mL of diesel fuel.

Results

The chromatogram in Figure 6.102 shows the GC-MS analysis in SIM mode of a diesel fuel contaminated water sample after the automated µSPE clean-up using a silica cartridge. The analytes VG, VX, and VX disulfide are shown with good selectivity and response. HD is here not eluted from the silica clean-up.

Conclusion

The advantage of the described automation is in the handling of only small amounts of sample, fast processing time, and the improved operator protection from hazardous compounds. Sample quantities as low as 1 mL are shown to be sufficient. The overall analysis cycle time including automated clean-up is short with approx. 15 minutes, depending on the GC temperature program used.

The described workflow can easily be extended on the robotic system by other sample preparation tasks, e.g. derivatization steps if necessary. The LLE sample preparation approach for water and soil samples complements the previously published DI-SPME method for the analysis of organophosphate agents in complex samples like foliage or grass [137] and provides the potential for the control of other related compounds under the CWC treaty. The described workflow is used in daily laboratory routine in the processing of chemical warfare agents and other analytes.

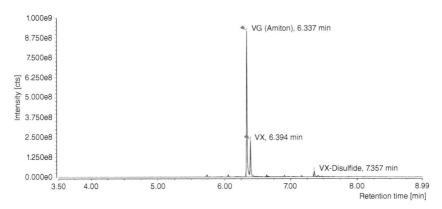

Figure 6.102 GC-MS chromatogram of a diesel fuel contaminated water sample after clean-up with silica cartridge (10 ppm spike, SIM mode). Source: Courtesy M.A. Althoff, private communication.

6.19 Shale Aldehydes in Beer

> The constituents of freshly bottled beer are not in chemical equilibrium.
> Vanderhaegen et al. [138]

Application

Aldehydes are widely found in nature. Natural sources for aldehydes are for instance the alcoholic fermentation or lipid oxidation. Aldehydes were also identified as by-products of drinking water disinfection, particularly ozonation, the currently introduced fourth treatment stage in WWTPs.

Some aldehydes are important flavor compounds, and in some cases causing unpleasant off-flavors. The comprehensive identification of flavor compounds using SPME Arrow extraction is accomplished by AMDIS deconvolution of the complex GC-MS full scan chromatograms with the important trace components [139]. The

Table 6.35 Flavor aldehydes derived from multiple chemical transformations with their flavor threshold.

Aldehyde	Generation	Boiling point	Flavor threshold [ppb]	Flavor description
(E)-2-Nonenal	Fatty acid oxidation	100–102 °C	0.03	Cardboard, papery, cucumber
Methional	Strecker degradation	165–166 °C	4.2	Cooked potatoes, worty
3-Methylbutanal	Strecker degradation	92–93 °C	56	Malty, chocolate, cherry, almond
2-Methylpropanal	Strecker degradation	64 °C	86	Grainy, varnish, fruity
Hexanal	Fatty acid oxidation	131 °C	88	Bitter, winey
Phenylacetaldehyde	Strecker degradation	195 °C	105	Hyacinth, flowery, roses
Benzaldehyde	Strecker degradation	179 °C	515	Almond, cherry stone
Acetaldehyd	Glycolysis byproduct	21 °C	1114	Green apple, fruity
Furfural	Maillard reaction	161.8 °C	15 157	Caramel, bready, cooked meat
5-Hydroxymethylfurfural	Maillard reaction	114–116 °C	35 784	Bready, caramel

Source: Adapted from Baert et al. [141].

6.19 Shale Aldehydes in Beer

formation of aldehydes is also responsible in the flavor development of beer during storage, the so-called "stale aldehydes" with very low odor thresholds. Sensory changes during storage is observed, for instance, bitterness decreases, sweetness grows, stale cardboardiness, and ribes called black currant/catty notes increase, requiring the control of the brewing process to avoid unpleasant staleness [140]. Therefore, the analysis of aldehydes as shown in Table 6.35, on the wider aspect of carbonyl compounds, is of high priority for quality control.

The low level of compounds requires a dynamic method to enrich the aldehydes for analysis at best directly from aqueous media. The analysis of such small molecules is best achieved by GC-MS after derivatization.

Scope and Principle of Operation

This method is directed to the extraction and concentration of the volatile aldehydes by SPME [142]. It is described for the quality control analysis of beer, but can be applied to many other beverages and foods. The polar aldehydes are derivatized on-fiber for GC analysis with O-(2,3,4,5,6-pentafluorobenzyl)hydroxylamine (PFBHA) to form pentafluorobenzyl oxime derivatives.

PFBHA reacts quantitatively, even with conjugated aliphatic aldehydes (Figure 6.103). The resulting oximes (E- and Z-isomers) do not decompose at elevated temperatures, neither do they require a time-consuming clean-up step and can easily be extracted by SPME and analyzed by GC with electron capture detector (ECD) or MS detection.

Solvents and Chemicals

- Derivatization reagent: PFBHA·HCl, purity >99%, CAS No. 57981-02-9, prepare a 60 mg/L solution.
- Aldehydes C2 to C12 as individual retention time standards.
- Suggested ISTD: 2,3,5,6-tetrafluorobenzaldehyd, CAS 19842-76-3.
- NaCl, CAS No. 7647-14-5, analytical quality.
- Water, CAS No. 7732-18-5, HPLC quality.

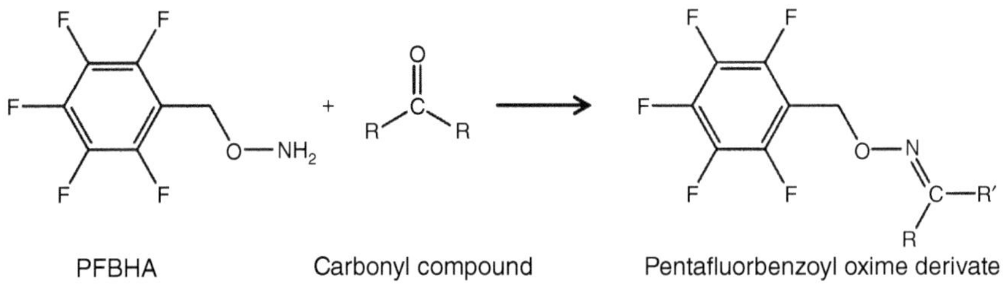

Figure 6.103 PFBHA derivatization reaction to form pentafluorobenzyl oxime derivatives.

Consumables

- 20 mL headspace screw-top clear vials with magnetic screw caps.
- SPME Arrow, 1.1 mm OD, PDMS, 100 µm, or PDMS/DVB, 65 µm.
- GC inlet liner, 1.3 or 1.7 mm ID.
- GC column: DB-5MS, 30 m × 0.25 mm × 0.25 µm, or equivalent.

System Configuration

The minimum hardware requirements for the robotic sampler is used as illustrated in Section 6.2.2 with Figure 6.12 describing the robotic sampler configuration for SPME derivatization workflows.

Analysis Parameter

The analytical parameters for the HS-SPME extraction and GC-MS analysis are listed in Table 6.36.

Table 6.36 Selected analysis parameters for the HS-SPME GC-MS analysis of aldehydes in beer.

SPME headspace extraction	
Pre-conditioning time	0.5 min
Pre-incubation dipped into reagent	10 min
Sample incubation temperature	60 °C
Incubation time	5 min
Agitator speed	500 rpm
Heatexer temperature, speed	60 °C, 1500 rpm
Extraction time	30 min
GC	
Inlet temperature	250 °C
Inlet mode	Splitless
Desorption time	5 min
Carrier gas	He, constant flow at 1 mL/min
Oven program	70 °C (1 min) → 5 °C/min → 280 °C
Transfer line temperature	280 °C
MS	
System, ionization	Single quadrupole, EI mode
Ion source temperature	250 °C
Scan mode	Full scan
Scan m/z	m/z 75–250
Scan time	0.25 s

Workflow

Initial Manual Steps

- 1.5 g of sodium chloride is placed in empty 20 mL screw cap headspace vials (optional).
- Remove the CO_2 bubbles of the fresh beer at room temperature before filling the HS vials.
- 2–5 mL of water or beer sample is added to the vial and closed with a magnetic cap.

Automated Workflow

- The automated workflow on the robotic sampler follows as outlined in Section 6.2.2. The sample vial is transferred by magnetic transport from the storage rack to the agitator and incubated at 60 °C for five minutes.
- After incubation time, the sample vial is transferred into the Heatex stirrer for extraction.
- After incubation time and before extraction, the SPME Arrow is dipped for one minute into the derivatization reagent and then applied to the headspace of the sample vial.
- For analysis, the loaded SPME Arrow is moved to the GC injection port and thermally desorbed at 250 °C for five minutes. During the desorption time, the injector split is kept closed.

Results

The quantitative calibration was performed as an external calibration without ISTD. The quantification of the C2 to C12 aldehydes was achieved with good precision in the range of 0.2–500 µg/L with a correlation factor R^2 better than 0.99 for all compounds. For the very nose sensitive E-2-nonenal, a very low LOD of 0.002 µg/L could be achieved.

Note: The derivatization reaction yields stereoisomers, which are separated chromatographically; hence, the doublets found in the chromatogram. Salting out

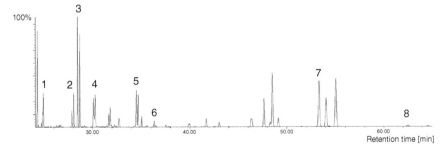

Figure 6.104 Beer analysis for aldehydes using automated SPME on-fiber derivatization GC-MS. Source: Courtesy CTC Analytics AG. 1 2-Methyl-propionaldehyde; 2 3-Methyl-butyraldehyde; 3 2-Methyl-butyraldehyde; 4 Valeraldehyde; 5 Caproaldehyde; 6 Furaldehyde; 7 Phenylacetaldehyde; and 8 E-2-Nonenal.

6 Solutions for Automated Analyses

Table 6.37 Aldehyde detection limits using SPME Arrow GC-MS with online derivatization.

Compounds	Detection Limits [µg/L]	Calibr. precision R^2
2-Methyl-propionaldehyde	0.05	0.992
3-Methyl-butyraldehyde	0.04	0.994
2-Methyl-butyraldehyde	0.02	0.993
Valeraldehyde	0.12	0.990
Caproaldehyde	0.07	0.991
Furaldehyde	10.0	0.997
Phenylacetaldehyde	0.02	0.997
E-2-Nonenal	0.002	0.990

Source: Wu et al. [143].

with NaCl showed a small increase of signals for the smaller aldehydes (C2–C4) only. However, since the effect was limited, a salt addition was kept optional for this method.

The chromatogram in Figure 6.104 shows the analysis of an actual brew sample from an SPME Arrow extraction with the elution of the aldehydes of Table 6.37, and the clear absence of the E-2-nonenal [143].

6.20 Phthalates in Polymers

Application

Phthalates are mostly used as plasticizers to make polymer materials more flexible (Figure 6.105). They also serve as solvents for instance for perfumes or pesticides. Due to the widespread use in plasticized polymer products, they are ubiquitously found in the environment, food, and even human body fluids. Measurable levels of many phthalate metabolites were found in the general population [144]. Several phthalates are considered harmful to human health, with a confirmed acting as endocrine disruptors.

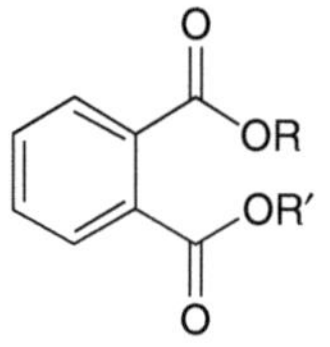

Figure 6.105 Phthalate general formula, with R, R' variety of linear and branched alkyl esters.

Several international regulations restrict the use of certain phthalates and set maximum concentration levels for control. In the European Union, the use of some phthalates is restricted for use in children's toys since 1999 [145]. The directive prohibits the use of certain categories of phthalates in the manufacture of toys and childcare articles intended for children. It applies to di(2-ethylhexyl) phthalate (DEHP), dibutyl phthalate (DBP), and butylbenzyl phthalate (BBP). DEHP, DBP, and BBP are restricted for all toys. Di-iso nonyl phthalate (DINP), di-*iso*-decyl phthalate (DIDP), and di-*n*-octyl phthalate (DNOP) are restricted only in toys for children under 3 years of age that can be expected to be taken into the mouth [146].

Scope and Principle of Operation

In general, the most often manually performed approach to phthalate analysis is the dissolution of the polymer in a suitable solvent to release the analytes of interest, and then precipitate the polymer using a solvent the material is not soluble in. After filtration or centrifugation, the extract is subjected to GC-MS analysis [147, 148].

Solvents and Chemicals

- Phthalates Certified reference material (CRM), e.g. NIST SRM 3074.
- Dibutyl phthalate (DBP), CAS No. 84-74-2.
- Di-(2-ethylhexyl) phthalate (DEHP), CAS No. 117-81-7.
- Benzyl butyl phthalate (BBP), CAS No. 85-68-7.
- Di-*n*-octyl phthalate (DNOP), CAS No. 117-84-0.
- Diisononyl phthalate (DINP), CAS No. 28553-12-0/68515-48-0.
- Diisodecyl phthalate (DIDP), CAS No. 26761-40-0/68515-49-1.
- Benzyl benzoate (BB), CAS No. 120-51-4 as optional ISTD.
- Tetrahydrofuran (THF), CAS No. 109-99-9, HPLC grade.
- Hexane, CAS No. 110-54-3, HPLC grade.

Consumables

- 10 mL screw-top clear vials with magnetic screw caps.
- 20 mL screw-top clear vials with magnetic screw caps.
- PTFE syringe filters, 0.45 μm.
- GC column: HP-5MS Column 30 m × 0.25 mm × 0.25 μm, or equivalent.
- GC inlet: Baffled inlet liner.
- Liquid syringe, 2500 μL, for liquid handling, needle gauge 22 or 23, point style conical.
- Liquid syringe, 10 μL, for GC injection, needle gauge 23s, point style conical.

System Configuration

The minimum hardware requirements of a robotic sampling system for the automated phthalate analysis shown in Figure 6.106 are:

System – Tool – Module	Task
PAL RTC System	x,y,z-Robotic system with automated tool change for sample preparation and liquid GC injection, 1200 mm rail length, alternatively a dual-head system (DHR) can be employed, avoiding frequent tool changes
1 pc Park Station	Park station for tools not in use during workflow (omitted with a DHR system)
1 pc D7/57 Liquid Tool	Tool for a 10 µL GC injection syringe
1 pc D8/57 Liquid Tool	Tool for a 2500 µL sample preparation syringe
1 pc Dilutor and Tool	For dispensing of solvents, 10 mL syringe
2 pc Tray Holder	Tray holder for 3× vial racks 10/20 mL and 2 mL vials
3 pc Rack 15× 10/20 mL	15 pos. racks for 10/20 mL vials
3 pc Rack 54× 2 mL	54 pos. racks for 2 mL vials for filtered extract
1 pc Rack 60× 10/20 mL vial	60 pos. rack for 10/20 mL vials (alternatively for higher sample numbers)
1 pc Filtration Station	40 pos. filtration tray for 4 mm PTFE filter
1 pc Solvent Module	For 3× 100 mL solvent/reagent reservoirs
1 pc Large Wash Module	Wash module for 2× 100 mL syringe wash solvents
1 pc Standard Wash Module	Wash module for up to 5× syringe wash solvents, or ISTD positions
1 pc Adapter	Standard Wash Station Adapter to place 2 mL vials instead of the 10 mL wash vial
1 pc Vortex Mixer	For intense vial mixing
1 pc Agitator/Incubator	Sample incubation of up to 6× 10 mL vials (optional)
1 set Vial inserts 10 mL	Inserts for the use of 10 mL vials in the Agitator
1 pc Mounting kit	Installs the x,y,z-robot GC instrument top
1 pc GC system	With Cooled Injection System (CIS) split/splitless, or PTV type injector
1 pc MS system	Single quadrupole analyzer, EI ion source

Analysis Parameter

The analysis parameters for the extraction of phthalates from polymers and GC-MS analysis are provided with Tables 6.38 and 6.39 [149].

6.20 Phthalates in Polymers

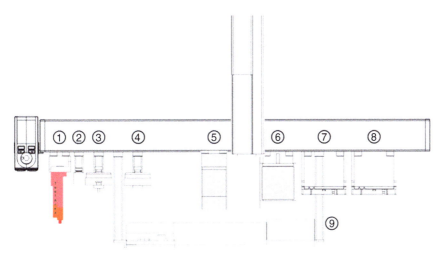

Figure 6.106 Robotic sampler configuration for the phthalate extraction workflow. 1 Tool park station; 2 Standard wash station; 3 Large wash station; 4 Solvent station; 5 Vortex mixer; 6 Incubator/Agitator module; 7 Syringe filter tray; 8 Sample tray; and 9 Instrument top mounting legs.

Table 6.38 Analysis parameter for the extraction and GC-MS analysis of phthalates.

Extraction	
Sample weight	50 mg
Extraction solvent, volume	THF, 5 mL
Extraction time	30 min (or more)
Precipitation solvent, volume	Hexane, 5 mL
Syringe filter	0.45 mm PTFE filter
Injection volume to GC	1.0 µL
GC	
GC inlet type	Cooled Injection System (CIS), PTV
Inlet temperature program	50 °C → 12 °C/s → 280 °C (3 min)
Inlet mode	Splitless for low concentrations, split (20 : 1) for high concentration levels
Carrier gas	He, constant flow at 1.0 mL/min
Oven program	50 °C (1 min) → 20 °C/min → 310 °C (5 min)
MS	
System, ionization	Single quadrupole, EI mode
Scan mode	SIM/scan
Scan m/z	See Table 6.39

Source: Adapted from GERSTEL [149].

Table 6.39 GC-MS SIM acquisition parameters for phthalate analysis (bold quantification masses).

Compound name	CAS#	M [g/mol]	SIM ions [m/z]
Benzyl benzoate (BB), ISTD	120-51-4	212	91, **105**, 194, 212
Benzyl butyl phthalate (BBP)	85-68-7	312	91, 149, **206**
Bis(2-ethylhexyl)phthalate (DEHP)	117-82-8	390	149, 167, **279**
Bis(2-n-butoxyethyl)phthalate (DBEP)	117-83-9	366	**149**, 176, 193
Dibutyl phthalate (DBP)	84-74-2	278	149, 167, 205, **223**
Diethyl phthalate (DEP)	84-66-2	222	**149**, 177, 222
Dihexyl phthalate (DHP)	84-75-3	334	**149**, 233, 251
Di-isononyl phthalate (DINP)	28553-12-0	418	149, 167, **293**
Dimethyl phthalate (DMP)	131-11-3	194	**163**, 194
Di-n-octyl phthalate (DNOP)	117-84-0	390	149, 167, 261, **279**

Source: Adapted from GERSTEL [149].

Workflow

Initial Manual Steps

- All samples to be extracted are cut into small pieces of <2 mm particle size. Alternatively, a cryogenic mill can be used to grind samples to powder.
- Weigh a 50 mg aliquot of a polymer sample exact into the sample vial.
- Cap and place the vial into the rack of the robotic sampler.

Automated Workflow

- The automated workflow is illustrated stepwise in Figure 6.107.
- Ensure the complete dissolution, ultrasound and/or mild heating in the incubator (optional configuration) to facilitate the dissolution process. Consider the BP of THF with 66 °C.
- If the sample has not been completely dissolved, the mixing period should be extended before continuing.
- Hexane precipitates the dissolved polymer. Add 10 mL hexane for each 5 mL of THF used during the dissolution step.
- Wait at least five minutes for the polymer material to precipitate.

Quantitative Calibration

Due to the ubiquitous presence of phthalates, the solvents in use must be tested for the presence of phthalate residues by performing a blank extraction. A series of standard dilutions with concentrations ranging from 50 to 1000 ng/mL and an upper range of 5–100 µg/mL are prepared for quantitative analyses.

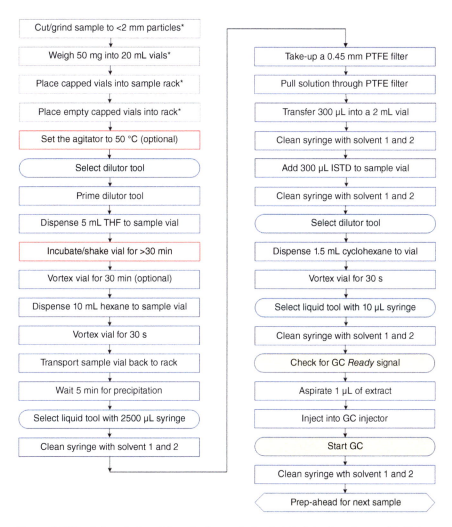

Figure 6.107 Automated workflow for the analysis of phthalates in polymers (*Initial manual pre-treatment steps).

Sample Measurements

The automated phthalate extraction procedure was found to be in good agreement with the manual sample preparation method [147]. The automated quantitative determination delivered accurate results with a precision of 1.9–5.5% RSD for DEHP, DBP, BBP, DINP, DIDP, and DNOP phthalates [149]. A chromatogram of the phthalate elution series is shown in Figure 6.108. Nonyl- and decylphthalates elute in unresolved humps due to the large number of differently branched isomeric structures [146].

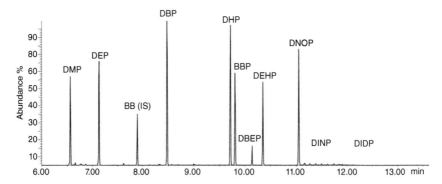

Figure 6.108 Chromatogram with the phthalate elution series (after, Source: Courtesy Agilent Technologies Inc. [148]).

Conclusion

The amount of sample and solvent required to perform the described automated method compared to the manual method was reduced significantly. The obtained results were in good agreement.

Using the prep-ahead function allows the sample preparation to be performed simultaneously with the GC-MS run for increased productivity.

References

1 ISO 11095:1996 (1996). *Linear calibration using reference materials*. Geneva, Switzerland: International Organization for Standardization.
2 Prichard, L. and Barwick, V. (2003). Preparation of Calibration Curves – A Guide to Best Practice. *Report #LGC/VAM/2003/032*. Teddington, UK: LGC Ltd.
3 SANTE/12682/2019 (Implemented by 01.01.2020) (2019). *Guidance document on analytical quality control and method validation procedures for pesticides residues analysis in food and feed*. European Commission Directorate-General for Health and Food Safety. https://doi.org/10.13140/RG.2.2.33021.77283.
4 Li, Y., Chen, X., Fan, C., and Pang, G. (2012). Compensation for matrix effects in the gas chromatography–mass spectrometry analysis of 186 pesticides in tea matrices using analyte protectants. *Journal of Chromatography A* 1266: 131–142.
5 Anastassiades, M., Maštovská, K., and Lehotay, S.J. (2003). Evaluation of analyte protectants to improve gas chromatographic analysis of pesticides. *Journal of Chromatography A* 1015: 163–184. https://doi.org/10.1016/S0021-9673(03)01208-1.
6 Rothmeier, S., M. Nestola, A. Hentschel, F. Fischer et al. (2018). Fully automated dilution workstation for pesticides working standards mixtures. *Presented at the European Pesticides Workshop 2018 in Munich, Germany*. Berlin, Germany: Institut Kirchhoff GmbH.

7 Donike, M. (1969). *N*-methyl-*N*-Trimethylsilyl-Trifluoracetamid ein neues Silylierungsmittel aus der Reihe der silylierten Amide. *Journal of Chromatography* 42: 103–104. https://doi.org/10.1016/S0021-9673(01)80592-6.

8 Donike, M. and Zimmermann, J. (1980). Zur Darstellung von Trimethylsilyl-, Triethylsilyl- und tert.-Butyldimethylsilyl-enoläthern von Ketosteroiden für gas-chromatographische und massenspektrometrische Untersuchungen. *Journal of Chromatography A* 202: 483–486. https://doi.org/10.1016/S0021-9673(00)91836-3.

9 Thevis, M. and Schänzer, W. (2007). Mass spectrometry in sports drug testing: structure characterization and analytical assays. *Mass Spectrometry Reviews* 26 (1): 79–107.

10 Fußhöller, G. and Schänzer, W. (2009). Improved steroid profiling using GC tandem mass spectrometry. *Poster at the Manfred Donike Workshop*, Cologne.

11 Neves Dias, A., Barnes Rodrigues, M., Cerqueira, R. et al. (2012). Optimization of a method for the simultaneous determination of glycerides, free and total glycerol in biodiesel ethyl esters from castor oil using gas chromatography. *Fuel* 94: 178–183. https://doi.org/10.1016/j.fuel.2011.10.037.

12 Alves, E.A., Agonia, A.S., Cravo, S.M. et al. (2016). GC-MS method for the analysis of thirteen opioids, cocaine and cocaethylene in whole blood based on a modified QuEChERS extraction. *Current Pharmaceutical Analysis* 12: 1–9. https://doi.org/10.2174/1573412912666160502163846.

13 Fluka (2005). *Silylating Agents*. Buchs SG: Fluka Chemie AG (today part of Sigma-Aldrich).

14 Gionfriddo, E., Passarini, A., and Pawliszyn, J. (2016). A facile and fully automated on-fiber derivatization protocol for direct analysis of short-chain aliphatic amines using a matrix compatible solid-phase microextraction coating. *Journal of Chromatography A* 1457: 22–28. https://doi.org/10.1016/j.chroma.2016.06.051.

15 Fiehn, O. (2005). Metabolite profiling in Arabidopsis. In: *Arabidopsis Protocols, Methods in Molecular Biology*, vol. 323 (eds. J. Salinas and J.J. Sanchez-Serrano), 439–447. Humana Press Inc. https://doi.org/https://doi.org/10.1385/1-59745-003-0:439.

16 Soma, Y., T. Yamashita, M. Takahash, K. Sugitate. et al. (2017). Automation of sample preparation for metabolomic analysis using robotic platform. *Poster #362 at the 13th International Conference of the Metabolomics Society*. Brisbane, Australia.

17 Fragner, L., Weckwerth, W., and Huebschmann, H.-J. (2012). Metabolomics strategies using GC-MS/MS technology. *Application Note 51999*. Austin TX, USA: Thermo Fisher Scientific.

18 Wikipedia (2020). 2,4,6-Tribromoanisole. https://en.wikipedia.org/wiki/2,4,6-Tribromoanisole (as of 30 March 2020).

19 Kotseridis, Y., Baumes, R.L., Bertrand, A. et al. (1999). Quantitative determination of beta-ionone in red wines and grapes of bordeaux using a stable isotope dilution assay. *Journal of Chromatography A* 848 (1–2): 317–325. https://doi.org/10.1016/s0021-9673(99)00422-7.

20 Langen, J., Wegmann-Herr, P., and Schmarr, H.G. (2016). Quantitative determination of α-ionone, β-ionone, and β-damascenone and enantiodifferentiation of α-ionone in wine for authenticity control using multidimensional gas chromatography with tandem mass spectrometric detection. *Analytical and Bioanalytical Chemistry* 408 (23): 6483–6496. https://doi.org/10.1007/s00216-016-9767-6.

21 Stenerson, K.K. (2011). TCA and precursors in red wine using in-matrix derivatization followed by SPME on the SLB TM-5ms. *Supelco Reporter* 29 (1): 14–15.

22 Kaziur, W., Salemi, A., Jochmann, M.A. et al. (2019). Automated determination of picogram-per-liter level of water taste and odor compounds using solid-phase microextraction arrow coupled with gas chromatography-mass spectrometry. *Analytical and Bioanalytical Chemistry* 411: 2653–2662. https://doi.org/10.1007/s00216-019-01711-7.

23 McGorrin, R.J. (2011). The significance of volatile sulfur compounds in food flavours – an overview. *ACS Symposium Series* 1068: 3–31. https://doi.org/10.1021/bk-2011-1068.ch001.

24 Cannon, R.J. and Ho, C.-T. (2018). Volatile sulfur compounds in tropical fruits. *Journal of Food and Drug Analysis*: 1–24. https://doi.org/10.1016/j.jfda.2018.01.014.

25 Apriyantono, A. and Septiana, E.E. (1995). Characterisation and identification of flavour of durian (Durio zibethinus Murr.) fruit using GC, GC-MS and GC-olfactometry. *Proceedings of EURO FOOD CHEM VII*, Volume 3, 74 ff.

26 Baldry, J., Dougan, J., and Howard, G.E. (1972). Volatile flavouring constituents of Durian. *Phytochemistry* 11: 2081–2084.

27 Moser, R., Düvel, D., and Greve, R. (1980). Volatile constituents and fatty acid composition of lipids in Durio zibethinus. *Phytochemistry* 19 (1): 79–81. https://doi.org/10.1016/0031-9422(80)85017-5.

28 Wong, K.C. and Tie, D.T. (1995). Volatile constituents of durian (Durio zibethinus Murr.). *Flavour and Fragrance Journal* 10: 79–83.

29 Weenen, H., Koolhaas, W.E., and Apriyantono, A. (1996). Sulfur-containing volatiles of durian fruits (Durio zibethinus Murr.). *Journal of Agriculture and Food Chemistry* 44: 3291–3293.

30 Koh, M. (2017). Guide to Durian: know the difference between Mao Shan Wang and D24. *Michelin Guide Singapore*, July 19, pp. 1–4. https://guide.michelin.com/sg/en/article/features/guide-to-durian-know-the-difference-between-mao-shan-wang-and-d24 (as of 17 April 2020).

31 Hashim, D.M.A.T. (2010). Unraveling the issue of alcohol for the Halal industry. *World Halal Research*. hdcglobal.com/upload-web/cms-editor-files/b08c8a04-c946-4ebe-99b9-2492bd32fcfc/file/11.%20En%20Dzulkifli%20Mat%20Hashim%20-%20WHR2010_Unraveling%20the%20Issue%20of%20Alcohol_Final.pdf.

32 Alzeer, J. and Hadeed, H.A. (2016). Ethanol and its Halal status in food industries. *Trends in Food Science and Technology* 58: 14–20. https://doi.org/10.1016/j.tifs.2016.10.018.

33 Zhang, X. (2017). CDS 7000C Purge and Trap performance with CTC PAL automation. *CDS Application Note #185*. Oxford, PA, USA: CDS Analytical LLC.

34 Bristow, R.L., Young, I.S., Pemberton, A. et al. (2019). An extensive review of the extraction techniques and detection methods for the taste and odour compound geosmin (trans-1,10-dimethyl-trans-9-decalol) in water. *TrAC Trends in Analytical Chemistry* 110: 233–248. https://doi.org/10.1016/J.TRAC.2018.10.032.

35 Marchal, R. and Waters, E.J. (2010). New directions in stabilization, clarification and fining of white wines. In: *Managing Wine Quality*, 188–225. Elsevier https://doi.org/10.1533/9781845699987.1.188.

36 Choi, J., S.-Y. Ahn, Y. Kim, I. Choi et al. (2016). Development of real time monitoring of off-flavour compounds using multi functionalized auto sampler with SPME-GC-MS/MS. *Poster at the 2016 Korea Society of Mass Spectrometry Conference, Water Analysis & Research Center, K-Water*, Dejoun, Korea.

37 Kong, Y. and Cao, Z. (2018). Determination of off-odor compounds in drinking water using an SPME device with gas chromatography and mass spectrometry. *Application Note 5994-0626EN*. Beijing, China: Agilent Technologies Inc.

38 Dettmer, K. and Engewald, W. (2003). Ambient air analysis of volatile organic compounds using adsorptive enrichment. *Chromatographia* 57: S339–S347. https://doi.org/10.1007/BF02492126.

39 Krebs, G., Schneider, E., and Schuhmann, A. (1991). Analytik flüchtiger aromatischer und halogenierter Kohlenwasserstoffe aus Bodenluft. *GIT Labor-Fachzeitschrift* 35: 19–22.

40 US EPA (2017). *Selected Analytical Methods for Environmental Remediation and Recovery (SAM 2017)*. Cincinnati, OH: U.S. Environmental Protection Agency (EPA), Office of Research and Development. https://cfpub.epa.gov/si/si_public_record_report.cfm?Lab=NHSRC&dirEntryId=339252.

41 US EPA (2014). *Method 8270E (SW-846): Semivolatile Organic Compounds by Gas Chromatography/ Mass Spectrometry (GC/MS)*. Washington, DC: U.S. Environmental Protection Agency (EPA).

42 WHO (2003). Polynuclear aromatic hydrocarbons in drinking-water – background document for development of WHO guidelines for drinking-water quality. In: *Guidelines for Drinking-Water Quality*, 2nd Ed. (eds. A. Boehncke and J. Kielhorn), 1–27. Geneva: World Health Organization.

43 Weis, L.M., Rummel, A.M., Masten, S.J. et al. (1998). Bay or baylike regions of polycyclic aromatic hydrocarbons were potent inhibitors of gap junctional intercellular communication. *Environmental Health Perspectives* 106 (1): 17–22. https://doi.org/10.1289/ehp.9810617.

44 Agency for Toxic Substances and Disease Registry (ATSDR) (2009). Case studies in environmental medicine – toxicity of polycyclic aromatic hydrocarbons (PAHs). *Course: WB 1519*. Atlanta, GA, USA: Agency for Toxic Substances and Disease Registry (ATSDR). https://www.atsdr.cdc.gov/csem/csem.asp?csem=13&po=10 (accessed 4 November 2020).

45 Lerda, D. (2010). Polycyclic aromatic hydrocarbons (PAHs) factsheet. *Joint Research Centre Technical Notes* 3: 1–25. https://doi.org/10.1016/S0167-7799(02)01943-1.

46 Coelho, E., Ferreira, C., and Almeida, C.M.M. (2008). Analysis of polynuclear aromatic hydrocarbons by SPME-GC-FID in environmental and tap waters. *Journal of the Brazilian Chemical Society* 19 (6): 1084–1097. https://doi.org/10.1590/S0103-50532008000600006.

47 Lim, G.S.Y., Preiswerk, T., Huebschmann, H.-J. et al. (2019). Robustness of SPME arrow immersion sampling: polycyclic aromatic hydrocarbons in drinking water. *Poster at the RAFA Conference 2019*, Prague, Czech Republic, CTC Analytics Asia Ltd. Pte.

48 The Commission of the European Communities (2006). COMMISSION REGULATION (EC) No 1881/2006 of 19 December 2006 Setting Maximum Levels for Certain Contaminants in Foodstuffs. *Official Journal of the European Union*, No. 2006R1881-EN-01.07.2009-003.001: 1-25.

49 Health Canada (2015). Benzo[a]pyrene in drinking water. Federal-Provincial-Territorial Committee on Drinking Water. https://www.canada.ca/content/dam/hc-sc/migration/hc-sc/ewh-semt/alt_formats/pdf/consult/_2015/bap/consult-eng.pdf

50 The Council of the European Union (1998). COUNCIL DIRECTIVE 98/83/EC of 3 November 1998 on the Quality of Water Intended for Human Consumption. *Official Journal of the European Union*, L 330/32-54.

51 China Official Method (2006). GB5749-2006 – Standards for Drinking Water Quality – 生活饮用水卫生标准生活饮用水标准检验方法, MOH, SAC, 2006-12-29.

52 Buchgraber, M., Ulberth, F., and Anklam, E. (2004). Interlaboratory evaluation of injection techniques for triglyceride analysis of cocoa butter by capillary gas chromatography. *Journal of Chromatography A* 1036 (2): 197–203. https://doi.org/10.1016/j.chroma.2004.03.011.

53 Christie, W.W. and Hutton, J. (2020). *High-Temperature Gas Chromatography of Triacylglycerols*. AOCS Lipid Library. https://lipidlibrary.aocs.org/lipid-analysis/selected-topics-in-the-analysis-of-lipids/high-temperature-gas-chromatography-of-triacylglycerols.

54 Restek Corporation (2020). Robust GC analysis of glycerides in edible oils. *Technical Resources*, Lit. Cat.# FFFA3175-UNV. Belefonte PA, USA: Restek Corporation.

55 Buchgraber, M., Androni, S., and Anklam, E. (2007). Determination of cocoa butter equivalents in milk chocolate by triacylglycerol profiling. *Journal of Agricultural and Food Chemistry* 55: 3284–3291. https://doi.org/10.1021/jf063350z.

56 Fontecha, J., Díaz, V., Fraga, M.J. et al. (1998). Triglyceride analysis by gas chromatography in assessment of authenticity of goat milk fat. *Journal of the American Oil Chemists Society* 75 (12): 1893–1896. https://doi.org/10.1007/s11746-998-0347-6.

57 Christie, W.W. (2011). *The Analysis of Fatty Acids, Preparation of Derivatives of Fatty Acids*. AOCS Lipid Library. http://lipidlibrary.aocs.org.

58 Kail, B.W., Link, D.D., and Morreale, B.D. (2012). Determination of free fatty acids and triglycerides by gas chromatography using selective esterification reactions. *Journal of Chromatographic Science* 50 (10): 934–939. https://doi.org/10.1093/chromsci/bms093.

59 AOAC International (2002). Fat (total, saturated and unsaturated) in foods. AOAC Official Method 996.01, 1996, Revised 2001.

60 AOAC International (2002). Fat (total, saturated, and unsaturated) in foods. AOAC Official Method 996.06, 1996, Revised 2001.

61 AOCS (2017). Saturated, *cis*-monounsaturated, and *cis*-polyunsaturated fatty acids in marine and other oils containing long chain polyunsaturated fatty acids (PUFAs) by capillary GLC. AOCS Official Method Ce 1i-07.

62 ISO 12966-2:2017 (2017). *Animal and vegetable fats and oils – gas chromatography of fatty acid methyl esters – Part 2: preparation of methyl esters of fatty acids ISO/EN 12966 method for animal and vegetable fats and oils*. International Standards Organization.

63 David, F., Sandra, P., and Vickers, A.K. (2005). Column selection for the analysis of fatty acid methyl esters. *Application Note 5989-3760EN*. Agilent Technologies Inc.

64 Tivanello, R., Capristo, M., Vicente, E. et al. (2020). Effects of deodorization temperature and time on the formation of 3-MCPD, 2-MCPD, and glycidyl esters and physicochemical changes of palm oil. *Journal of Food Science* https://doi.org/10.1111/1750-3841.15304.

65 Basle, Q. (2020). Glycidyl esters: pivotal role of analytics. *Food Analysis* 1: 6–10. www.food-analysis.org.

66 EFSA (2018). *Revised Safe Intake for 3-MCPD in Vegetable Oils and Food*. EFSA Media Relations Office.

67 Fett, G.A. (2012). *Ergänzende Hinweise zu den DGF-Einheitsmethoden C-VI 17 (10) Und C-VI 18 (10) zur Bestimmung der 3-MCPD- und Glycidyl-Ester*. Frankfurt: Deutsche Gesellschaft für Fettwissenschaft e.V. http://www.dgfett.de/methods/hinweise.pdf.

68 European Food Safety Authority (2013). Analysis of occurrence of 3-monochloropropane-1,2-diol (3-MCPD) in food in Europe in the years 2009–2011 and preliminary exposure assessment. *EFSA Journal* 11 (9): 3381. https://doi.org/10.2903/j.efsa.2013.3381.

69 Velisek, J., Davidek, J., Hajšlová, J. et al. (1978). Chlorohydrins in protein hydrolysates. *Zeitschrift für Lebensmittel-Untersuchung und -Forschung* 167: 241–244.

70 Lee, B.Q. and Khor, S.M. (2015). 3-Chloropropane-1,2-diol (3-MCPD) in soy sauce: a review on the formation, reduction, and detection of this potential carcinogen. *Comprehensive Reviews in Food Science and Food Safety* 14 (1): 48–66. https://doi.org/10.1111/1541-4337.12120.

71 Svejkovska, B., Novotny, O., Divinova, V. et al. (2004). Formation and decomposition of 3-chloropropane-1,2-diol esters in models simulating processed foods. *Czechoslovak Journal of Food Sciences* 22: 190–196.

72 The European Commission. 2018. COMMISSION REGULATION (EU) 2018/290 of 26 February 2018 amending Regulation (EC) No 1881/2006 as regards maximum levels of glycidyl fatty acid esters in vegetable oils and fats, infant formula, follow-on formula and foods for special medical purposes intended for infants and young children. *Official Journal of the European Union* L 55/27 (27.2.2018). https://doi.org/10.2903/j.efsa.2016.4426.

73 Zwagerman, R. and Overman, P. (2016). A novel method for the automatic sample preparation and analysis of 3-MCPD-, 2-MCPD-, and glycidylesters in edible oils and fats. *European Journal of Lipid Science and Technology* 118 (7): 997–1006. https://doi.org/10.1002/ejlt.201500358.

74 Zwagerman, R. and Overman, P. (2019). Optimized analysis of MCPD- and glycidyl esters in edible oils and fats using fast alkaline transesterification and 13C-correction for glycidol overestimation: validation including interlaboratory comparison. *European Journal of Lipid Science and Technology* 1800395, 1–12. https://doi.org/10.1002/ejlt.201800395.

75 Wagner, C., Neukom, H.-P., and Grob, K. (2001). Mineral paraffins in vegetable oils and refinery by-products for animal. *Mitteilungen aus Lebensmitteluntersuchung und Hygiene* 92: 499–514.

76 Biedermann, M., Fiselier, K., and Grob, K. (2009). Aromatic hydrocarbons of mineral oil origin in foods: method for determining the total concentration and first results. *Journal of Agricultural and Food Chemistry* 57 (19): 8711–8721. https://doi.org/10.1021/jf901375e.

77 Bundesinstitut für Risikobewertung (2009). Übergänge von Mineralöl aus Verpackungsmaterialien auf Lebensmittel. *Stellungnahme Nr. 008/2010 des BfR vom 09*. Berlin, Germany: German Federal Institute for Risk Assessment (BfR).

78 EFSA Panel on Contaminants in the Food Chain (CONTAM) (2012). Scientific opinion on mineral oil hydrocarbons in food. *EFSA Journal* 10 (6): 2704. https://doi.org/10.2903/j.efsa.2012.2704.

79 Arcella, D., Baert, K., and Binaglia, M. (2019). Rapid risk assessment on the possible risk for public health due to the contamination of infant formula and follow-on formula by mineral oil aromatic hydrocarbons (MOAH). *EFSA Supporting Publication*, EN-1741. European Food Safety Authority. https://doi.org/10.2903/sp.efsa.2019.EN-1741.

80 Bratinova, S. and Hoekstra, E. (eds.) (2019). Guidance on sampling, analysis and data reporting for the monitoring of mineral oil hydrocarbons in food and food contact materials. In the Frame of Commission Recommendation (EU) 2017/84. European Commission, Joint Research Center (JRC). https://doi.org/10.2760/208879.

81 Nestola, M. and Schmidt, T.C. (2017). Determination of mineral oil aromatic hydrocarbons in edible oils and fats by online liquid chromatography–gas chromatography–flame ionization detection – evaluation of automated removal strategies for biogenic olefins. *Journal of Chromatography A* 1505: 69–76. https://doi.org/10.1016/j.chroma.2017.05.035.

82 Fiselier, K., Fiorini, D., and Grob, K. (2009). Activated aluminum oxide selectively retaining long chain n-alkanes. Part I. Description of the retention properties. *Analytica Chimica Acta* 634 (1): 96–101. https://doi.org/10.1016/j.aca.2008.12.007.

83 Fiselier, K., Fiorini, D., and Grob, K. (2009). Activated aluminum oxide selectively retaining long chain n-alkanes: Part II. Integration into an

on-line high performance liquid chromatography-liquid chromatography-gas chromatography-flame ionization detection method to remove plant paraffins for the determination of mineral paraffins in foods and environmental samples. *Analytica Chimica Acta* 634 (1): 102–109. https://doi.org/10.1016/j.aca.2008.12.011.

84 Weber, S., Schrag, K., Mildau, G. et al. (2018). Analytical methods for the determination of mineral oil saturated hydrocarbons (MOSH) and mineral oil aromatic hydrocarbons (MOAH) – a short review. *Analytical Chemistry Insights* 13: 1–16. https://doi.org/10.1177/1177390118777757.

85 Restek (2020). MOSH/MOAH Standard (9 components). https://www.restek.com/catalog/view/39665 (as of 13 April 2020).

86 DIN EN 16995 (2017). *Vegetable oils and foodstuff on basis of vegetable oils – determination of mineral oil saturated hydrocarbons (MOSH) and mineral oil aromatic hydrocarbons (MOAH) with on-line LC-GC-FID analysis*. DIN EN 16995:2017-08. German Institute for Standardisation (Deutsches Institut für Normung).

87 Niederer, M. (2016). *Lippenpflegeprodukte/Mineralparaffine (MOSH/MOAH) und Allergene Duftstoffe*. Switzerland: Cantonal Laboratory of Basel (Kantonales Laboratorium).

88 Anastassiades, M., Lehotay, S.J., Stajnbaher, D. et al. (2003). Fast and easy multiresidue method employing acetonitrile extraction/partitioning and "dispersive solid-phase extraction" for the determination of pesticide residues in produce. *Journal of AOAC International* 86: 412–431.

89 EN 15662 (2018). *Foods of plant origin – multimethod for the determination of pesticide residues using GC- and LC-based analysis following acetonitrile extraction/partitioning and clean-up by dispersive SPE – modular QuEChERS-method*. DIN EN 15662:2018. German Institute for Standardisation (Deutsches Institut für Normung).

90 AOAC (2007). Pesticide residues in foods by acetonitrile extraction and partitioning with magnesium sulfate. Method Number 2007.01. AOAC. http://www.eoma.aoac.org/methods/info.asp?ID=48938 (accessed 4 November 2020).

91 Hübschmann, H.-J. and Chong, C.M. (2018). Automated clean-up of QuEChERS extracts for GC-MS and LC-MS. *Poster at the Food Safety Analysis Conference*, Singapore (27–28 November 2018).

92 Morris, B.D. and Schriner, R.B. (2015). Development of an automated column solid-phase extraction cleanup of QuEChERS extracts, using a zirconia-based sorbent, for pesticide residue analyses by LC-MS/MS. *Journal of Agricultural and Food Chemistry* 63: 5107–5119. https://doi.org/10.1021/jf505539e.

93 Han, L., Sapozhnikova, Y., and Nuñez, A. (2019). Analysis and occurrence of organophosphate esters in meats and fish consumed in the United States. *Journal of Agricultural and Food Chemistry* 67 (46): 12652–12662. https://doi.org/10.1021/acs.jafc.9b01548.

94 Han, L. and Sapozhnikova, Y. (2020). Semi-automated high-throughput method for residual analysis of 302 pesticides and environmental contaminants in

catfish by fast low-pressure GC-MS/MS and UHPLC-MS/MS. *Food Chemistry* 319: 126592. https://doi.org/10.1016/j.foodchem.2020.126592.

95 Lehotay, S.J., Han, L., and Sapozhnikova, Y. (2016). Automated mini-column solid-phase extraction cleanup for high- throughput analysis of chemical contaminants in foods by low-pressure gas chromatography – tandem mass spectrometry. *Chromatographia* 79: 1113–1130. https://doi.org/10.1007/s10337-016-3116-y.

96 Maštovská, K., Lehotay, S.J., and Anastassiades, M. (2005). Combination of analyte protectants to overcome matrix effects in routine GC analysis of pesticide residues in food matrixes. *Analytical Chemistry* 77 (24): 8129–8137.

97 Sánchez-Brunete, C., Albero, B., Martin, G., and Tadeo, J.L. (2005). Determination of pesticide residues by GC-MS using analyte protectants to counteract the matrix effect. *Analytical Science* 21 (11): 1291–1296. https://doi.org/10.2116/analsci.21.1291.

98 Giddings, J.C. (1962). Theory of minimum time operation in gas chromatography. *Analytical Chemistry* 34 (3): 314–319. https://doi.org/10.1021/ac60183a005.

99 de Zeeuw, J., Peene, J., De Jong, R. et al. (2000). A simple way to speed up separations by GCMS using short 0.53 mm columns and vacuum outlet conditions. *Journal of High Resolution Chromatography* 23 (12): 677–680.

100 Peene, J., De Zeeuw, J., and De Jong, R. (2000). Low-pressure gas chromatography: fast analysis with high sensitivity. *International Laboratory*: 41–44.

101 Maštovská, K. and Lehotay, S.J. (2003). Practical approaches to fast gas chromatography–mass spectrometry. *Journal of Chromatography A* 1000 (1–2): 153–180. https://doi.org/10.1016/S0021-9673(03)00448-5.

102 Maštovská, K., Lehotay, S.J., and Hajšlová, J. (2001). Optimization and evaluation of low-pressure gas chromatography-mass spectrometry for the fast analysis of multiple pesticide residues in a food commodity. *Journal of Chromatography A* 926 (2): 291–308. https://doi.org/10.1016/S0021-9673(01)01054-8.

103 Sapozhnikova, Y. and Lehotay, S.J. (2015). Review of recent developments and applications in low-pressure (vacuum outlet) gas chromatography. *Analytica Chimica Acta* 899: 13–22. https://doi.org/10.1016/j.aca.2015.10.003.

104 Lehotay, S.J., De Zeeuw, J., Sapozhnikova, Y. et al. (2020). There is no time to waste: low-pressure gas chromatographymass spectrometry (LPGC-MS) is a proven solution for fast, sensitive, and robust GC-MS analysis. *LCGC North America* 38 (8): 457–466.

105 Monteiro, S.H., Lehotay, S.J., Sapozhnikova, Y. et al. (2020). High-throughput mega-method for the analysis of pesticides, veterinary drugs, and environmental contaminants by UHPLC- MS/MS and robotic mini-SPE cleanup + LPGC-MS/MS, Part 1: beef. *Journal of Agricultural and Food Chemistry* https://doi.org/10.1021/acs.jafc.0c00710.

106 Goon, A., Shinde, R., Ghosh, B. et al. (2019). Application of automated mini–solid-phase extraction cleanup for the analysis of pesticides in complex spice matrixes by GC-MS/MS. *Journal of AOAC International* 10: 1–6. https://doi.org/10.5740/jaoacint.19-0202.

107 Anastassiades, M. (2013). Variation of QuEChERS-method for avocado. Community Reference Laboratory for Single Residue Methods. Germany: CVUA Stuttgart.

108 Sun, L., Q. Guo, C.C. Jacob, C.P.B. Martins et al. (2020). Multi-pesticide residues analyses of QuEChERS extracts using an automated online µSPE clean-up coupled to LC-MS/MS. *Application Note 65684*. San Jose, CA, USA: Thermo Fisher Scientific.

109 European Food Safety Authority (2019). Review of the existing maximum residue levels for glyphosate according to article 12 of regulation (EC) No 396/2005 – revised version to take into account omitted data. *EFSA Journal* 17 (10) https://doi.org/10.2903/j.efsa.2019.5862.

110 Alferness, P.L. and Iwata, Y. (1994). Determination of glyphosate and (aminomehtyl) phosphonic acid in soil, plant and animal matrices, and water by capillary gas chromatography with mass selective detection. *Journal of Agricultural and Food Chemistry* 42: 2751–2759.

111 Anastassiades, M., D. I. Kolberg, E. Eichhorn, A.-K. Wachtler et al. (2020). Quick method for the analysis of numerous highly polar pesticides in foods of plant origin via LC-MS/MS involving simultaneous extraction with methanol (QuPPe-Method). *EU Reference Laboratory for Pesticides Requiring Single Residue Methods (EURL-SRM)*, Fellbach, Germany. https://www.eurl-pesticides.eu/docs/public/tmplt_article.asp?CntID=887&LabID=200&Lang=EN (accessed 4 November 2020).

112 Antec Scientific (2020). Glyphosate and AMPA. *ALEXYS Application Note #216_007_03*. Zoeterwoude, The Netherlands.

113 Helle, N. and Chmelka, F. (2012). Glyphosate/AMPA a global presence, *GERSTEL Newsletter*. Mühlheim: GERSTEL GmbH & Co KG.

114 Institut Kirchhoff Berlin (2016). Bestimmung von Glyphosat, AMPA und Glufosinat in Lebensmitteln mittels Online SPE-LC-MS/MS nach FMOC-Derivatisierung. Web presentation. https://www.institut-kirchhoff.de/dienstleistungen/analytik/pflanzenschutzmittel/glyphosat-ampa-und-glufosinat/ (as of 20 March 2020).

115 Helle, N. and Chmelka, F. (2013). A global presence: speaking of glyphosate. *GERSTEL Solutions Worldwide* 13: 6–9.

116 GERSTEL (2020). *MAESTRO PrepBuilder*, Product Information. GERSTEL GmbH & Co KG. http://www.gerstel.com/en/maestro-prepbuilder.htm.

117 US EPA (2020). National Primary Drinking Water Regulations. https://www.epa.gov/ground-water-and-drinking-water/national-primary-drinking-water-regulations#Organic (as of 22 March 2020).

118 Cabrices, O.G. and Schreiber, A. (2013). *Automated Derivatization, SPE Cleanup and LC/MS/MS Determination of Glyphosate and Other Polar Pesticides*. Framingham, MA, USA: Global Analytical Solution. https://sciex.com/Documents/brochures/GlyphosateQTRAP4500_GERSTEL_8013813.pdf.

119 Häkkinen, V.M.A. (1991). Analysis of chemical warfare agents in water by solid phase extraction and two-channel capillary gas chromatography. *Journal of Separation Science* 14: 811–815.

120 Meyer, M.T., K.A. Loftin, E.A. Lee, G.H. Hinshaw et al. (2009). Determination of glyphosate, its degradation product aminomethylphosphonic acid, and glufosinate in water by isotope dilution and online solid-phase extraction and liquid chromatography/tandem mass spectrometry. U.S. Geological Survey Techniques and Methods, Book 5, 32. Reston, Virginia, USA: U.S. Geological Survey.

121 U.S. EPA (2015). *Method 543. Determination of Selected Organic Chemicals in Drinking Water by On-Line Solid Phase Extraction Liquid Chromatography/Tandem Mass Spectrometry (On-Line SPE-LC/MS/MS)*. Cincinatti, OH: U.S. EPA, Office of Research and Development, National Exposure Research Laboratory.

122 U.S. EPA (2018). *Method Development for Unregulated Contaminants in Drinking Water: Public Meeting and Webinar*. Washington, DC: U.S. EPA, Office of Ground Water and Drinking Water. https://www.epa.gov/sites/production/files/2018-07/documents/method-development-unregulated-contaminants-drinking-water-meeting-materials-june2018.pdf.

123 Gusmaroli, L., Insa, S., and Petrovic, M. (2018). Development of an online SPE-UHPLC-MS/MS method for the multiresidue analysis of the 17 compounds from the EU 'Watch List'. *Analytical and Bioanalytical Chemistry* 410: 4165–4176. https://doi.org/10.1007/s00216-018-1069-8.

124 U.S. EPA (2007). *Method 1694: Pharmaceuticals and Personal Care Products in Water, Soil, Sediment, and Biosolids by HPLC/MS/MS*. Washington, DC: U.S. EPA, Office of Water Office of Science and Technology Engineering and Analysis Division.

125 Ramirez, C.E., Wang, C., and Gardinali, P.R. (2014). Fully automated trace level determination of parent and alkylated PAHs in environmental waters by online SPE-LC-APPI-MS/MS. *Analytical and Bioanalytical Chemistry* 406: 329–344. https://doi.org/10.1007/s00216-013-7436-6.

126 Jang, J., Kim, Y., and Choi, J. (2012). Fast and accurate determination of algal toxins in water using online preconcentration and UPLC-orbitrap mass spectometry. *Journal of Korean Society on Water Environment* 28 (6): 843–850.

127 McHale, K. and Sanders, M. (2010). Quantification of EPA 1694 pharmaceuticals and personal care products in water at the ng/L level utilizing online sample preparation with LC-MS/MS. *Application Note 508*. Somerset, NJ, USA: Thermo Fisher Scientific.

128 Beck, J.R. and Yang, C.T. (2014). EPA draft method 543 quantitation of organic pesticides in drinking water using online pre-concentration/solid phase extraction and tandem mass spectrometry. White Paper PN-64151-EN-0614S. San Jose CA, USA: Thermo Fisher Scientific.

129 Beck, J., Yamaguchi, M., and Saito, K. (2006). Quantitation enhanced data-dependent (QED) scanning of drinking water samples using equan for pesticide analysis on a triple stage quadrupole. *Application Note #378*. San Jose, CA USA: Thermo Scientific.

130 The United States Pharmacopeial Convention Inc (2019). <467> Residual Solvents. Interim Revision Announcement, C185876-M99226-GCCA2015, rev. 00 20190927. https://doi.org/C185876-M99226-GCCA2015,rev.00 20190927.

131 Belsky, J.L., Ashley, A.J., Bhatt, P.A. et al. (2010). Optimization of the water-insoluble procedures for USP general chapter residual solvents <467>. *AAPS PharmSciTech* 11 (2): 994–1004. https://doi.org/10.1208/s12249-010-9460-6.

132 Organisation for the Prohibition of Chemical Weapons (OPCW) (2020). Convention on the prohibition of the development, production, stockpiling and use of chemical weapons and on their destruction (the Chemical Weapons Convention or CWC). https://www.opcw.org/chemical-weapons-convention (as of 03 April 2020).

133 Vanninen, P. (ed.) (2017). *Recommended Operating Procedures for Analysis in the Verification of Chemical Disarmament*, 2017e. Finland: University of Helsinki. http://www.helsinki.fi/verifin/bluebook.

134 Finnish Institute for Verification of the Chemical Weapons Convention (VERFIN) (2020). http://www.helsinki.fi/verifin/VERIFIN/english/verification/designated.htm (as of 11 March 2020).

135 Althoff, M.A., Bertsch, A., and Metzulat, M. (2019). Automation of μ-SPE (smart-SPE) and liquid–liquid extraction applied for the analysis of chemical warfare agents. *Separations* 6 (4) https://doi.org/10.3390/separations6040049.

136 LABC-Labortechnik (2017). *Dispersive Liquid–Liquid-Microextraction (DLLME) mit Bilimex*, Product Information. Hennef: LABC-Labortechnik Zilliger KG. www.LABC.de.

137 Althoff, M.A., Bertsch, A., Metzulat, M. et al. (2017). Application of headspace and direct immersion solid-phase microextraction in the analysis of organothiophosphates related to the chemical weapons convention from water and complex matrices. *Talanta* 174: 295–300. https://doi.org/10.1016/j.talanta.2017.05.024.

138 Vanderhaegen, B., Neven, H., Verachtert, H. et al. (2006). The chemistry of beer aging – a critical review. *Food Chemistry* 95 (3): 357–381. https://doi.org/10.1016/j.foodchem.2005.01.006.

139 Kawamura, K. (2020). Qualitative analysis of aroma components in beer using peak deconvolution. *Application Note M290*. Kyoto, Japan: Shimadzu Corp.

140 Vanderhaegen, B., Delvaux, F. et al. (2007). Aging characteristics of different beer types. *Food Chemistry* 103 (2): 404–412. https://doi.org/10.1016/j.foodchem.2006.07.062.

141 Baert, J.J., De Clippeleer, J., Hughes, P.S. et al. (2012). On the origin of free and bound staling aldehydes in beer. *Journal of Agricultural and Food Chemistry* 60 (46): 11449–11472. https://doi.org/10.1021/jf303670z.

142 Tsai, S.W. and Chang, C.M. (2003). Analysis of aldehydes in water by solid-phase microextraction with on-fiber derivatization. *Journal of Chromatography A* 1015: 143.

143 Wu, Q., Chen, H., and Yang, Z. (2016). GC-MS determination of stale aldehydes in beer by SPME on-fibre derivatization. *Ingenious News* 06, GC/MS Application Note. Zwingen, Switzerland: CTC Analytics AG.

144 Center of Disease Control and Prevention (2020). Phthalates Factsheet. CDC website https://www.cdc.gov/biomonitoring/Phthalates_FactSheet.html (as of 11 March 2020).

145 The Commission of the European Communities (1999). COMMISSION DECISION of 7 December 1999 adopting measures prohibiting the placing on the market of toys and childcare articles intended to be placed in the mouth by children under three years of age made of soft PVC containing one or more of the substances. *Official Journal of the European Union*, No. 9. 12. 1999, L 315/46-49.

146 COWI, IOM Consulting and AMEC (2012). Evaluation of New Scientific Evidence Concerning the Restrictions On DINP and DIDP Contained in Entry 52 of Annex XVII to Regulation (EC) No 1907/2006 (REACH). https://echa.europa.eu/documents/10162/a35fa99b-ed8f-4451-a4d5-f012e9ba69c7 (as of 22 March 2020).

147 US Consumer Product Safety Commission (2010). *Test Method: CPSC-CH-C1001-09.3 Standard Operating Procedure for Determination of Phthalates*. Gaithersburg, MD: US Consumer Products Safety Commission, Directorate for Laboratory Services.

148 Zou, Y. and Cai, M. (2013). Determination of phthalate concentration in toys and children's products. *Application Note 5990-4863EN*. Shanghai, China: Agilent Technologies Inc.

149 GERSTEL (2014). Phthalates – easy extraction of plasticizers from toys. *GERSTEL Solutions Worldwide* 14: 12–14.

Appendix

A.1 Robotic System Control

The programming, load, and execution of x,y,z-robot workflows and system control can be achieved in several ways using a variety of software solutions. This can include the individual programming of workflows using the proprietary programming language tools of the manufacturer, as well as graphical object supported interfaces with icons for the available workflow steps (graphical user interface (GUI)). The execution of workflows is often integrated into the chromatography data systems (CDSs) of the leading gas chromatography-mass spectrometry (GC-MS) and liquid chromatography-mass spectrometry (LC-MS) manufacturer by providing dedicated system drivers. Integrated solutions offer the benefit of working from one sample sequence table for sample preparation and analysis with a seamless tracking of all operations. Also, independent programming and workflow execution software solutions are in use offering the extended integration of additionally needed sample preparation functions like weighing, powder dosing, sonication, freezing, and other controlled external devices. This allows highly customized robotic workflows. Furthermore, the local execution of workflows mainly used for stand-alone installations can often be accomplished by operation of a loaded workflow script right away via the system terminals. This independent solution working autonomously from chromatography data systems offers a suitable choice for routine sample preparation tasks aside of online integrated analyses.

A summary of the most used x,y,z-robotic system control and workflow software solutions is provided in the following sections A1.1. to A1.6 without the pretension of a comprehensive introduction.

A.1.1 Maestro Software

The Maestro® named software package developed by the GERSTEL GmbH & Co KG, Mülheim an der Ruhr, Germany, offers seamless integration with the CDSs of Agilent Technologies, Inc. [1]. One sequence table runs the complete system integrated with sample preparation and online analysis via GC-MS or LC-MS. The sample preparation can be accomplished in stand-alone mode as well. Several proprietary sample preparation modules are integrated for extended workflow functionality.

Automated Sample Preparation: Methods for GC-MS and LC-MS,
First Edition. Hans-Joachim Hübschmann.
© 2022 WILEY-VCH GmbH. Published 2022 by WILEY-VCH GmbH.

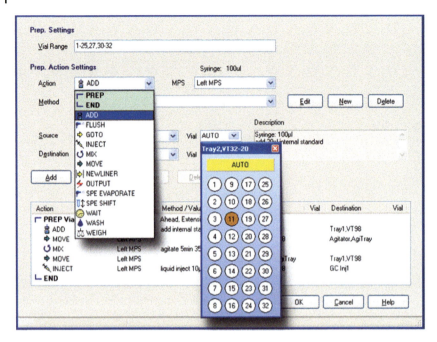

Figure A.1 Maestro PrepBuilder user interface. Source: Courtesy GERSTEL GmbH & Co KG.

Additional features allow the introduction of priority samples at any point in the analysis sequence, logging of the system operation providing traceability, sending a user email notification in case of system events, as well as a built-in maintenance planning function.

Workflows are created using the graphical Maestro PrepBuilder for the automated sample preparation and sample introduction to GC, GC-MS and LC, LC-MS.

The workflow steps are prepared for easy access as icons on the screen for standard addition, derivatization, sample transfer, incubation, agitation, extraction, mixing, conditioning, dilution, liner exchange, weighing, solid-phase extraction (SPE), washing, evaporation, and sample introduction (Figure A.1). The prep-ahead functionality is provided by a built-in scheduler for all sample preparation steps.

A.1.2 Chronos Software

The Chronos® named software package developed by the Axel Semrau GmbH & Co KG, Sprockhövel, Germany, organizes sample preparation processes for time-efficient and parallel sample handling also integrating external devices into workflows [2]. The sample preparation workflows and analyses are controlled from one sequence table for the integrated analytical systems of the major instrument manufacturer. Stand-alone as well as online analysis installations are both addressed. A large number of third-party instruments are integrated via dedicated interface functions with, for instance, the weighing of vials with balances providing the standard serial interface handshake. Weighing results are registered and used for further processing in the sample preparation process allowing conditional decisions

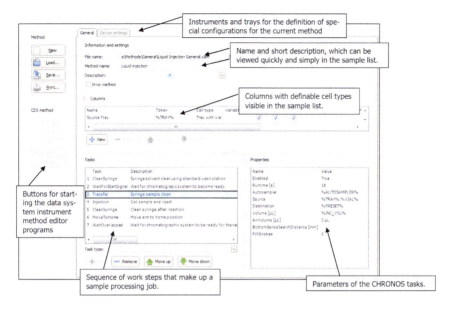

Figure A.2 Chronos method editor. Source: Courtesy Axel Semrau GmbH & Co KG.

within workflows. External functions also include online SPE, GC liner exchange, sonication, refractometer, dilutors, stirrer plates, LIMS connectivity, and more.

The workflow programming uses a comprehensive method editor as shown in Figure A.2. A list of tasks is provided, which are selected to generate the stepwise workflow. Each task offers the related parameter set with default values for customizing. Multistage sample preparation procedures are executed in parallel through an integrated algorithm for time efficiency. A graphical representation of the programmed sample processing monitors the workflow progress in real time and informs about the overall sequence duration. Additional functionality comprises the management of multiple sample lists, priority samples, xlsx/CSV export and import, LIMS integration, SMS messaging, or the log file generation during operation. Communication with CDSs includes those from leading instrument manufacturers like Agilent, Bruker, DataApex, GL Sciences, Sciex, Shimadzu, Thermo Fisher Scientific, or Waters.

A.1.3 Graphical Workflow Programming

The most popular method developing tools in a lab are those offering a GUI for ease of use. Programs are available with the "Sampling Workflow Editor Software" by Thermo Fisher Scientific [3], or the "PAL Method Composer" by CTC Analytics AG, Zwingen, Switzerland [4], or similar software packages.

Methods are developed and tested while being directly connected with the x,y,z-robotic system. This ensures that only activities for tools and modules are shown on the screen which are needed and compatible with the current system configuration in use. This kind of built-in user guidance makes workflow programming

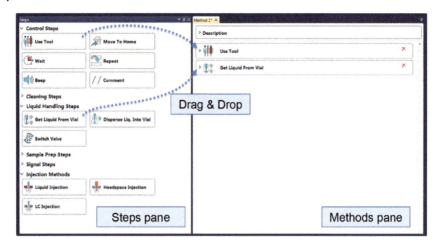

Figure A.3 Graphical user interface for methods creation by Drag & Drop. Source: Courtesy CTC Analytics AG.

straightforward and avoids potential mistakes. Each activity icon carries a set of operating parameters with suggested default, and minimum/maximum parameters for customization. All operations and movements can be checked and adjusted stepwise on the connected instrument during building the workflow. Completed methods can be imported from the CDS in use or directly loaded in the stand-alone mode.

The GUI allows the creation methods by Drag & Drop (Figure A.3). Workflows are built with several steps pulled into the desired sequence on a method pane. Comprehensive tasks like solid-phase micro-extraction (SPME), µSPE clean-up, or injections to GC and LC are provided comprising already the necessary operation steps. Parameters for each workflow step are taken from the default setting for customization. Tooltips can be used and edited by the operator in local languages. The GUI guides access to the workflow activities grouped into the basic functionalities:

- Control steps
- Cleaning steps
- Liquid handling steps
- Sample prep steps
- Signal steps
- Injection methods

A.1.4 Sample Control Software

The "PAL Sample Control" software is a programming tool provided by CTC Analytics AG, Zwingen, Switzerland, for the daily routine jobs [5]. Workflows are created using a user interface for setting up the task list from the available system activities with appropriate analytical parameters, also useful for complex

Figure A.4 PAL Sample Control window with task and activity list. Source: Courtesy CTC Analytics AG.

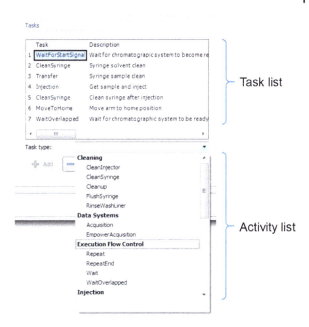

workflows (Figure A.4). It optimizes automatically the timing of various steps in a sample preparation process and generates a schedule that minimizes the runtimes of sequences. The access to the trayholder and rack index positions allows the set up of sample list which can be exported to a CDS in use, so that the system operation is executed from one combined sequence table. The CDSs of the major chromatographic instrument providers are supported.

A.1.5 Local System Control

Workflow programs can be executed on many robotic samplers also locally without being connected or integrated to a CDS. Stand-alone installations in routine environments are typically operated via a local terminal control. Also, in case of the operational hyphenation with a nonintegrated analysis system, a synchronized workflow can be achieved by simple *Start/Ready* interfacing. In this installation, the robot checks a *Ready* signal from a served sampling or analysis system and provides the *Start* upon completing the programmed workflow or sample transfer. For programming local workflows, a method composer/editor (see Section A.1.3) or a script control language programming is required (see Section A.1.6).

Using local scripts, it is possible to create, edit, and run sample lists (job queues) using different methods that are based on templates or imported custom scripts. Often standard workflows are already preinstalled for liquid injection to GC or LC, static and dynamic headspace, SPME extractions, wash or mixing cycles. Additional workflows can usually be imported from a USB memory stick. The local system control provides an easy-to-use system operation for customized standard workflows in a routine environment.

A.1.6 Script Control Language

The detailed programming of complex workflows requires access to a scripting language as specified by the robot manufacturer. Such "scripts" called workflows are executable programs usually resembling a BASIC-like annotation with access to all robotic system operation parameters. A particular script editor, executor, and system language documentation are required from the robotic system manufacturer for advanced user programming tasks.

A.2 System Maintenance

A.2.1 Syringes

A.2.1.1 Manual Syringe Handling

Avoid moving a plunger in a dry syringe barrel. This will result in unnecessary wear of the plunger. When using a syringe manually for replacement, injection, or dispensing, grasp only the syringe flange and plunger button. Variations in liquid measurement due to heating the syringe barrel from body temperature should be avoided [6].

Note: Steel plungers cannot be replaced or swapped between syringes due to the individual manufacturing process. Polymer tip plungers, e.g. fitted with a polytetrafluoroethylene (PTFE) tip, are available for exchange from several syringe manufacturers.

A.2.1.2 Syringe Cleaning

Cleaning the syringes in use on a regular preventive maintenance schedule is a must, assures the reliable performance and extends the syringe lifetime.

Syringes are best cleaned with solvents of different polarity during the workflow and for manual maintenance. Solvents must be suitable for the processed samples and analytes. Alcohols, acetone, water, or ethyl acetate, hexane in high quality are most used as cleaning solvents. Alkaline and detergent-based agents are not recommended for cleaning procedures within workflows. Potential carryover between samples can be eliminated by flushing the syringe with solvents repeatedly during the sample preparation process. Depending on the analyte, matrix and application of 3 and more cleaning cycles (plunger strokes) are observed. Dedicated cleaning cycles have to be introduced in workflows before the use of a syringe or syringe tool, after each sample, standard or reagent dispensing, and after and before starting the next work cycle.

Some syringe manufacturers offer cleaning agents. The cleaning solution usually is a neutral Extran™, a special detergent preparation for cleaning of alkali-sensitive metals, glass or quartz equipment comprising of anionic, and nonionic surfactants, and low concentrations of excipients [7]. Avoid soaking the syringe barrel into solvents. Some solvents will gradually dissolve the glue used to hold the back flange and the front flange (fixed needle or removable needle thread). Do not immerse the front flange below the level of the solvent. After extensive cleaning, rinse the syringe with solvent, finally acetone for drying. Then, take the plunger off and allow the syringe to dry at room temperature. Removable and fixed needle all-metal plunger microliter syringes can be heated in an oven of up to 120 °C. Plungers should be removed before heating [8].

A.2.1.3 Plunger Cleaning

Cleaning the plunger regularly is essential in a preventive maintenance schedule. The wetted plunger collects dust during continuous automated operation when pulled up outside of the syringe barrel during sample aspiration, and in particular, during full volume cleaning steps. Dust and potentially also sample residue deposits on the upper part of the plunger can build up resistance for a free movement, even blockage with freezing of the plunger movement can occur. Risk of fatal plunger zigzag is caused by missing appropriate syringe maintenance. Remove the plunger from the syringe barrel and wipe it several times with a solvent-wetted lint-free tissue, see Figures A.5 and A.6. Re-insert the plunger into the barrel and pump a solvent e.g. acetone through the needle and syringe. When reinserting a PTFE-tipped plunger into a syringe barrel, lubricate the tip by wetting it with a sample compatible solvent [6, 8].

A.2.1.4 Needle Cleaning

In case of a blocked needle, e.g. from septum piercing, and solvent wash does not eliminate the blocking, mechanical cleaning can be the only remedy. Forcing liquid through the syringe can cause the glass barrel to split with excessive pressure.

The needle cleaning kits offered commercially usually contain a cleaning solution concentrate (mostly Extran™) and a set of tungsten cleaning wires for different needle gauges as shown in Table A.1. The needle cleaning wires are extremely thin in diameter and require a pair of duckbill or flat nose pliers to feed the wire into the needle. Feed the wire slowly into the clogged needle. If the wire stops at the obstruction, do not force the wire, since the small wires bend easily. Obtain a smaller cleaning wire and try to work your way around or through the obstruction. Avoid bending the wire [9].

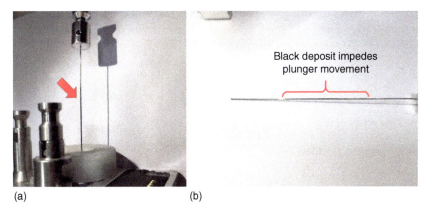

Figure A.5 Black-contaminated syringe plunger impeding even blocking the movement. (a) Plunger up from syringe barrel and (b) plunger removed.

Figure A.6 Wiping the syringe plunger of Figure A.5 with a methanol-soaked tissue.

Table A.1 Dimension of cleaning wires (after [9]).

Needle gauge	Cleaning wire OD	
22, 23 (and larger)	0.01207"	0.307 mm
24–26	0.00815"	0.207 mm
27	0.00659"	0.167 mm
25s, 22s, 28–30	0.00497"	0.126 mm
26s, 31–33	0.00350"	0.089 mm

Also, heated syringe needle cleaners are available, but applicable only for removal of persistent organic residues from needles, especially for trace analysis of high-boiling compounds, to prevent a potential carryover.

A.2.1.5 Confirming the Dispensed Volume of a Syringe

The precision of a syringe can be determined by accurately weighing several vials and dispensing a defined volume of water. Typical specifications of an automated liquid handling system for microliter syringes of 10 and 100 µL volume with 8 and 100 µL of water dispensed are given in Table A.2.

A.2.1.6 Sterilization

For sterilization, syringes can be autoclaved or gassed using ethylene oxide. Plungers need to be removed before. Set the autoclave to a maximum temperature of 100 °C

Table A.2 Precision of automated syringe dispensing, gravimetric measurements, n = 7 [10].

Syringe and needle type	Precision specification
10 µL, Gauge 22s	RSD ≤ 1%
10 µL, Gauge 26s	RSD ≤ 1%
100 µL, Gauge 22	RSD ≤ 0.1%
100 µL, Gauge 26s	RSD ≤ 0.1%

for removable needle syringes and 70 °C for fixed needle and Luer tip syringes. Autoclaving by heating a syringe too high and cooling down quickly by purging the system results in breaking the barrel. Allow for a slow cooldown. Inserting the plunger back into the barrel only at room temperature [8, 9].

A.2.2 Pipettes

A.2.2.1 Calibration

Pipette volume calibrations are performed according to ISO 8655 [11] by weighing defined volumes of water into pre-weight vials using the known general formula of Eq. (A.1): volume calculation from dispensed solvent weight

$$V = m \cdot \rho^{-1} \cdot 10^{-6} \, [\text{mL}] \tag{A.1}$$

with V = volume dispensed
m = weight of the pipetted volume [g]
ρ = specific mass [g/L]

Procedure for the calibration of a pipetting tool with the liquid class "Water" on x,y,z-robotic systems [12]:

1. Make a copy of the liquid class "Water."
2. In the copied liquid class, set "Correction factor" to "1" and "Volume offset" to "0."
3. Execute a single dispense script/method using different volumes of water into pre-weighed target vials.
4. Weigh back the target vials to retrieve the weight of the pipetted water.
5. Convert the weights to the corresponding volumes with respect to the temperature.
6. Prepare a diagram plot of the measured values against the set values and calculate slope and offset for a linear regression curve.
7. Set the new "Correction factor" of the liquid class "Water" to the determined slope value and the "Volume offset" to the offset on the y-axis of the linear regression curve.

A.2.2.2 Pipette Parts Maintenance

The sealing parts of a pipette are consumables. The typical lifetime is >100 000 strokes. Refer to the manufacturer guidelines with the use of a spare parts kit typically comprising plunger grease, brush and seals. Remove any grease from the plunger and the tip adapter, in particular on the face side with the seal, using a lint-free cloth (Figure A.7). Make sure to lubricate both parts with some grease after cleaning or autoclaving.

After cleaning or autoclaving the pipette tool, a tightness check is recommended. The easiest way to check the reassembled tool for tightness is done by executing a routine test method. Use a workflow activity to fill the tip with water. If no water

Figure A.7 Autoclavable parts of a pipetting tool for automated operation. Source: Courtesy CTC Analytics AG.

Tip ejector   Tip adapter   Plunger with stop nut

has leaked after 30-second wait time, all critical parts of the tool can be considered as tight.

The pipette tool precision performance can be confirmed using the liquid class "Water" as well. Set Correction factor = 1, and Volume Offset = 0 in single dispense mode. Transfer at least seven times 50% of the pipette tip volume of water. Weigh the target vials before and after dispensing. Determine the weights of the transferred volumes and calculate the pipetting volumes as of Eq. (A.1). The pipetting precision from the measurement series is expressed as the relative standard deviation (%RSD). In case of poor precision results, check the tool for tightness with the correct mount and quality of the plunger seal, and the fitting of the pipette tips. If another sample than water has to be tested or optimized, adjust the aspirate flow rate and dispense flow rate accordingly. An airgap on the rear is always required to blow the sample out of the tip if single dispense is used.

The accuracy of a pipette tool is verified by transfer of 20, 40, 60, 80, and 100% of pipette tip volume of water and weighing the target vials before and after dispensing water. The effectively transferred volume is calculated by Eq. (A.1).

The plot of a graph with x = used volume setting, y = calculated/transferred volume V provides both correction factor and offset value, as outlined in Section A.2.2.1. From the linear regression, the slope is determined and used as the correction factor. The y-offset is entered as offset value in the liquid class.

A.2.3 System Hardware Maintenance Schedule

Regular maintenance procedures help to ensure the accuracy and precision of the robotic sample preparation system. Recommended schedules are

Daily:
- Clean wash and waste solvent reservoirs.
- Check for bubbles in solvent lines.
- Check for potential syringe needle bending.
- Empty the waste solvent collection reservoir.

Weekly:
- Clean syringes and syringe plungers, depending on the application daily or after each long sample series.
- Check the integrity of the dilutor syringe.
- Clean wash and waste solvent reservoirs.
- Check solvent frits for contamination.

Monthly:
- Replace metal plunger syringes, depending on sample throughput.
- Replace polymer tip plungers, depending on sample throughput.
- Replace septa for wash and waste vials.
- Replace needle seal in the LC injection port, depending on sample throughput.
- Check syringe use count, replace as needed.
- Check the glass liner of the SPME conditioning station, replace if needed.
- Check liners of the fast wash station and clean if necessary.

Annual:
- Preventive maintenance schedule by a service engineer.
- Use a manufacturer-provided PM kit.
- Clean and apply a thin layer of grease on rails and rollers.
- Check valves for proper function.
- Check pump flow rates, as of manufacturer specs.
- Check and clean mechanical devices, e.g. drawer rails.
- Check and clean the pipette tool, as of manufacturer guides.
- Check the glass liner of the fast wash station and replace if needed.
- Calibrate the pipette tool.
- Replace the dilutor syringe.
- Replace solvent lines in case of deposit or discoloration.
- For belt-driven devices, replace driving belts, e.g. agitator.
- Clean electrical contacts in head and exchangeable tools.
- Check tool coupling count for connector replacements, as of manufacturer specs.

A.3 Syringe Needle Gauge

The syringe needle gauge describes the needle inner and outer diameter. Higher gauge numbers correspond with smaller diameters. The gauge value for needles also provides information on the resulting wall thickness. The "s"-type needles show at the same outer diameter an approx. double wall thickness. With their improved stability, "s"-type needles are recommended for septum piercing in routine operations.

Table A.3 Nominal needle gauge dimensions [13–16].

Gauge	Outer diameter [inch]	[mm]	Inner diameter [inch]	[mm]	Wall thickness [inch]	[mm]	Volume [μL/in.]	[μL/mm]	[μL/57 mm]
10	0.1330–0.1350	3.404	0.1040–0.1080	2.693	0.014	0.356	144.641	5.695	324.6
11	0.1190–0.1210	3.048	0.0920–0.0960	2.388	0.013	0.330	113.728	4.477	255.2
12	0.1080–0.1100	2.769	0.0830–0.0870	2.159	0.012	0.305	93.000	3.661	208.7
13	0.0940–0.0960	2.413	0.0690–0.0730	1.804	0.012	0.305	64.895	2.555	145.6
14	0.0820–0.0840	2.109	0.0610–0.0650	1.600	0.010	0.254	51.076	2.011	114.6
15	0.0715–0.0725	1.829	0.0525–0.0555	1.372	0.009	0.229	37.529	1.478	84.2
16	0.0645–0.0655	1.651	0.0455–0.0485	1.194	0.009	0.229	28.444	1.120	63.8
17	0.0575–0.0585	1.473	0.0405–0.0435	1.067	0.008	0.203	22.715	0.894	51.0
18	0.0495–0.0505	1.270	0.0315–0.0345	0.838	0.009	0.216	14.011	0.552	31.4
19	0.0415–0.0425	1.067	0.0255–0.0285	0.686	0.008	0.191	9.389	0.370	21.1
20	0.0355–0.0360	0.908	0.0230–0.0245	0.603	0.006	0.152	7.255	0.286	16.3
21	0.0320–0.0325	0.819	0.0195–0.0210	0.514	0.006	0.152	5.271	0.208	11.8
22	0.0280–0.0285	0.718	0.0155–0.0170	0.413	0.006	0.152	3.403	0.134	7.6
22s	0.0280–0.0285	0.718	0.0055–0.0077	0.168	0.011	0.279	0.563	0.022	1.3
23	0.0250–0.0255	0.642	0.0125–0.0140	0.337	0.006	0.152	2.266	0.089	5.1
23s	0.0250–0.0255	0.642	0.0040–0.0051	0.116	0.011	0.267	0.268	0.011	0.6
24	0.0220–0.0225	0.566	0.0115–0.0130	0.311	0.005	0.127	1.930	0.076	4.3
25	0.0200–0.0205	0.515	0.0095–0.0110	0.260	0.005	0.127	1.349	0.053	3.0
25s	0.0200–0.0205	0.515	0.0055–0.0065	0.153	0.007	0.178	0.464	0.018	1.0
26	0.0180–0.0185	0.464	0.0095–0.0110	0.260	0.004	0.102	1.349	0.053	3.0
26s	0.0184–0.0189	0.474	0.0045–0.0055	0.127	0.007	0.178	0.322	0.013	0.7
27	0.0160–0.0165	0.413	0.0075–0.0090	0.210	0.004	0.102	0.876	0.034	2.0
28	0.0140–0.0145	0.362	0.0065–0.0080	0.184	0.004	0.089	0.675	0.027	1.5
29	0.0130–0.0135	0.337	0.0065–0.0080	0.184	0.003	0.076	0.675	0.027	1.5
30	0.0120–0.0125	0.312	0.0055–0.0070	0.159	0.003	0.076	0.504	0.020	1.1
31	0.0100–0.0105	0.261	0.0045–0.0060	0.133	0.003	0.064	0.353	0.014	0.8
32	0.0090–0.0095	0.235	0.0035–0.0050	0.108	0.003	0.064	0.233	0.009	0.5
33	0.0080–0.0085	0.210	0.0035–0.0050	0.108	0.002	0.051	0.233	0.009	0.5
34	0.0060–0.0065	0.159	0.0015–0.0025	0.051	0.002	0.051	0.052	0.002	0.1

The "stronger" s-type needle gauges are highlighted. The needle volume for the most used 57 mm needle length is provided in a separate column.

Depending on the needle length and inner diameter, the resulting needle volumes vary significantly. Using injection syringes for hot vaporizing GC injectors, the needle volume need to be considered to assess the correct sample injection volume. Table A.3 informs besides outer and inner diameter and the wall thickness of different needle gauge measures about the volume of the most used 57 mm long GC injection needle.

A.4 Pressure Units Conversion

Table A.4 Pressure units conversion table.

	Pa	kPa	bar	Torr	psi	at	atm
Pa	—	1×10^{-3}	1×10^{-5}	7.5×10^{-3}	1.45×10^{-4}	1.02×10^5	9.87×10^{-6}
kPa	1×10^3	—	1×10^{-2}	7.5	0.145	1.02×10^8	9.87×10^{-3}
bar	1×10^5	100	—	750	14.514	1.02	0.987
Torr	133	0.133	1.33×10^{-3}	—	1.94×10^{-2}	1.36×10^{-3}	1.32×10^{-3}
psi	6.89×10^3	6.89	6.89×10^{-2}	51.67	—	7.03×10^{-2}	6.80×10^{-2}
at	9.81×10^4	98.1	0.981	736	14.224	—	0.968
atm	1.0133×10^5	1.0133×10^2	1.0133	760	14.706	1.033	—

A.5 Solvents

A.5.1 Solvent Miscibility

Solubility and miscibility describe similar phenomena, but they are not the same. Solvents may combine poorly or very well. It is possible to describe solutions as partially soluble and quantify the effect. Miscibility, however, is an absolute property that is not expressed in degrees. When substances are miscible, there are no layering effects, precipitates, partial mixing, or separation. Miscible means the substances mix completely. The most common determination of miscibility of solvents is by visual evaluation. If two substances form a layer, then they are immiscible [17]. For further solvent property information, see the book on "Solvents and Solvent Effects in Organic Chemistry" by C. Reichard [18], also W.M. Jackson and J.S. Drury "Miscibility of Organic Solvent Pairs" [19].

The miscibility of common solvents with boiling point (BP) and polarity most often used in sample preparation workflows is provided in Table A.5.

Table A.5 Miscibility of solvents, boiling point (BP), and polarity of often used in sample preparation workflows.

Solvent	Formula	BP (°C)	Polarity index	Acetic acid	Acetone	Acetonitrile	Benzene	Butanol (n-)	Butyl acetate (n-)	Carbon tetrachloride	Chloroform	Cyclohexane	Dichloroethane (1,2-)	Dichloromethane	Diethyl ether	Diisopropyl ether	Dimethyl ether	DMF	DMSO	Dioxane	Ethanol	Ethyl acetate	Heptane	Hexane	Methanol	MEK (butanone)	MTBE	Octane (iso-)	Pentane	Propanol (n-)	Propanol (iso-)	THF	Toluene	Trichloro ethylene	Water	Xylene
Acetic acid	$C_2H_4O_2$	118	6.2																										No								
Acetone	C_3H_6O	56	5.1																																		
Acetonitrile	C_2H_3N	82	5.8																				No	No				No	No								
Benzene	C_6H_6	80	2.7																															No		No	
Butanol (n-)	$C_4H_{10}O$	125	4.0																															No		No	
Butyl acetate (n-)	$C_6H_{12}O_2$	126	3.9																															No		No	
Carbon tetrachloride	CCl_4	77	1.6																															No		No	
Chloroform	$CHCl_3$	61	4.1																															No		No	
Cyclohexane	C_6H_{12}	81	0.2															No	No						No									No		No	
Dichloroethane (1,2-)	$C_2H_4Cl_2$	84	3.5																															No		No	
Dichloromethane	CH_2Cl_2	40	3.1																															No		No	
Diethyl ether	$C_4H_{10}O$	35	2.8															No							No									No		No	
Diisopropyl ether	$C_6H_{14}O$	68	2.2																No															No		No	
Dimethyl ether	C_2H_6O	80	4.7																															No		No	
DMF	C_3H_7NO	155	6.4																				No	No			No	No									
DMSO	C_2H_6OS	189	7.2																				No	No			No	No									
Dioxane	$C_4H_8O_2$	101	4.8																																		
Ethanol	C_2H_6O	78	5.2																																		
Ethyl acetate	$C_4H_8O_2$	77	4.4																															No		No	
Heptane	C_7H_{16}	98	0			No												No	No															No		No	

Table A.5 (Continued)

Solvent	Formula	BP (°C)	Polarity index	Acetic acid	Acetone	Acetonitrile	Benzene	Butanol (n-)	Butyl acetate (n-)	Carbon tetrachloride	Chloroform	Cyclohexane	Dichloromethane	Dichloroethane (1,2-)	Diethyl ether	Diisopropyl ether	Dimethyl ether	DMF	DMSO	Dioxane	Ethanol	Ethyl acetate	Heptane	Hexane	Methanol	MEK (butanone)	MTBE	Octane (iso-)	Pentane (iso-)	Propanol (n-)	Propanol (iso-)	THF	Toluene	Trichloro ethylene	Water	Xylene
Hexane	C_6H_{14}	69	0	No		No												No																	No	
Methanol	CH_4O	65	5.1							No		No No																	No No					No	No No		No
MEK (butanone)	C_4H_8O	80	4.7																																		
MTBE	$C_5H_{12}O$	55	2.5																																		
Octane (iso-)	C_8H_{18}	99	0		No													No No																	No	
Pentane	C_5H_{12}	36	0		No	No												No No							No									No	
Propanol (n-)	C_3H_8O	97	4.0																																		
Propanol (iso-)	C_3H_8O	82	3.9																																		
THF	C_4H_8O	65	4.0																																		
Toluene	C_7H_8	111	2.4																																	No	
Trichloroethylene	C_2HCl_3	87	1.0							No No No No No No No No No																									No No	
Water	H_2O	100	9.0														No No																		No No		No No
Xylene	C_8H_{10}	139	2.5																																	No No	

Blanks: solvents are miscible in any ratio. Red "No": not miscible.

A.5.2 Solvent Stability

Many organic reagents need to be stabilized. The knowledge about the stabilizer used may be of importance. Certain solvents will degrade over time requiring special handling and storage considerations. Also, the products of certain degradation processes can pose a potential safety risk if present at sufficiently high levels. This also imposes an important safety concern with storage and evaporation on robotic systems. For these types of solvents, small amounts of stabilizing chemicals are added to slow down or stop degradation [20]. Stabilizers can interfere with spectroscopic detection or chromatographic analysis and concentrate in methods for trace analysis. Always refer to the manufacturer documentation for details.

A.5.2.1 Halogenated Solvents

Halogenated solvents degrade with time via mechanisms that are catalyzed by light, heat, and oxygen. Degradation products include phosgene ($COCl_2$) and hydrochloric acid (HCl). Organic halides as chloroform, deuterated chloroform, iodobutane are decomposed by light or heat and can become acidic. Free halogen (Cl_2, I_2) and radical halogens (Cl·, I·), halogenic acids (HCl, HI), and free halides as Cl- and I- are formed. Decomposition is minimized if bottles are stored refrigerated and in the dark. Silver and copper foils act as a radical scavenger and can be used as stabilizer. Silver and copper forms insoluble, stable metal halides (CuX and AgX) that trap the generated halogen radicals, neutral halogens, and ionic halides, stabilizing the organic halides.

Chloroform Chloroform is unstable and is combined with a variety of stabilizers to enhance shelf life like ethanol or amylene. Ethanol must be added to chloroform at relatively high concentrations of about 1% to be effective. This will increase the polarity of the solvent and potentially impact certain applications. Amylene (2-methyl-2-butene) is added to chloroform to scavenge free radicals, effective at levels of approximately 100 ppm.

Dichloromethane (DCM, methylene chloride) Dichloromethane degrades with time and requires a stabilizer to prevent the formation of HCl. Without a stabilizer, HCl will form and injection of acidic DCM will cause inlet liners and columns to become reactive. Two classes of stabilizers are applied. Stabilizers keep HCl from forming and eliminate HCl upon formation. Methanol is used to stop HCl from forming. It is not recommended to extract aqueous samples, or solid samples that contain water, using methanol-stabilized DCM. The methanol will partition into the water and could leave an unstabilized extract [21]. Alkanes such as cyclohexane, cyclohexene, or amylene (2-methyl-2-butene) scavenge HCl after its formation and are typically employed at levels of 100 ppm.

Due to the popularity of DCM as a solvent in analytical labs, it should be noted that over time, alkene stabilizers produce chlorinated by-products that may interfere with some GC analyses. This is particularly true in the case of cyclohexene since these by-products with relatively high boiling point will elute further away from the solvent front and may interfere with target analytes.

A.5.2.2 Ethers

Organic peroxides are formed in ethers catalyzed by heat, light, and oxygen. A high risk for potential explosion occurs at concentrations above 100 ppm. In particular, low levels of peroxides may be present and cause a safety hazard during heating and during ether extract evaporation, also concentrating peroxides in the solvent residue. Due to their high reactivity, ether peroxides can interfere with a given method.

Ethyl Ether A common stabilizer system for ethyl ether is butylhydroxytoluene (BHT). BHT is a very effective suppressor of peroxide formation. It is typically added to ethyl ether at low ppm concentrations and scavenges the free radical species responsible for peroxide formation. BHT is incompatible with methods requiring high optical purity due to strong UV absorbance resulting from the aromatic functionality of this molecule.

Ethanol is another common stabilizer, which is added at much higher concentrations of 1–2% than BHT. Due to this relatively high stabilizer concentration, the presence of ethanol significantly increases solvent polarity and may affect certain applications.

Tetrahydrofuran (THF) Tetrahydrofuran as ether forms organic peroxides upon storage like ethyl ether. BHT is typically used as a stabilizer in THF at levels of 100–300 ppm. Like in ethyl ether, BHT interferes in methods utilizing UV detection causing a high background level.

A.5.3 Solvent Viscosity

The specific dynamic density is given in g/mL at 25 °C (298.15 K), surface tension in dynes/cm at 25 °C (298.15 K) [22], and viscosity in cP (=mPa·s) at 25 °C (298.15 K) [23, 24]. Table A.6 provides the viscosity, density, and surface tension of common solvents relative to water. The solvent viscosity needs to be considered when programming syringe aspirations speeds to prevent bubble formation by cavitation.

A.6 Material Resistance

A wide variety of different materials employed in robotic sample preparation systems are in contact with the sample, solvents, and reagents. Syringes, pumps, tubes, sealings and for sure vials, septa and closures, even racks and holders can be affected. All materials of an automated system and consumables in use must be compatible within a selected workflow. Concerning heat resistance at certain operation temperatures, most of the sample preparation steps are run at room temperature. Exceptions are those devices where incubation, derivatization, conditioning, desorption, hot water cleaning, or similar steps at elevated temperatures are involved. A consideration for resistance is also the residence time of solvents and chemicals with certain devices. A short exposure to a steel needle might be

Table A.6 Solvent viscosity, density, and surface tension, sorted by increasing viscosity relative to water.

Compound	Formula	CAS#	M [g/mol]	Spec. density [g/mL]	Surface tension [dynes/cm]	Viscosity [mPa·s], [cP]	Viscosity relative to H_2O
Diethyl ether	$C_4H_{10}O$	60-29-7	74.12	0.71	16.7	0.22	0.25
Hexane	C_6H_{14}	110-54-3	86.18	0.66	17.9	0.30	0.34
Acetone	C_3H_6O	67-64-1	58.08	0.79	23	0.31	0.35
Methyl t-butyl ether (MTBE)	$C_5H_{12}O$	1634-04-4	88.15	0.74	72.5	0.36	0.40
Acetonitrile	C_2H_3N	75-05-8	41.05	0.78	28.7	0.37	0.42
n-Heptane	C_7H_{16}	142-82-5	100.2	0.68	19.8	0.39	0.44
Dichloromethane (DCM)	CH_2Cl_2	75-09-2	84.93	1.32	27.8	0.41	0.46
Ethyl acetate	$C_4H_8O_2$	141-78-6	88.11	0.89	23.2	0.42	0.47
Tetrahydrofuran	C_4H_8O	109-99-9	72.11	0.88	26.7	0.46	0.52
Chloroform	$CHCl_3$	67-66-3	119.4	1.48	26.7	0.54	0.61
Methanol	CH_4O	67-56-1	32.04	0.79	22.1	0.54	0.61
Trichloroethylene (TCE)	C_2HCl_3	79-01-6	131.4	1.46	28.7	0.55	0.62
Methyl isobutyl ketone	$C_6H_{12}O$	108-10-1	100.2	0.80	23.5	0.55	0.62
Toluene	C_7H_8	108-88-3	92.14	0.87	27.9	0.56	0.63
Benzene	C_6H_6	71-43-2	78.11	0.87	28.2	0.60	0.67
p-Xylene	C_8H_{10}	106-42-3	106.2	0.86	27.9	0.60	0.67
Styrene	C_8H_8	100-42-5	104.2	0.90	32	0.70	0.79
o-Xylene	C_8H_{10}	95-47-6	106.2	0.88	29.6	0.76	0.85
1,2-Dichloroethane	$C_2H_4Cl_2$	107-06-2	98.96	1.25	32.6	0.78	0.88
1,1,1-Trichloroethane	$C_2H_3Cl_3$	71-55-6	133.4	1.33	25	0.79	0.89

Name	Formula	CAS	MW				
N,N-Dimethylformamide (DMF)	C$_3$H$_7$NO	68-12-2	73.09	0.95	34.4	0.79	0.89
Pyridine	C$_5$H$_5$N	110-86-1	79.1	0.98	36.7	0.88	0.99
Cyclohexane	C$_6$H$_{12}$	110-82-7	84.16	0.77	24.7	0.89	1.00
Water	**H$_2$O**	**7732-18-5**	**18.02**	**1.00**	**72.7**	**0.89**	**1.00**
Tetrachloromethane	CCl$_4$	56-23-5	153.8	1.58	26.3	0.91	1.02
Acetic acid	C$_2$H$_4$O$_2$	64-19-7	60.05	1.04	27	1.06	1.19
Ethanol	C$_2$H$_6$O	64-17-5	46.07	0.79	22	1.07	1.20
1,4-Dioxane	C$_4$H$_8$O$_2$	123-91-1	88.11	1.03	32.9	1.18	1.33
Hydrogen peroxide	H$_2$O$_2$	7722-84-1	34.02	1.45	74	1.25	1.40
Formic acid	CH$_2$O$_2$	64-18-6	46.03	1.21	37.7	1.61	1.81
n-Propanol	C$_3$H$_8$O	71-23-8	60.1	0.80	20.9	1.95	2.19
Dimethyl sulfoxide (DMSO)	C$_2$H$_6$OS	67-68-5	78.13	1.10	42.9	1.99	2.24
iso-Propanol	C$_3$H$_8$O	67-63-0	60.1	0.78	23.3	2.04	2.29
n-Butanol	C$_4$H$_{10}$O	71-36-3	74.12	0.81	25	2.54	2.85
2-Butanol	C$_4$H$_{10}$O	78-92-2	74.12	0.81	22.6	3.10	3.48
Benzyl alcohol	C$_7$H$_8$O	100-51-6	108.1	1.04	36.8	5.47	6.15
Ethylene glycol	C$_2$H$_6$O$_2$	107-21-1	62.07	1.11	48.4	16.1	18.1
Dibutyl phthalate	C$_{16}$H$_{22}$O$_4$	84-74-2	278.3	1.04	37.4	16.6	18.7
Diethylene glycol	C$_4$H$_{10}$O$_3$	111-46-6	106.1	1.11	55.1	30.2	33.9
Propylene glycol	C$_3$H$_8$O$_2$	57-55-6	76.1	1.03	45.6	40.4	45.4
Cyclohexanol	C$_6$H$_{12}$O	108-93-0	100.2	0.96	33.4	57.5	64.6
Triethanolamine	C$_6$H$_{15}$NO$_3$	102-71-6	149.2	1.12	51.5	609	684
Glycerol	C$_3$H$_8$O$_3$	56-81-5	92.09	1.26	76.2	934	1049

tolerable and manageable by a preventive maintenance plan, while a permanent exposure in a pump head creates a different resistance scenario.

Materials are often characterized in the industry according to their resistance in different classes [25]:

Class	Temperature range	Resistant Temperature range
A	at 20 °C	Little or no damage after 30 d
B	at 20 °C	Shows some effect after 7 d
C	at 20–50 °C	Immediate damage may occur
D	at 20–50 °C	Little or no damage after 30 d
E	at 50 °C	Some effect after 7 d
F	at 50 °C	Immediate damage may occur

In cases material incompatibility of classes B or C (or similar letter code) is identified for particular chemicals or solvents, it is advised to replace the same with compatible ones or check for a replacement of applied solvents. If it is unavoidable and crucial for a specific preparation method, the regular preventive maintenance plan must cover the affected devices with a schedule for maximum usage until mandatory replacement. The use of material incompatible chemicals or solvents classes C, D, E, and F must be avoided to prevent failure and serious damage.

A.6.1 Glass

Laboratory glassware, for instance, the typical autosampler vials, but also other glass products as frits or syringes, are mostly manufactured from borosilicate glass made of silicon dioxide and >8% boron trioxide as the main components. It is internationally known with many brand names like Duran®, Pyrex®, and Schott®, just to mention a few of them.

The outstanding quality of borosilicate glass, in contrast to soda–lime glass, is the low thermal expansion, making products resistant against rapid temperature changes, and its high chemical resistance and durability. The temperature differential that borosilicate glass can withstand before fracturing is about 165 °C [26]. Borosilicate glass material is resistant to a wide variety of acids and bases at room to moderate temperatures. Also at high temperatures, the alkali surface erosion is low [27].

A.6.2 Polymers

A wide variety of polymer materials found use in sample preparation instrumentation. Here, the resistance against solvents and chemicals is of importance for reliable routine operation and need to be carefully checked in the design of every workflow.

Most affected, by organic solvents or reagents, are tubing materials but also less visible parts for instance inside of solvent pumps or valves. Materials used in automated sample preparation systems and consumables, if not from glass or stainless steel, were mostly comprising of different polymer materials. The explanation of the commonly used acronyms with material description, typical use, and melting point (MP) is provided as follows:

HDPE: Polyethylene of high density (PEHD) is a thermoplastic polymer produced from the monomer ethylene, MP of 131 °C.

FEP: Fluorinated ethylene propylene is a copolymer of hexafluoropropylene and tetrafluoroethylene, MP of 260 °C, and typically used with solvent tubings, waste tubings, O-rings, etc.

FFKM: Poly(tetrafluoro ethylene) is a perfluoroelastomer with cross-linkable monomer structure, maximum operation temperature of 325 °C, and often used with solvent pumps, O-rings, sealings, gaskets, ferrules.

LDPE: Low-density polyethylene is a thermoplastic made from the monomer ethylene, has more branching (about 2% of the carbon atoms) than HDPE, MP typically in the range of 120–130 °C, and typically used with solvent tubings, waste tubings, etc.

PEEK: Polyether ether ketone is a polyaryletherketone, machinable, MP of 343 °C, and manifold use for tubes, micro-pump body, housings, fittings, needle seals, etc.

PFA: Perfluoroalkoxy alkane is a fluoropolymer, copolymers of tetrafluoroethylene and perfluoroethers, MP of 315 °C, maximum operation temperature of 260 °C, and typically used with solvent tubings, waste tubings, etc.

PP: Polypropylene is a polyolefin thermoplastic polymer with properties similar to polyethylene, MP of 130–171 °C and widely used for pipette tips, vials, tubes, MTPs, DWPs, etc.

PPS: Polyphenylene sulfide is a thermoplastic polymer consisting of aromatic rings linked by sulfides, maximum operation temperature of 218 °C, does not dissolve in any solvent at room temperatures and used in solvent pumps.

PVDF: Polyvinylidene difluoride is a highly nonreactive thermoplastic fluoropolymer, stable of up to 375 °C and used for housings, plates, tubings, etc.

The typically found polymer materials in instruments show very different solvent resistance. In the design of sample preparation methods with contact of solvents to polymer materials in form of tubings, sealings, pipette tips, or vials, just to mention a few typical areas, the compatibility of the materials need to be considered. Table A.7 informs about the compatibility of the most found polymer materials in analytical sample preparation instrumentation with often applied solvents and reagents.

A.6.3 Stainless Steel

Iron-based alloys with a minimum of 11% chromium are termed "stainless steel", which prevents the steel from rusting. A wide variety of stainless steel alloys are known with different material properties and recommended for usage. Types of stainless steel are named with a three-digit number in the American Iron and Steel Institute (AISI) designation. The most common stainless steel grades are 304 and 316 with both good formability and corrosion resistance. The steel needles of syringes are usually manufactured from AISI 304 stainless steel [35]. The steel

Table A.7 Resistance of the most used polymer materials with typical solvents or reagents.

Chemical	FEP 20°C	FEP 60°C	FEP 100°C	FFKM	PEEK	PFA 20°C	PFA 60°C	PFA 100°C	PE LDPE	PE HDPE	PP
Acetaldehyde	R	R	R	R	R	R	R	R	A, E	A, E	R
Acetic Acid 10%	R	R	R	R	R	R	R	R	D	D	R
Acetic acid (glac./anh.)	R	R	R	R	—	R	R	R	A, F	A	R
Acetic anhydride	R	R	R	R	—	R	R	R	A	A	A
Acetone	R	R	R	R	R	R	R	R	F	F	R
Acetonitrile	R	R	R	R	R	R	R	R	R	R	R
Alcohols: Allyl	R	R	R	—	—	R	R	R	D	D	A
Alcohols: Benzyl	—	—	—	R	R	—	—	—	F	B, F	R
Alcohols: Ethyl	R	R	R	R	R	R	R	R	D	D	R
Alcohols: Isobutyl	R	R	R	R	—	R	R	R	D	D	R
Alcohols: Isopropyl	R	R	R	R	R	R	R	R	D	D	R
Alcohols: Methyl	R	R	R	R	R	R	R	R	D	D	R
Ammonia, anhydrous	R	R	R	R	R	A	R	R	D	D	R
Ammonia, aqueous	R	R	R	R	—	R	R	R	R	R	R
Amyl Acetate	R	R	R	R	R	R	R	R	A, E	A	A
Aniline	R	R	R	R	R	R	R	R	A	B, E	R
Aqua Regia (80% HCl$_3$), 20% HNO	R	R	R	R	NR	A	R	R	C	C	A
Benzaldehyde	R	R	R	R	R	R	R	R	R	R	C
Benzene	R	R	R	R	R	R	R	R	C	C	C
Boric Acid	R	R	R	R	R	R	R	R	D	D	R
Bromine	R	R	R	R	NR	R	R	R	F	B, F	C
Bromine Water, sat. aqueous	R	R	R	R	—	R	R	R	C	C	—
Butyl Acetate	R	R	R	R	—	R	R	R	R	R	A
Carbon Disulfide	R	R	R	R	NR	R	R	R	—	—	C
Carbon Tetrachloride	R	R	R	R	R	R	R	R	B, F	A, B, E	C
Carbonic Acid	R	R	R	R	R	R	R	R	R	R	R
Chlorine, wet	R	R	R	B	NR	R	R	R	A, F	A	C
Chloroacetic Acid	R	R	R	R	R	R	R	R	R	R	B
Chlorobenzene (mono)	R	R	R	R	R	R	R	R	C	C	B

Table A.7 (Continued)

Chemical	FEP 20°C	FEP 60°C	FEP 100°C	FFKM	PEEK	PFA 20°C	PFA 60°C	PFA 100°C	PE LDPE	PE HDPE	PP
Chloroform	R	R	R	R	R	R	R	R	B, F	B, F	B
Chlorosulphonic Acid	R	R	R	R	R	R	R	R	C	C	C
Chromic acid 80%	R	R	R	R	NR	R	R	R	A	A	C
Citric Acid	R	R	R	R	R	R	R	R	A	A	R
Cyclohexane	R	R	R	R	R	R	R	R	B, F	B, F	C
Cyclohexanone	—	—	—	NR	R	—	—	—	B	B	C
Diethyl Ketone	—	—	—	—	—	—	—	—	C	C	—
Dimethyl Formamide (DMF)	—	—	—	R	R	—	—	—	B	B	R
Dimethylsulfoxide (DMSO)	—	—	—	—	—	—	—	—	D	D	—
Diethyl Ether	R	R	R	R	R	R	R	R	C	C	R
1,4 Dioxane	—	—	—	—	—	—	—	—	A, E	A	—
Ethyl Acetate	—	—	—	R	R	—	—	—	D	D	R
Ethyl Benzene	—	—	—	—	—	—	—	—	C	B, F	—
Ethylene Glycol	—	—	—	R	R	—	—	—	D	D	R
Fluorinated Refrigerants	R	R	R	B	R	R	R	R	—	—	—
Fluorine, dry	R	R	NR	B	NR	R	R	NR	B	A	C
Formaldehyde 40%	R	R	R	R	R	R	R	R	D	D	R
Formic Acid	R	R	R	—	C	R	R	R	R	R	R
Glycerol	R	R	R	R	R	R	R	R	D	D	R
Glycols	R	R	R	R	—	R	R	R	R	R	—
Glycol, ethylene	R	R	R	—	—	R	R	R	R	R	R
Heptane	—	—	—	R	R	—	—	—	R	R	B
Hexane	—	—	—	R	R	—	—	—	NR	D, E	A
Hexamine	R	R	R	—	—	R	R	R	—	—	—
Hydrazine	R	R	R	R	R	R	R	R	R	R	B
Hydrobromic Acid	R	R	R	B	NR	R	R	R	R	R	B
Hydrochloric Acid	R	R	R	R	R	R	R	R	R	R	B
Hydrofluoric Acid	R	R	R	R	NR	R	R	R	D	D	B
Hydrogen Peroxide (30 — 90%)	R	R	R	R	R	R	R	R	D	D	A
Hydrogen Sulfide	R	R	R	R	R	R	R	R	R	R	R
Isooctane	—	—	—	R	R	—	—	—	B	B	R
Isopropyl Ether	—	—	—	R	R	—	—	—	C	C	A
Kerosene	—	—	—	R	R	—	—	—	B, F	B, F	A
Ketones	R	R	R	—	—	R	R	R	—	—	B
Methyl Ethyl Ketone (MEK)	—	—	—	R	R	—	—	—	C	C	A

(Continued)

Table A.7 (Continued)

Chemical	FEP 20°C	FEP 60°C	FEP 100°C	FFKM	PEEK	PFA 20°C	PFA 60°C	PFA 100°C	PE LDPE	PE HDPE	PP
Methyl Isobutyl Ketone (MIBK)	—	—	—	R	R	—	—	—	C	C	R
Methylene Chloride	R	R	R	R	R	R	R	R	C	B, F	A
Monoethanolamine	R	R	NR	R	—	R	R	NR	—	—	A
Nitric Acid (50%)	R	R	R	—	—	R	R	R	A, E	B, F	A
Nitric Acid (90%)	R	R	R	R	NR	R	R	R	B, F	B, F	C
Nitrobenzene	R	R	R	R	R	R	R	R	C	C	A
n—Octane	—	—	—	—	—	—	—	—	D	D	—
Oils, mineral	R	R	R	—	—	R	R	R	A, F	A	R
Oleic Acid	—	—	—	R	R	—	—	—	C	D	A
Oxalic Acid (cold)	R	R	R	R	R	R	R	R	B, D	D	R
Ozone	R	R	R	R	R	R	R	R	A, F	A, F	A
Pentane	—	—	—	R	R	—	—	—	—	—	C
Perchloric Acid	R	R	R	R	R	R	R	R	A, F	A, F	B
Perchloroethylene	—	—	—	R	R	—	—	—	C	C	C
Phenol (Carbolic Acid)	R	R	R	R	NR	R	R	R	C	C	A
Phosphoric Acid (95%)	R	R	R	R	R	R	R	R	A, F	A	R
Potassium Hydroxide conc.	—	—	—	—	—	—	—	—	D	D	R
Propylene Glycol	—	—	—	—	—	—	—	—	D	D	R
Pyridine	R	R	R	R	R	R	R	R	C	C	R
Sodium Hydroxide (50%)	—	—	—	—	—	—	—	—	D	D	R
Sodium Hypochlorite (15%)	—	—	—	—	—	—	—	—	A, E	D	R
Sulfur Dioxide	R	R	R	R	R	R	R	R	—	—	R
Sulphuric Acid (<50%)	R	R	R	R	—	R	R	R	D	D	R
Sulphuric Acid (95%)	R	R	R	R	NR	R	R	R	D	B, E	B
Tetrachloroethane	—	—	—	R	—	—	—	—	C	C	B
Tetrachloroethylene	—	—	—	R	R	—	—	—	C	C	C
Tetrahydrofuran (THF)	—	—	—	R	R	—	—	—	B, F	B, F	B
Toluene (Toluol)	—	—	—	R	R	—	—	—	B, F	C	B
Trichloroacetic Acid	—	—	—	R	—	—	—	—	B, F	C	R
Trichloroethane	—	—	—	R	R	—	—	—	C	C	B
Trichlorethylene	R	R	R	R	R	R	R	R	C	C	B
Turpentine	—	—	—	R	R	—	—	—	B, F	B, F	C
Urea (30%)	R	R	R	R	R	R	R	R	D	D	R
Xylene	—	—	—	—	R	—	—	—	B, F	B, F	A

R, resistant; A, at 20 °C little or no damage after 30 days; B, at 20 °C shows some effect after seven days; C, at 20–50 °C immediate damage may occur; D, at 20–50 °C little or no damage after 30 days; E, at 50 °C some effect after seven days; F, at 50 °C immediate damage may occur; NR, not recommended; —, no data.

contains both chromium, between 18% and 20%, and nickel between 8% and 10.5% [36]. It is also named "18/8 stainless steel," embossed in many tools, also known from household cutlery. The 316 steel grade contains an additional 2–3% of molybdenum for improved corrosion resistance in chloride-containing environments. 316 steel is also used in the manufacture of medical surgical instruments.

While stainless steel is known to be highly corrosion-resistant, some agents are not compatible in permanent exposure, in particular in chemical applications. A known limitation is the contact with HCl. Also, 304 steel is not resistant to permanent high chloride exposure. Solutions with as little as 25 ppm of sodium chloride can begin to have a corrosive effect [37].

The following selected chemicals for sample preparation purpose are listed for a critical chemical behavior with 304 steel, categorized as "severe effect applies for a 48 hour exposure period" [38]:

- Acetic acid
- Aqua regia
- Benzonitrile
- Bromine
- Chloric acid
- Chlorosulfonic acid
- Hydrobromic acid
- Hydrochloric acid
- Iodine
- Phosphoric acid
- Sodium hypochlorite
- Sulfur dioxide
- Sulfuric acid
- Trichloroacetic acid

A short-term exposure of 304 steel syringe needle material to the listed critical chemicals, as it is typical in sample preparation procedures, will not exclude its usage but induce chemical "wear and tear" with a reduced useful lifetime. Depending on the application, a maintenance plan for the early and regular exchange should be considered.

References

1 GERSTEL (2013). Maestro software. Product information, GERSTEL GmbH & Co.KG Mülheim an der Ruhr, Germany. https://www.gerstel.com/en/maestro-en.htm (accessed 11 April 2020).
2 Axel Semrau (2018). CHRONOS 4.9 product information. Axel Semrau GmbH & Co KG, Sprockhövel, Germany. https://www.axelsemrau.de/produktdetails/chronos (accessed 11 April 2020).
3 Thermo Fisher Scientific (2018). Automate sample handling workflow. Product brochure BR10670_E 11/18M, Thermo Fisher Scientific Inc.

4 CTC Analytics (2020). PAL Method Composer. https://www.palsystem.com/index.php?id=850 (accessed 11 April 2020).
5 CTC Analytics (2019). PAL sample control user manual. Version March 15, 2019, CTC Analytics AG, Zwingen, Switzerland.
6 Restek (2018). Syringe Basics. http://www.restek.com/techtips/Syringe-Basics (accessed 9 August 2018).
7 Merck (2016). Extran detergents. Product information, Merck KGaA, Darmstadt, Germany.
8 Trajan Scientific (2018). SGE syringe – instructions for syringe use, care and maintenance 5 μL to 500 μL syringes. Document No. MN-0673-S Rev.C, 08/2018, Trajan Scientific Australia Pty. Ltd., Victoria, Australia.
9 Hamilton Bonaduz (2014). Needle cleaning kit. Document No. 69136 Rev. C, 11/2014, Hamilton Bonaduz AG, Bonaduz, GR, Switzerland.
10 CTC Analytics AG (2017). PAL RTC and RSI specifications (LC & GC). Specifications Rev 7 – October 2017.
11 International Organization for Standardization (2002). ISO 8655-1:2002(en) Piston-operated volumetric apparatus – Part 1: terminology, general requirements and user recommendations. https://www.iso.org/obp/ui/#iso:std:iso:8655:-1:ed-1:v1:en (accessed 22 March 2020).
12 CTC (2019). User manual PAL system. Rev. 3.1.0, CTC Analytics AG, Zwingen, Switzerland.
13 Hamilton Bonaduz. (2019). "Needle Gauge Chart". https://www.hamiltoncompany.com/laboratory-products/needles-knowledge/needle-gauge-chart (accessed 10 November 2019).
14 Wikipedia. (2019). "Birmingham gauge". https://en.wikipedia.org/wiki/Birmingham_gauge (accessed 10 November 2019).
15 Merck. (2019). "Syringe Needle Gauge Chart". https://www.sigmaaldrich.com/chemistry/stockroom-reagents/learning-center/technical-library/needle-gauge-chart.html (accessed 10 November 2019).
16 Medical Tube Technology. (2004). "Hypodermic Needle Gauge Chart". http://www.medtube.com/hypo_chrt.htm (accessed 10 November 2019).
17 Merck. (2019). "Solvent Miscibility Table" https://www.sigmaaldrich.com/chemistry/solvents/solvent-miscibility-table.html (accessed 10 November 2019).
18 Reichardt, C. (2003). *Solvents and Solvent Effects in Organic Chemistry*, 3rd Ed. Wiley-VCH Publishers.
19 Jackson, W.M. and Drury, J.S. (1959). Miscibility of organic solvent pairs. *Industrial & Engineering Chemistry* 51 (12): 1491–1493. https://doi.org/10.1021/ie50600a039.
20 Merck. (2019). "Solvent Stabilizer Systems". https://www.sigmaaldrich.com/chemistry/solvents/learning-center/stabilizer-systems.html (accessed 16 July 2019).
21 Restek (2014). Analysis of Halogenated Environmental Pollutants Using Electron Capture Detection. In: Bellefonte, PA, USA: Restek Corp.
22 Yaw, C.L. and Andrew, W. (2008). *Thermophysical Properties of Chemicals and Hydrocarbons*. Norwich, NY: Elsevier.

23 Lide, D.R. (ed.) (2004). *CRC Handbook of Chemistry and Physics*, 85th Ed. Boca Raton, FL, USA: CRC Press.
24 Diversified Enterprises. (2019). Viscosity, surface tension, specific density and molecular weight of selected liquids. https://www.accudynetest.com/visc_table.html?sortby=sort_centipoise%20ASC (accessed 10 November 2019).
25 Thermo Fisher Scientific (2000). Labware chemical resistance table. Application Note. Thermo Fisher Scientific. https://tools.thermofisher.com/content/sfs/brochures/D20480.pdf (accessed 22 March 2020).
26 Wikipedia (2020). Borosilicate glass. https://en.wikipedia.org/wiki/Borosilicate_glass (accessed 22 March 2020).
27 Duran-Group (2020). Chemical resistance of Duran®. The surface erosion after 3 hours boiling in a mixture of equal volume fractions of sodium hydroxide solution (concentration 1 mol/L) and sodium carbonate solution (concentration 0.5 mol/L) is only 134 mg/100 cm^2. https://www.duran-group.com/en/about-duran/duran-properties/chemical-properties.html (accessed 22 March 2020).
28 Polyfluor Plastics. (2019). Chemical resistance PFA. https://www.polyfluor.nl/en/chemical-resistance/pfa/ (accessed 16 July 2019).
29 CP Lab Safety. (2019). Polyetherether ketone (PEEK) chemical compatibility chart. https://www.calpaclab.com/polyetherether-ketone-peek-chemical-compatibility-chart/ (accessed 16 July 2019).
30 DuPont. (2019). Kalrez® parts chemical resistance. https://www.dupont.com/knowledge/chemical-resistance-kalrez-parts.html (accessed 16 July 2019).
31 DuPont. (2019). Kalrez® chemical resistance and fluid compatibility. www.parrinst.com/wp-content/uploads/downloads/2011/07/Parr_DuPont-Kalrez-O-ring-Materials-Corrosion-Info.pdf (accessed 16 July 2019).
32 Polyfluor Plastics. (2019). Chemical resistance FEP. https://www.polyfluor.nl/en/chemical-resistance/fep/ (accessed 16 July 2019).
33 CP Lab Safety. (2019). Chemical charts. https://www.calpaclab.com/chemical-compatibility-charts/ (accessed 16 July 2019).
34 BRAND. (2015). Technical information. Product information GK900_07_Technical Info_e, BRAND GMBH + CO KG, Wertheim, Germany.
35 Hamilton Bonaduz (2020). Syringe needles. Hamilton Bonaduz AG, Bonaduz, GR, Switzerland. https://www.hamiltoncompany.com/laboratory-products/needles#top (accessed 28 March 2020).
36 Iron Boar Labs (2018). AISI 304 (S30400) stainless steel. Iron Boar Labs Ltd. https://www.makeitfrom.com/material-properties/AISI-304-S30400-Stainless-Steel (accessed 22 March 2020).
37 Reliance-Foundry. (2020). 304 Stainless steel versus 316 stainless steel. https://www.reliance-foundry.com/blog/304-vs-316-stainless-steel (accessed 20 March 2020).
38 Thomas & Betts (2006). *304 Stainless Steel Corrosion Compatibility Chart*. Memphis, TN, USA: Thomas & Betts Corporation www.tnb.com.

Glossary

#

24/7 operation	A system operation 24 h the day over 7 d the week

A

AISI	American Iron and Steel Institute
aka	"also known as"
AMPA	Aminomethylphosphonic acid. Metabolite of the pesticide glyphosate
AOCS	American Oil Chemists' Society, international professional organization, founded in 1909 with approx. 4000 members in 90 countries
AOAC	Founded in 1884 as the Association of Official Agricultural Chemists, in 1965, the name was changed to the Association of Official Analytical Chemists, later to the Association of Analytical Communities, since 1991 AOAC International. See https://en.wikipedia.org/wiki/AOAC_International
API	Atmospheric pressure ionization, see also ESI
APPI	Atmospheric pressure photo ionization
AR	Analytical reagent grade, pro-analysis grade
ASE™	Accelerated Solvent Extraction™ by Thermo Fisher Scientific

B

BHT	Buthylhydroxytoluene. Stabilizer, e.g. in ethyl ether
BIN	β-Ionone, odorous compound
Bottom sensing	A special optional feature of some automated systems to detect the vial bottom or valve seal in a defined height distance, works with syringes and pipet tips likewise
	In case of a syringe use, the needle pulls back less than a millimeter to aspirate liquid media. Typically, three injections of 1 μL are possible from a 5 μL sample. It works with syringe needle gauges 26 and stronger. Tapered bottom vials or microvials are recommended to prevent bending of the needle toward the sidewall
BP	Boiling point
BSA	*N,O*-Bis(trimethylsilyl)acetamide. Silylation reagent

Automated Sample Preparation: Methods for GC-MS and LC-MS,
First Edition. Hans-Joachim Hübschmann.
© 2022 WILEY-VCH GmbH. Published 2022 by WILEY-VCH GmbH.

BSTFA	N,O-Bis(trimethylsilyl)trifluoroacetamide. Silylation reagent

C

C18	Length of an alkyl chain, e.g. in bonded LC phases, length of an n-alkane chain. In general, such abbreviations are used as Cn, n giving the number of C atoms in the chain, for instance, C18 or C40 with a C chain of 18 resp. 40 carbon atoms
CA	Cellulose acetate
Cartesian robot	x,y,z-Robots are classified as Cartesian robots. Every point in the Cartesian coordinate system is defined by its x, y, and z values. The principal working axes are in right angles to each other and controlled linear. Every point within this working space can be addressed or excluded. As a naming convention, the x-axis is the width of the unit from left to right, the y-axis described the depth from a front point to the back, and the z-axis is the the vertical axis
Cavitation	Bubble formation in a tube or syringe caused by rapid aspiration and liquid viscosity
CDC	Center of Disease Control
CDS	Chromatography Data System
CI	Chemical ionization. Ionization mode in GC-MS with low ionization energy for reduced molecular ion fragmentation, mediated by a CI gas introduced into the specific ion source. Commonly applied for structural identification purposes requiring the information of the molecular mass of an unknown analyte, and also for quantification
CIS	Cooled Injection System. Multifunctional and temperature programmable GC injector (by GERSTEL GmbH & Co KG, Mülheim, Germany)
CRM	Certified reference material. Manufactured under traceable conditions and certified quality control
CSR	Concurrent solvent recondensation. Large volume liquid injection technique in GC

D

DAG	Diacylglycerol
DBDI	Dielectric barrier discharge ion source. Ion source for atmospheric pressure ionization, used for automated direct SPME-MS analysis
DCM	Dichloromethane
DHA	Docosahexaenoic acid. 22:6(n-3), an essential omega-3 fatty acid
DHS	Dynamic headspace
DHS-VTT	Dynamic headspace vacuum transfer in trap extraction, applies vacuum during the HS-SPME or HS-ITEX process
DI-SPME	Direct immersion SPME. The SPME Fiber or Arrow is immersed into the (aqueous) sample for extraction of semivolatile analytes. Strong shaking is required during the extraction phase
Dielectric constant	The measure k of a solvent's ability to insulate opposite charges from one another, most common measure for solvent polarity classification
Dilution factor	The total number of equal volumes in which one volume, e.g. of a standard solution, is diluted. For instance, the dilution factor of 10 hence is 1 volume of standard in a total of 10 equal volumes, corresponds to the pipetting of 1 + 9 volumes

Dilution ratio	The dilution ratio describes the ratio of equal volumes used for dilution, e.g. for a ratio of 1:9 it is 1 volume of standard plus 9 volumes of solvent, to achieve a dilution factor of 10
DIN	Deutsche Industrie Norm. National standards organization of Germany
DLLME	Dispersive liquid–liquid microextraction
DMF	Dimethylformamide
DNS-Cl	Dansylchloride, 5-dimethylaminonaphthalene-1-sulfonyl chloride. Derivatization agent for LC fluorescence detection
Duty cycle	The instrument operation time which delivers analytical data. Standby, equilibration or maintenance times are reducing the duty cycle of an instrument
DWP	Deep well plate

E

ECD	Electron capture detector. A selective, highly sensitive GC detector for electrophilic compounds, for instance, organic halogen, organic metal, diketone analytes
ECTFE	Halar™, ethylene-chlorotrifluoroethylene copolymer
EDCs	Endocrine disrupting chemicals
EI	Electron ionization. Most used ionization mode in GC-MS with a standardized ionization energy of 70 eV, commonly applied for the registration of full scan mass spectra for library search identification of the compound-specific fragmentation pattern and quantification
Endcapped	Endcapping refers to the replacement of accessible silanol groups of silica stationary phase materials by trimethylsilyl groups and prevents retention of polar groups. See https://en.wikipedia.org/wiki/Endcapping
ESI	Electrospray ionization. The prevailing ionization mode in LC-MS for positive (ESI+) and negative (ESI−) ions
ETFE	Tefzel™, ethylene-tetrafluoroethylene
EPA	Environmental Protection Agency
EPA	Eicosapentaenoic acid (also icosapentaenoic acid). 20:5(n-3), an essential omega-3 fatty acid
EPC	Electronic pressure control. Programmable carrier gas regulation of the GC used for column, split, or backflush gas flows
Extran™	Laboratory cleaning agent. Aqueous solution contains, anionic and nonionic surfactants, benzenesulfonic acid (1–3%), n-alkylbenzenesulfonic acid, and sodium salts (1–2.5%). See Merck, Extran, Safety Data Sheet, Revision Date 15.06.2017 Version 2.5

F

FAMEs	Fatty acid methyl esters
FC	Freeze concentration
FD	Fluorescence detection
FDA	Food and Drug Administration
FEP	Teflon FEP™, copolymer of hexafluoropropylene and tetrafluoroethylene
FFA	Free fatty acid

FID	Flame ionization detector
FLPE	Nalgene™, fluorinated high density polyethylene
FMOC-Cl	Fluorenylmethyloxycarbonyl chloride. Derivatization agent for LC
Full Scan MS	Detection mode in GC-MS or LC-MS for the registration of complete (full) mass spectra, typically used for compound identification via library search

G

GAC	Green analytical chemistry
Gantry robot	Resembles the portal or gantry cranes with rails on two sides of the moving bar. Gantry robots are usually enclosed, benchtop installed, and mostly offline operating
Gauge	The measure of the syringe needle dimensions OD and ID, see the Needle Gauge Chart in Appendix Table A.3
g force	The centrifugation force of a centrifuge is measured in multiples of the earth acceleration, g is positively related to the radius and angular speed of the centrifuge rotor
GC	Gas chromatograph or Gas chromatography
General Unknown	The "General Unknown" analyses tries to identify the majority of (harmful) analytes in a sample. Typically full scan mass spectrometry is used, combined with MS/MS and HR/AM analysis for compound identification
GF	Glass fiber
GFC	Gel filtration chromatography. See also SEC
GPC	Gel permeation chromatography. See also SEC
GSM	Geosmin, off-odor compound
GUI	Graphical user interface

H

HD	Mustard gas, 1-chloro-2-[(2-chloroethyl)sulfanyl]ethane, CAS# 505-60-2
HDMS	Hexamethyldisilazane. Derivatization agent
HDPE	High-density polyethylene
HFBA	Heptafluorobutyric anhydride. Derivatization agent
HR/AM-MS	High-resolution accurate-mass mass spectrometry. Accurate mass determination of analytes typically requires a sufficiently high mass resolving analyzer of $R > 30\,000$, provided by the Orbitrap, magnetic sector, or Time-of-Flight (ToF) technology
HSGC	Headspace gas chromatography. Coupling of the headspace extraction with GC analysis
HS-SPME	Headspace SPME. The SPME Fiber or Arrow is kept during extraction in the headspace above a sample. Agitation of the sample is required
HSSE	Headspace sorptive extraction. Headspace extraction using a coated stir bar positioned in the sample headspace during extraction

I

ID	Inner diameter
IBMP	3-Isobutyl-2-methoxypyrazine, off-odor compound
IL	Ionic liquid
IPMP	3-Isopropyl-2-methoxypyrazine, off-odor compound
ISO	International Organization for Standardization
ISO/EN	ISO document/European Standard
ISTD	Internal Standard
ITEX DHS	Acronym for the dynamic headspace analysis using an in-tube extraction method. A syringe-based dynamic headspace purge and trap process for the analysis of VOCs. The analytes are collected on sorbent material (e.g. Tenax™). The sorbent material is located in the syringe needle. Multiple plunger strokes control the sensitivity of the method during shaking of the sample vial

L

LC	Liquid chromatograph or liquid chromatography
LDPE	Low-density polyethylene
LLE	Liquid–liquid extraction
LOD	Limit of detection. Defined in the signal domain, for instance, using a S/N ratio of 1:3
LOQ	Limit of quantification. Statistically defined in the concentration domain
Luer Lock™	The Luer lock fitting, named after the nineteenth-century German Wülfing Luer, is a standardized conical syringe/needle joint for exchangeable single-use needles (ISO 594, DIN and EN standard 1707:1996 and 20594-1:1993). A male-taper fitting on the syringe side connects to a female Luer slip tip at the needle. Luer lock fittings are held together by friction or screwed into the syringe joint. Luer Lock fittings are common for medical equipment, e.g. for hypodermic needles
LVI	Large volume injection in GC. Commonly as LVI, the injection of volumes larger than 10 µL of up to several hundreds of µL is understood

M

µSPE	Micro-SPE. Miniaturized SPE cartridges for wash and load volumes in the µL scale. Elution flow rates range in the µL/s range
MAG	Monoacylglycerol
MBTFA	N-methyl-bis(trifluoroacetamide). Derivatization agent
MCE	Mixed cellulose ester
MCPD	Monochloropropane-1,2-diol, or 3-chloropropane-1,2-diol, occurring as 2-MCPD and 3-MCPD
MDL	Method detection limit

MeOx-TMS	Two-step derivatization process by methoximation followed by silylation applied for metabolomic profiling
MEPS™	Microextraction by Packed Sorbent™, by Trajan Scientific and Medical
MHE	Multiple headspace extraction. Headspace method for matrix-independent quantification and high sample throughput
MIB	2-Methylisoborneol. Off-odor compound
MS/MS	Hyphenation of two mass spectrometer stages. Most popular in the targeted analysis is the usage of triple quadrupole technology with two quadrupole analyzers connected by a collision cell
MSTFA	N-Trimethylsilyl-N-methyl trifluoroacetamide. Derivatization agent
MTBE	Methyl tert-butyl ether
MTP	Microtiter Plate™. In general named microplate, a flat plate with multiple "wells" used as small test tubes. A microplate typically has 6, 12, 24, 48, 96, 384, or 1536 sample wells arranged in a 2:3 rectangular matrix with a standardized dimension of 5″ × 3.33″. See also https://en.wikipedia.org/wiki/Microplate
MHE	Multiple headspace extraction. Method for quantitative headspace analysis by repeated equilibration and measurement from the same vial
MMSE	Monolithic material sorptive extraction
MOH	Mineral oil hydrocarbons. The term includes saturated and aromatic hydrocarbons
MOAH	Mineral oil aromatic hydrocarbons
MOSH	Mineral oil saturated hydrocarbons
MRM	Multiple reaction monitoring. MS/MS operation mode for targeted analysis with the detection of multiple product ions after a specific precursor ion has passed the collision cell of a triple quadrupole MS. See also SRM
MTBSTFA	N-tert-Butyldimethylsilyl-N-methyltrifluoroacetamide. Derivatization agent
MUFA	Monounsaturated fatty acids

N

NC	Nitrocellulose
NCI	Negative chemical ionization. Selective ionization method in mass spectrometry producing negative ions of the analyte
Needle termination	Terminations are called the connection point and style of the syringe barrel with the needle. They are available in different configurations optimized for various applications. Termination codes can vary with manufacturer: (N) Cemented Needle (F) Fixed Needle (RN) Removable Needle (R) Removable Needle (ASN) Autosampler Cemented Needle (ASRN) Autosampler Removable Needle (TLL) PTFE Luer Lock (KH) Knurled Hub (LT) Luer Tip (LTN) Luer Tip Cemented Needle (LL) Luer Lock

Needle Gauge	It describes the different needle sizes of OD and ID, hence wall thickness and volume, which are available for syringe needles, measured in "gauge." Large numbers correspond with a smaller OD. The small "s" (e.g. 23s) after the gauge size specifies a small internal volume in the needle (ID); thus, a higher wall thickness at the same OD provides increased ruggedness for automated applications. See Gauge and the Needle Gauge Chart in Appendix Table A.3
Ninhydrin	Reagent for amino acid derivatization, generates a dark purple color of the reaction products, absorbance max. at 570 nm
NT	Needle trap
NTD	Needle trap device
NTME	Needle trap micro-extraction

O

OCPs	Organochlorine pesticides
OD	Outer diameter
OPCW	Organization for the Prohibition of Chemical Weapons

P

p.a. (PA)	Pro-analysis grade, analytical reagent grade. The chemical purity grade for analytical purposes
PA	Polyamide, trade name e.g. Nylon™
PAH	Polyaromatic hydrocarbon
PC	Polycarbonate
PCPs	Personal care products
PEEK	Polyetheretherketone
PEI	Polyetherimide
PES	Polyethersulfone
PETG	Polyethylene terephthalate copolymer
PFA	Polyfluoroalkoxy alkane polymer
PFBAY	Pentafluorobenzaldehyde
PFBB	Pentafluorobenzyl bromide. Derivatization agent
PFBCI	Pentafluorobenzoyl chloride. Derivatization agent
PFBHA·HCl	O-(2,3,4,5,6-Pentafluorobenzyl)hydroxylamine hydrochloride. Derivatization agent
PFE	Pressurized fluid extraction
PFPA	Pentafluoropropionic anhydride. Derivatization agent
PFPOH	Pentafluoro-propanol. Derivatization agent
PM	Preventive maintenance
PMMA	Polymethyl methacrylate
PMP	Polymethylpentene
POPs	Persistent organic pollutants. Covering the groups of polychlorinated/brominated dioxins and furans, organochlorine pesticides, and similar persistent compounds
POSH	Polyolefin oligomeric saturated hydrocarbons
PP	Polypropylene

PPCO	Polypropylene copolymer
PPCPs	Pharmaceuticals and personal care products, e.g. as of EPA Method 1694
Priming	Prepare a device like a syringe or a dilutor for operation using a suitable solvent to commence the intended workflow. This step can be a security part of an automated workflow for instance for rinsing, cleaning, or air removal after installation or some idle time of the tool applied
PS	Polystyrene
PSA	Primary secondary amines. Sorbent material
PSF	Polysulfone
PT	Proficiency test
PTFE	Polytetrafluoroethylene, trade name e.g. Teflon™
PTV	Programmed temperature vaporizer. Temperature programmable GC injector for liquid band injection and low discrimination of high-boiling compounds
PUFAs	Polyunsaturated fatty acids, namely EPA and DHA
PVC	Polyvinyl chloride
PVDF	Polyvinylidene fluoride

Q

QA	Quality assurance
Quantification	The term meaning includes both quantitation and identification
QuEChERS	Quick Easy Cheap Efficient Rugged Safe. Acronym for the pH-buffered acetonitrile extraction known for pesticides, drugs, and similar residues in food, soil, and other complex matrices

R

RC	Regenerated cellulose
RCF	The "relative centrifugal force" of a centrifuge is measured in multiples of the standard earth acceleration g
RFID	Radio-frequency identification. RFID tags contain electronically stored information which is collected from a reader. The RFID tag can be embedded inside of an object for unequivocal identification
Rinsing	The term "rinsing" is used for priming a syringe with a liquid sample. This step reduces remaining solvents and may cover active sites within the wetted syringe surface. Rinsing with sample is a recommended security step in automated workflows
RM	Reference material, not certified, commercial, external, or in-house prepared
RT	Retention time
RTL	Retention time locking

S

Sample	The "sample" is the original collected sample. In general it is used as well for the comminuted representative test portion used for the sample processing steps

SAN	Styrene acrylonitrile
SANCO	Directorate-General for Health and Consumers. Correct DG SANCO stands for the French words Santé (Health) & Consommateurs (Consumers)
SBS	Society for Biomolecular Screening, since 2010 the Society for Laboratory Automation and Screening (SLAS). Microplates use the formerly by SBS recommended format, today the ANSI SLAS 1-2004 (R2012) Standard (American National Standards Institute, Inc)
SEC	Size exclusion chromatography. Compound separation by molecular size, large molecules are early eluting. Also termed GPC, GFC
SFA	Saturated fatty acids
SFC	Supercritical fluid chromatography
SFE	Supercritical fluid extraction
SICRIT™	Soft ionization by chemical reaction in transfer. Ion source for the direct MS coupling of GC, LC, SPME, SFE or sample headspace into MS, trademark of Plasmion GmbH Augsburg, Germany
SIM	Selected ion monitoring. Detection of selected ions from target analytes in GC-MS or LC-MS
Smart-SPE	See µSPE
Solvent miscibility	Solvents form a homogenous solution without layers
Solvent polarity	Solvent classification by the dielectric constant k. Solvents with $k > 15$ are considered polar, $k < 15$ nonpolar
SOP	Standard operation procedure. A written document of step-by-step instructions to set up and carry out routine analyses. SOPs aim to achieve efficiency, quality output, and uniformity of performance, while reducing miscommunication and failure to comply with regulations (Wikipedia)
SPE	Solid-phase extraction
SPDE	Solid-phase dynamic extraction. In-tube extraction using an inside coated fused silica tube
SPME	Solid-phase micro-extraction
SRM	Selected reaction monitoring. MS/MS operation mode for targeted analysis with the detection of selected product ions after a specific precursor ion has passed the collision cell of a triple quadrupole MS. See also MRM
SVE	Solvent vapor exit. A heated valve used for the evaporated solvent vapor used in online LC-GC connection
SVOC	Semi-volatile organic compounds

T

TAA	Group of tetraalkylammonium salts. Derivatization agents
TAG	Triacylglycerides
Targeted Analysis	The "Targeted Analysis" monitors a limited number of compounds, for instance, provided by regulated methods. In GC-MS or LC-MS typically SIM, MS/MS or HR/AM techniques are used
TBA	2,4,6-Tribromoanisole, off-odor compound
TBAC	Tetrabutylammonium chloride. Derivatization agent

TBAH	Tetrabutylammonium hydroxide. Derivatization agent
TBAHS	Tetrabutylammonium hydrogen sulfate. Derivatization agent
TBH	Tetrabutylammonium hydroxide. Derivatization agent
TCA	2,4,6-Trichloroanisole, off-odor compound
TCE	Trichloroethylene
Tenax TA™	A low bleeding polymer resin, poly(2,6-diphenylphenylene oxide (PPPO) with high thermal stability of up to 350 °C. "Tenax TA" is the product name, "TA" stands for "trapping agent"
Tenax GR™	The black composite material of Tenax TA and 30% graphite with higher breakthrough volume for most volatile organics
Test portion	The "test portion" is the comminuted part of the sample which is analyzed
Tetra	Tetrachloromethane
TFE	Tetrafluoroethylene, e.g. Teflon™
TIC	Total ion current. The reconstructed chromatogram (ion current vs. time) of a mass spectrometer, summing up all ion intensities of each mass spectrum in the scanned mass range
TFAA	Trifluoroacetic anhydride. Derivatization reagent
TFME	Thin-film micro-extraction
TMAH	Tetramethylammonium hydroxide. Derivatization agent
TMSH	Trimethylsulfonium hydroxide. Derivatization agent
TTBB	4-(Trifluoromethyl)-2,3,5,6-tetrafluorobenzyl bromide. Derivatization agent

U

µSPE	Acronym for micro-SPE. Miniaturized SPE cartridges with up to 100 mg of sorbent material are used for full automation of the SPE procedure on x,y,z-robotic autosamplers
USP	United States Pharmacopeia, https://www.usp.org. The USP is a compendium of drug information for the United States, published annually by the United States Pharmacopeial Convention (Wikipedia)

V

Vac-HSSPME	Vacuum-assisted headspace solid-phase microextraction
VASE™	Vacuum-Assisted Sorbent Extraction. Headspace extraction under reduced pressure, patented by Entech Instruments
VG	V-series nerve agent of the organophosphorus class, O,O-Diethyl S-[2-(diethylamino)ethyl] phosphorothioate, CAS# 78-53-5. Also called Amiton or Tetram.
VX	Short for "venomous agent X," nerve agent of the organophosphorus class, Ethyl ({2-[bis(propan-2-yl)amino]ethyl}sulfanyl)(methyl) phosphinate, CAS# 50782-69-9

W

WWTP	Wastewater treatment plant

X

XLPE   Cross-linked high-density polyethylene

Z

Z-Sep   Zirconium-based sorbent material

Additional definitions of scientific terms, particular analytical method-related nomenclature, and terminology can be found in several IUPAC and common publications [1–5].

References

1 Majors, R. and Hinshaw, J. (2020). Chromatography & sample preparation terminology guide. *LCGC North America* 38 (February): 84.
2 Ettre, L. (1993). International Union of Pure Commission on Analytical Nomenclature, Nomenclature for chromatography (IUPAC Recommendations 1993). *Pure and Applied Chemistry* 65 (4): 819–872.
3 Smith, R.M. (1993). Nomenclature for supercritical fluid chromatography and extraction (IUPAC Recommendations 1993). *Pure and Applied Chemistry* 65 (11): 2397–2403.
4 Murray, K.K., Boyd, R.K., Eberlin, M.N. et al. (2013). Standard definitions of terms relating to mass spectrometry (IUPAC Recommendations 2013). *Pure and Applied Chemistry* 85 (7): 1515–1609.
5 Vessman, J., van Staden, S.B.I. et al. (2001). Selectivity in analytical chemistry (IUPAC Recommendations 2001). *Pure and Applied Chemistry* 73 (N (July)): 1381–1386. https://doi.org/10.1351/pac200173081381.

Index

a

acetylation   141, 157
acylating reagents   161
agitation   78, 79, 150, 151, 163, 164, 167–168, 197, 266
air displacement pipettes   68–69
alcohol   65, 107, 108, 125, 149, 155, 156, 168, 203, 281, 284, 390
alkylation reactions   160
analyte protectants (APs)   27, 64, 222, 248, 252, 348, 353, 358
aromatic acid chlorides   154–155
aspiration speed   62–63
automated liner exchange (ALEX)   226–230
automated solid-phase extraction   131
automatic tool changer (ATC)   42

b

barcodes   38–40, 219
bay region   315
benzo(a)pyrene   313, 314, 318–319
Birmingham Wire Gauge   58

c

carousel autosamplers   26
Cartesian robots   28–32
cellulose acetate (CA)   125
centrifugation   38, 49, 97, 99, 101–104, 125, 148–150, 348, 357, 383, 385–386, 395
charcoal
  application   301
  consumables   301
  scope and principle   301–302
  solvents and chemicals   301
  system configuration   302
  workflow   303
chemical analysis   1–3, 8, 20, 231
chemical filtration   129, 133, 139, 348
Chemical Weapons Convention (CWC)
  analysis parameter   385, 386
  application   382–383
  consumables   383–384
  recovery   388
  results   389
  sample measurements   388–389
  scope and principle   383
  solvents and chemicals   383
  system configuration   384–385
  workflow   385–387
chloroform   99, 103, 151, 158
clean-up procedures   124–148
  centrifugation   148–150
  filtration   124–129
  gel permeation chromatography   143–148
  solid phase extraction   129–143
collaborative robots   33–34
cycloidal mixing   169–170

d

dansylchloride   155
decapping   50–52

Automated Sample Preparation: Methods for GC-MS and LC-MS,
First Edition. Hans-Joachim Hübschmann.
© 2022 WILEY-VCH GmbH. Published 2022 by WILEY-VCH GmbH.

derivatization 153–163
  acetylation 157
  aromatic acid chlorides 154–155
  dansylchloride 155
  fluorinating reagents 158–159
  FMOC derivatization 155
  GC and GC-MS 156–159
  in-port derivatization 159–163
  LC and LC-MS 154–155
  methoxyamination 158
  methylation 157–158
  ninhydrin reaction 155
  silylation 156–157, 260–262
  SPME on-fiber 262–266
detection force 77
dichloromethane (DCM) 11, 91, 276, 342, 383
dilution pipetting mode 71
dilutions 248
  for calibration curves 251–256
  geometric 248–251
  working standards preparation 256–259
dilutor/dispenser operation 82–84
dispensing speed 27, 55–57, 63
dispersive liquid/liquid micro-extraction (DLLME) 66, 84, 92, 100–104, 203, 384
drop force 77
drops and droplets 55–57
dynamic headspace analysis 84, 168, 192, 201–210, 279
dynamic headspace analysis with in-tube extraction (ITEX DHS) 124, 169, 203–207, 278, 281
Dynamic Headspace Vacuum Transfer in Trap (DHS-VTT) 206, 207

e

economical aspects 15–16
EPA Method 8270E 304
EQuan™ 20
ethanol 103, 193–195, 284–288, 342, 344
European standards (EN) 2, 13, 21, 62, 130, 322, 347

evaporation 22, 66, 90–91, 94, 103, 104, 106, 124, 129, 132–135, 137–139, 146, 150–153, 163, 194, 216, 225–226, 233, 260, 262, 313–314, 332–333, 341, 348, 368, 369, 459
extraction 90–123
  dispersive liquid/liquid micro-extraction 100–104
  liquid 91–92
  liquid-liquid extractions 97–100
  pressurized fluid extraction 92–97
  sorptive extraction methods 104–123

f

fatty acid methyl ester (FAMEs) 21
  analytical parameter 326
  application 321
  consumables 324
  derivatization method 323
  health risk 324
  limitations 323
  method description 322–323
  sample measurements 328
  scope and principle 321–322
  solvents and chemicals 323–324
  system configuration 325–326
  workflow 326–327
filtration 124
  filter material 125–126
  filter vials 127–129
  syringe 126–127
flow cell sampling 84–85
fluorinating reagents 158–159
forward pipetting mode 69

g

gas chromatography (GC)
  automated liner exchange 226–230
  hot needle injection 222–224
  liquid band injection 224–226
  liquid injection 222–230
  sandwich injection 222
gastight syringes 61
GC volatiles analysis 191
  dynamic headspace analysis 201–210

needle trap micro-extraction 208–210
purge & trap 202–204
using sorbent tubes 207
with in-tube extraction 204–207
multiple headspace quantification 197–201
static headspace 192–194
analyte sensitivity 195
injection technique 195–197
overcoming matrix effects 194
tube adsorption 210–222
gel permeation chromatography (GPC) 91, 143–148, 231
GPC-GC online coupling 146–147
micro-GPC-GC online coupling 147–148
standard methods 145
workflow and instrument configuration 145–146
geometric dilution 248–251
geosmin
analytic parameters 296–297
application 295
consumables 296
quantitative calibration 297–299
recovery and precision 299
regulations 300
sample measurements 300
scope and principle 295
solvents and chemicals 295
system configuration 296
workflow 297
glass fiber (GF) 125
glycidol 329–339
glyphosate
analysis parameter 365–366
applications 362
consumales 364
sample measurements 367–368
scope and principle 362–363
sensitivity 366–367
solvents and chemicals 364
system configuration 364–365
workflow 365–366

green analytical chemistry (GAC) 4, 10–12, 19, 20, 33, 100, 147, 374, 459
gripper transport 48–50, 89, 92

h

halal food
analytical parameters 286–287
application 284
consumables 285
results 288
scope and principle 284–285
solvents and chemicals 285
system configuration 285–286
workflow 286, 287
height in rack 77
height on tool 77
hot needle injection 222–224
human performance 2–4

i

incubation overlapping 36, 163–164
injection-port 38, 119, 159, 233–237, 275, 297, 371, 393
instrumental workflows 9–10
data quality 11
green analytical chemistry 11–12
productivity 12–13
Turkey operation 11

j

just enough clean-up concept 124

l

laboratory logistics 7–16
LC injection 233–234
dynamic load and wash 234–235
pipette tool 235–237
LC-GC on-line injection 230–233
liquid band injection 224–226
liquid-liquid extractions (LLE) 11, 24, 27, 66, 82, 84, 92–94, 97–100, 122, 130, 134, 142, 148, 155, 163, 167, 170, 294, 304, 306–311, 312–314, 322, 340, 362, 364, 383–389
low retention tips 81–82

m

Maestro® 365
magnetic transport 46–48, 163, 251, 273, 323, 393
metal plunger 60, 301
methoximation reaction (MeOx) 158, 266–271
methylation 21, 157–158, 160, 321–324, 328
micro-extraction in packed sorbent (MEPS) 141–142
micro-SPE (μSPE) 50–51, 84, 124, 137, 142–143, 148, 163, 216, 229, 348–351, 353–362, 383–389
micropipet 68, 73, 78
mineral oil hydrocarbons (MOH)
  analysis parameter 343–344
  application 339
  consumables 342–343
  quantitative calibration 346
  sample measurements 346–347
  scope and principle 340–342
  solvents and chemicals 342
  system configuration 343
  workflow 344–346
mini-SPE 137, 349
miniaturization 2, 4, 10–12, 28, 90, 92, 94, 96, 130, 134, 137, 142, 146, 150
mixed cellulose ester (MCE) 125
mixing
  agitation 167–168
  cycloidal 169–170
  spinning 169
  syringes 169
  vortexing 166–167
3-Monochloropropane-1,2-diol (3-MCPD) 328, 329
  analysis parameter 334–336
  analytical method 332–333
  consumables 333
  recovery and precision 337–338
  sample measurements 336
  scope and principle 329–332
  system configuration 334
  workflow 335, 336

multiple axis robots 32–33
multiple headspace analysis (MHE) 163, 194, 197–201, 286, 325

n

needle transport 50, 138
needle trap device (NTD) 208–209
needle-trap micro-extraction (NTME) 208–210
ninhydrin reaction 155
nitrocellulose (NC) 125

o

on-line SPE 134–137
online-SPE water analysis
  analysis parameter 373–375
  application 368–369
  consumables 370
  sample preparation 369
  scope and principle 369
  solvents and chemicals 369–370
  system configuration 370–371
  workflow 371–373

p

peltier devices 165
pesticides analysis
  application 347–348
  consumables 353
  sample measurements 358–361
  sample preparation cycles 350–352
  scope and principle 348–350
  setup 354–357
  solvents and chemicals 352–353
  system configuration 354
  workflow 357–358
phthalates
  analysis parameter 396–398
  application 393–395
  consumables 395
  quantitative calibration 398
  sample measurements 399, 400
  scope and principle 395
  solvents and chemicals 395

system configuration 396, 397
workflow 398
pick up force 76
pipette
  disposable pipette extraction 78–79
  filter tips 78
  protein de-salting 79–81
pipettes
  air displacement pipettes 68–69
  aspiration 71, 73
  dispensing 73–74
  functional tips 78–81
  liquid classes 75
  liquid level detection 74–75
  materials 81–82
  modes 69–72
  positive displacement pipettes 69
  tips 75–77
polyamide (PA) 125
polyaromatic hydrocarbons (PAHs)
  analysis parameter 317–319
  application 315
  consumables 316
  regulations 318–319
  sample measurements 318, 320, 321
  scope and principle 315–316
  solvents and chemicals 316
  system configuration 316–317
  workflow 317, 320
polyethersulfone (PES) 125, 128
polymer tip plungers 61
polypropylene (PP) 81, 125, 353
polytetrafluoroethylene (PTFE) 51, 61, 125–126, 128, 249, 301, 353, 370, 376, 396, 397, 399
polyvinylidene difluoride (PVDF) 126, 128
pressurized fluid extraction (PFE) 92–97
priming 24, 59–61, 63–65, 82, 290
purge and trap technique (P&T) 202–204, 206, 289–291, 294

q

quantitative headspace analyses 197
Queen of the Fruits 276

r

radio-frequency identification (RFID)
  chips 39–40, 87, 219
regenerated cellulose (RC) 126
relative centrifugal force (RCF) 149
residual solvents
  analytical parameter 378, 380
  application 375
  consumables 376
  sample measurements 380–382
  scope and principle 376
  solvents and chemicals 376–378
  system configuration 378, 379
  workflow 379
reverse pipetting 69–70

s

sample preparation process 1, 12, 32, 129, 367–368
sandwich injection 27, 160, 162, 222
scavenging mode 129, 133, 139
selective compliance articulated robot arm (SCARA) 28, 29
semivolatile organic compounds
  analytical parameter 306–311
  application 304
  consumables 305–306
  quantitative calibration 312–313
  recovery 314
  regulations 314
  sample measurements 313–314
  scope and principle 304
  solvents and chemicals 304–305
  system configuration 306
  workflow 312
shale aldehydes
  analytical parameter 392
  application 390–391
  consumables 392
  results 393, 394
  scope and principle 391

shale aldehydes (contd.)
  solvents and chemicals 391
  system configuration 392
  workflow 393
silylation 118–119, 141, 156–159, 163, 260–262, 266–271
smart syringe concept 45
solid materials handling 85–88
  automated powder dispensing 86–88
  extraction 90–123
  powder dispensing 86–88
  weighing 88–89
  workflow 86
solid phase extraction (SPE) 129, 137, 138, 141
  general clean-up procedure 133
  micro-SPE clean-up 137–140
  on-line 134–137
  sample preparation process 129
  scavenging mode 129
  syringe micro-SPE 140–143
sorptive extraction methods 104
spinning 169
stainless steel 60, 94, 146, 220, 342
standard operating procedure (SOP) 11, 13–15, 46, 460
standardization 2, 4, 10, 59, 459
static headspace analysis 192–194
  analyte sensitivity 195
  injection technique 195–197
  overcoming matrix effects 194
Stir Bar Sorptive Extraction (SBSE) 108, 119, 121–123, 195, 201, 217
strip force 77
Stubs Iron Wire Gauge 58
sulfur compounds 276–283
syringe rinsing 22, 64
syringe washing 64–65, 235, 249
syringes 56–65
  large volume 57
  needles 58–59
  needles point styles 59–60
  operational parameters 62–64
  plunger types 60–61
  precision and accuracy 57–58
  termination 61–62

t

taste and odour compounds trace analysis
  analytical parameters 273
  application 271–272
  consumables 272–273
  quantitative calibration 275
  recovery and precision 275
  sample measurements 275–276
  scope and principle 272
  solvents and chemicals 272
  system configuration 273
  workflow 273–275
temperature control 163–166
  cooling 163–166
  heating 163–164
Tenax 201–202, 204–207, 209, 213, 216, 220
tetrahydrofuran (THF) 395
thermal desorption unit (TDU) 40, 122, 207–208, 217
thin-film micro-extraction (TFME) 123
traditional sample extraction 97
transesterification 21–22, 332
transferring standard methods 20
transport vial 46

u

United States Environmental Protection Agency (US EPA) 13, 40, 91, 97, 202, 204, 289

v

vacuum-assisted sorbent extraction (VASE) 220
vial bottom sensing 66–67
volatile organic compounds (VOCs) 375
  in drinking water
    analytic parameters 292, 293
    application 289
    consumables 290
    quantitative calibration 294
    regulations 294

scope and principle 289–290
solvents and chemicals 290
system configuration 290, 291
workflow 293
volatile sulfur compounds (VSCs) 207, 276, 278, 279, 281
volumetric dosing 88
vortex mixer 99, 166–167, 169, 251, 255, 268, 301, 306, 323, 325, 334, 371, 397

w

weighing 88–89
workflow concepts
  instrumental concepts
    Cartesian robots 28–32
    collaborative robots 33–34
    multiple axis robots 32–33
    selective compliance articulated robots 28
    tray autosamplers 26–28
    workstations 25–26

object transport 46
  grippers 48–50
  magnetic 46, 48
  needle 50
sample preparation 19–20
  online/offline configuration 25
  transferring standard methods 20–21
sample preparation process 35
  batch processing 36–37
  identification 38–40
  incubation overlapping 36, 37
  parallel processing 38
  prep-ahead mode 35–36
  sequential sample preparation 35
tool change 41
  automatic 42–44
  identification 44–46
  manual 41–42
vial decapping 50–52
workstation 25–26, 88, 335